T0183915

Lecture Notes in Computer Science 11785

More information about this series at http://www.springer.com/series/7407

David Parker · Verena Wolf (Eds.)

Quantitative Evaluation of Systems

16th International Conference, QEST 2019
Glasgow, UK, September 10–12, 2019
Proceedings

 Springer

Editors
David Parker (iD)
University of Birmingham
Birmingham, UK

Verena Wolf
Saarland University
Saarbrücken, Germany

ISSN 0302-9743   ISSN 1611-3349 (electronic)
Lecture Notes in Computer Science
ISBN 978-3-030-30280-1   ISBN 978-3-030-30281-8 (eBook)
https://doi.org/10.1007/978-3-030-30281-8

LNCS Sublibrary: SL1 – Theoretical Computer Science and General Issues

This Springer imprint is published by the registered company Springer Nature Switzerland AG
The registered company address is: Gewerbestrasse 11, 6330 Cham, Switzerland

Preface

This volume contains the papers presented at QEST 2019: 16th International Conference on Quantitative Evaluation of Systems held during September 10–12, 2019, in Glasgow, UK. QEST is a leading forum on quantitative evaluation and verification of computer systems and networks, through stochastic models and measurements. It was held as a standalone event this year.

QEST covers a broad range of topics, including quantitative specification methods, stochastic and nondeterministic models, and metrics for performance, reliability, safety, correctness, and security. QEST welcomes a diversity of modeling formalisms, programming languages, and methodologies that incorporate quantitative aspects such as probabilities, temporal properties, and other forms of nondeterminism. Papers may advance empirical, simulation, and analytic methods. QEST also has a tradition of presenting case studies, highlighting the role of quantitative evaluation in the design of systems, where the notion of system is broad. Systems of interest include computer hardware and software architectures, communication systems, embedded systems, infrastructural systems, and biological systems. Moreover, tools for supporting the practical application of research results in all of the aforementioned areas are also of interest to QEST. In short, QEST aims to encourage all aspects of work centered around creating a sound methodological basis for assessing and designing systems using quantitative means.

This year's edition of QEST featured three keynote speakers. Andrew Gordon (Microsoft Research) gave a talk on "End-User Probabilistic Programming," Piet Van Mieghem (Delft University of Technology) spoke about "Epidemic Spread on Networks," and Andre Platzer (Carnegie Mellon University) presented "The Logical Path to Autonomous Cyber-Physical Systems." This year, QEST also included two invited tutorials, integrated within the main program. Barbara Fila (Kordy) from INSA Rennes gave a tutorial titled "Twenty Years of Quantitative Evaluation of Security with Attack Trees" and Jan Křetínský (Technical University of Munich) spoke about "Learning in Probabilistic Verification and Synthesis."

The Program Committee (PC) consisted of 33 experts and we received a total of 40 submissions. Each submission was reviewed by either PC members or external reviewers (at least three) in a single-blind fashion. Based on the reviews and the PC discussion phase, 19 papers were selected for the conference program (17 full-length papers and 2 tool demonstration papers).

Our thanks go to the QEST community for making this an interesting and lively event; in particular, we acknowledge the hard work of the PC members and the additional reviewers for sharing their valued expertise with the rest of the community. The collection and selection of papers was organized through the EasyChair conference system and the proceedings volume was prepared and published with the help of Springer, in particular Alfred Hofmann, Aliaksandr Birukou, and Anna Kramer; we thank them all for their assistance. We would also like to thank Springer for kindly sponsoring the conference and SICSA (The Scottish Informatics & Computer Science Alliance) for sponsoring PhD Studentships for Scotland-based students. Finally, we would also like to thank the local organization team, especially Gethin Norman, the general chair, for his dedication and excellent work, and the QEST Steering Committee for their support and guidance.

We hope that you find the conference proceedings rewarding and will consider submitting papers to QEST 2020.

July 2019   David Parker
                                                       Verena Wolf

Organization

Program Committee

Alessandro Abate	University of Oxford, UK
Ezio Bartocci	Vienna University of Technology, Austria
Luca Bortolussi	University of Trieste, Italy
Doina Bucur	University of Twente, The Netherlands
Milan Ceska	Brno University of Technology, Germany
Krishnendu Chatterjee	Institute of Science and Technology (IST), Austria
Pedro R. D'Argenio	Universidad Nacional de Córdoba, CONICET, Argentina
Yuxin Deng	East China Normal University, China
Susanna Donatelli	Universita' di Torino, Italy
Lu Feng	University of Virginia, USA
Jean-Michel Fourneau	DAVID, Universite de Versailles St Quentin, France
Martin Fränzle	Carl von Ossietzky Universität Oldenburg, Germany
Marco Gribaudo	Politecnico di Milano, Italy
Arnd Hartmanns	University of Twente, The Netherlands
Nils Jansen	Radboud University, The Netherlands
Benjamin Lucien Kaminski	RWTH Aachen University, Germany
Rupak Majumdar	MPI-SWS, Germany
Radu Mardare	Aalborg University, Denmark
Andrea Marin	University of Venice, Italy
Marco Paolieri	University of Southern California, USA
David Parker	University of Birmingham, UK
Pavithra Prabhakar	Kansas State University, USA
Anne Remke	WWU Münster, Germany
Kristin Yvonne Rozier	Iowa State University, USA
Guido Sanguinetti	The University of Edinburgh, UK
Michele Sevegnani	University of Glasgow, UK
Markus Siegle	Bundeswehr University Munich, Germany
Jeremy Sproston	University of Turin, Italy
Miklos Telek	Budapest University of Technology and Economics, Hungary
Mirco Tribastone	IMT School for Advanced Studies Lucca, Italy
Benny Van Houdt	University of Antwerp, Belgium
Verena Wolf	Saarland University, Germany
Lijun Zhang	Institute of Software, Chinese Academy of Sciences, China

Additional Reviewers

Bacci, Giorgio
Bacci, Giovanni
Backenköhler, Michael
Balbo, Gianfranco
Batz, Kevin
Cairoli, Francesca
Clerc, Florence
Cubuktepe, Murat
Demasi, Ramiro
Dureja, Rohit
Fu, Chen
Goharshady, Amir Kafshadr
Hasanbeig, Hosein
Horvath, Illes
Horváth, András
Kovács, Péter
Kröger, Paul
Lal, Ratan
Lamp, Josephine
Li, Jianwen

Linard, Alexis
Liu, Jiaxiang
Ma, Meiyi
Meszaros, Andras
Michaelides, Michalis
Mikučionis, Marius
Milios, Dimitrios
Molyneux, Gareth
Pakdamanian, Erfan
Pilch, Carina
Salamati, Mahmoud
Sheng, Shili
Spel, Jip
Strathmann, Thomas
Taghian Dinani, Soudabeh
Turrini, Andrea
van der Laan, Hans
Wijesuriya, Viraj Brian
Xue, Bai
Zhang, Chenyi

Contents

Security

Probabilistic Modelling and Abstraction

Applications and Tools

Invited Talks

End-User Probabilistic Programming

Judith Borghouts, Andrew D. Gordon, Advait Sarkar$^{(\boxtimes)}$, and Neil Toronto

Microsoft Research, Cambridge, UK
advait@microsoft.com

Abstract. Probabilistic programming aims to help users make decisions under uncertainty. The user writes code representing a probabilistic model, and receives outcomes as distributions or summary statistics. We consider probabilistic programming for end-users, in particular spreadsheet users, estimated to number in tens to hundreds of millions. We examine the sources of uncertainty actually encountered by spreadsheet users, and their coping mechanisms, via an interview study. We examine spreadsheet-based interfaces and technology to help reason under uncertainty, via probabilistic and other means. We show how uncertain values can propagate uncertainty through spreadsheets, and how sheet-defined functions can be applied to handle uncertainty. Hence, we draw conclusions about the promise and limitations of probabilistic programming for end-users.

1 Introduction

In this paper, we discuss the potential of bringing together two rather distinct approaches to decision making under uncertainty: spreadsheets and probabilistic programming. We start by introducing these two approaches.

1.1 Background: Spreadsheets and End-User Programming

The spreadsheet is the first "killer app" of the personal computer era, starting in 1979 with Dan Bricklin and Bob Frankston's VisiCalc for the Apple II [15]. The primary interface of a spreadsheet—then and now, four decades later—is the *grid*, a two-dimensional array of cells. Each cell may hold a literal data value, or a formula that computes a data value (and may depend on the data in other cells, which may themselves be computed by formulas). Spreadsheets help democratise computing by allowing computer users to create their own customised calculations based on their own data. They are highly flexible and general-purpose, capable of performing a huge variety of jobs for a great many users in their working or personal lives.

Spreadsheet formulas comprise calls to a wide collection of built-in algorithms, encapsulated in functions known as *worksheet functions*. Formulas typically act on strings, numbers, two-dimensional arrays, and can treat fragments of the grid as arrays. Formulas may consist of complex, nested expressions, including conditionals and other forms of control flow. For these and other reasons,

© Springer Nature Switzerland AG 2019
D. Parker and V. Wolf (Eds.): QEST 2019, LNCS 11785, pp. 3–24, 2019.
https://doi.org/10.1007/978-3-030-30281-8_1

spreadsheets can be viewed as code [16], and spreadsheet users are canonical examples of *end-user programmers* [18]: people who write code primarily for their own use. Even though they write code, end-user programmers are usually not professional developers. An end-user programmer often has little intrinsic interest or education in computing but instead wishes to get some job done with the spreadsheet. They are "business professionals or scientists or other kinds of domain specialists whose jobs involve computational tasks" [24].

Spreadsheets often contain uncertain data: for example, academics may deal with noise and missing data in their data sets, managers may have to make business decisions based on projected sales data, and project leaders have to adapt schedules based on estimated workload. Some of the core affordances of spreadsheets are mechanisms to deal with uncertainty: for instance, uncertainty about future events can be modelled simply by trying out different parameters and immediately seeing an updated model. Due to their flexibility, ubiquity, and low knowledge barriers, spreadsheets are acknowledged to be a "breakthrough technology for practical modeling" [28]. Still, this paper considers some proposed additions to spreadsheets to propagate uncertain values through calculations and models.

1.2 Background: Probabilistic Programming

Let's turn to another approach to decision making under uncertainty: statistical models. The purpose of a statistical model is to infer insights from observed data. Much expertise is needed to write, and interpret the results of, statistical inference algorithms, such as randomised Monte Carlo methods or deterministic message-passing. The aim of *probabilistic programming* [12] is to empower domain experts and expert statisticians to get the benefits of statistical modelling and machine learning, without needing expertise in writing inference algorithms. The idea is that the user specifies a statistical model by writing a piece of code, and delegates the difficulty of statistical inference to an automatic compiler.

Probabilistic programming languages typically comprise a deterministic core language, plus (1) operations to sample from probability distributions, (2) operations to condition on observations, and (3) operations to infer properties of the resulting probability distributions. BUGS [9] is the first probabilistic programming language, first developed in 1989 [14], and used extensively in several textbooks for statisticians and social scientists [8,20,25]. Infer.NET [23], developed since 2004, is used at scale in Microsoft. BUGS and Infer.NET only support certain classes of graphical models. Church [10] introduced the idea of a *universal* probabilistic programming language, that can express any model written in a Turing-complete programming language (although efficient inference in general remains a challenge). More recently, Stan [6] and PyMC3 [21] have also gained wide popularity, and there is a wide range of research languages, including Figaro [27], Anglican [38], and many others. Probabilistic programming environments with graphical representations have also been developed, to aid the understanding of programmers new to the paradigm [13].

Table 1. Estimated users of probabilistic programming and of spreadsheets.

programming (writing formulas) in spreadsheets
(tens to hundreds of millions of people)

>>>

probabilistic programming (without conditioning) in spreadsheets
(hundreds of thousands of people)

>

probabilistic programming in probabilistic programming languages
(tens of thousands of people)

(Published estimates of spreadsheet users range from tens [32] to hundreds of millions https://irishtechnews.ie/seven-reasons-why-excel-is-still-used-by-half-a-billion-people-worldwide/. Palisade, the maker of @Risk, claims use by over 150,000 decision makers. See https://www.palisade.com/about/about.asp. RStan has about 20 K downloads per month and the Stan website has about 15 K unique visitors per month (personal communication, Matthijs Vákár, May 2019). See also https://discourse.mc-stan.org/t/estimating-popularity-of-stan-and-related-packages/8768.)

1.3 Bringing Probabilistic Programming to the Spreadsheet

Why might we want to enable probabilistic programming in spreadsheets? As we have already discussed, a substantial amount of decision making around the world is supported by data in spreadsheets. Many models of uncertain situations such as financial plans, events, scientific experiments, and so on, are built using spreadsheet formulas by end-user programmers.

Thus, the direction seems inevitable: let's take probabilistic programming to the data, to the spreadsheet! These observations have led researchers on probabilistic programming languages (including one of the authors) to design probabilistic programming systems aimed towards spreadsheet users. Examples include Tabular [11] and Invrea's Scenarios [40].

In fact, probabilistic modelling and even aspects of probabilistic programming have existed in spreadsheets from early on, before the interest in probabilistic programming for statistics and machine learning.

The formula RAND() draws at random from the uniform distribution on the unit interval. The formula NORM.INV(RAND(),0,1) draws at random from the standard normal distribution. Writing Monte Carlo simulations using such randomized spreadsheet formulas is a simple form of probabilistic modelling, based on repeated sampling. Monte Carlo simulations can be implemented, for example, by arranging a randomised computation in a row of a sheet, and then replicating the row many times.[1] Books on spreadsheet modelling devote whole chapters to this idiom [28,39]. Savage [31] advocates probabilistic modelling using features such as Excel's data tables.[2]

A key aspect of probabilistic programming is that the user writes a model, and the system handles inference. Writing Monte Carlo simulations by

---

[1] See https://www.youtube.com/watch?v=BQv2Uyea8i4&t=27s, for example.
[2] See https://www.probabilitymanagement.org/.

replicating some formulas is not probabilistic programming as the user is expressing the inference algorithm directly. Still, there are well-established add-ins that do support probabilistic programming without tedious replication by the user. These include @Risk (pronounced 'at risk', first released in 1987) or Crystal Ball. It seems to be uncommon, but these tools also support forms of Bayesian conditioning via rejection sampling. (Probabilistic programming languages support conditioning via inference techniques that are far more efficient than rejection sampling.)

Intriguingly, we can reason that there are more users of probabilistic programming in spreadsheets via these add-ins than in actual probabilistic programming languages such as BUGS or Stan. Table 1 shows rough estimates of orders of magnitude of users today. We cannot be certain, of course, because we have only rough estimates of usage numbers. End-user probabilistic programming does appear to be a relatively miniscule subset of all end-user programming: the use of formulas for probabilistic modelling is probably a tiny fraction of the use of formulas in general.

1.4   How Would Probabilistic Programming Help Spreadsheet Users?

To understand how better support for probabilistic programming might help end users, we conducted an interview study of how spreadsheet users manage uncertainty. The study used thematic analysis [5], a qualitative method, common in psychology, in which transcripts of the interviews are *coded* (that is, labelled by researchers) to mark significant phenomena, and the results aggregated.

This paper reports some technical background, the interview study itself, and the design implications of the study.

We begin in Sect. 2 by describing two different existing proposals for spreadsheet extensions that can deal with uncertainty. First, we describe *uncertain values*, that can be used like ordinary certain values, and that propagate uncertainty through calculations. We describe three formalisms for uncertain values: qualitative, possibilistic, and probabilistic. Second, we describe *sheet-defined functions*, and how they can be applied to model uncertainty. Section 3 describes our interview study and its findings, including a categorisation of the types of uncertainty encountered by spreadsheet users in spreadsheets they had constructed, and also a categorisation of the coping strategies adopted by the spreadsheet users. Section 4 describes design implications of the interview study, and explores how the formalisms of Sect. 2 apply to the categorisations of uncertainty faced by users in Sect. 3.

To the best of our knowledge, this work is the first study of how end-users deal with uncertain values in spreadsheets, and the first to discuss how various candidate spreadsheet extensions might match the potential needs of end-users. We augment Streit's framework with the idea of using arrays of possible scenarios or probabilistic samples as a representation of uncertain values.

2 Spreadsheet Extensions for Uncertainty

We consider how spreadsheets can be extended to better handle different types of uncertainty. We consider two extensions. First, Streit [36] proposed to store different sorts of uncertain value in cells, and have these uncertain values propagate through calculations. Second, Peyton Jones, Blackwell, and Burnett [26] proposed sheet-defined functions as a general-purpose mechanism for end-users to define new worksheet functions by an example calculation in a sheet. We explain how both these mechanisms can help users manage uncertainty. We do so with a running example, which we present next.

2.1 Example: Clara's Budget

Let us consider Clara,[3] a fictitious character who represents a common category of spreadsheet end-user. She is a confident computer user. She does not identify as a programmer, but does use spreadsheets in her work and personal life. She is confident to model with simple formulas including arithmetic and common functions like SUM, but is not highly confident with more complex formulas.

We join Clara as she is preparing a spreadsheet to help decide whether she should purchase a sofa, given her budget for this month. The equations below show her assignments of data and formulas to cells.

```
E1 = "Budget for Jan 2018"
E3 = "Income"; F4 = 2000
E5 = "Expenses"
E6 = "Rent"; F6 = 1100  // A certain cost
E7 = "Commute"; F7 = 85 // Another certain cost
E8 = "Sofa!"; F8 = 700  // Clara is deciding whether to buy this item
E9 = "Utilities"; F9 = 100  // But she is uncertain about her utilities bill
E10 = "Total expenses"; F10 = SUM(F6:F9) // value: 1985
E12 = "Balance"; F12 = F4−F10 // value: 15
```

(Instead of the standard view of the spreadsheet, which shows the values of formulas but not the formulas themselves, we use a textual notation for spreadsheets known as Calculation View [29], and show values of formulas in comments.)

Clara lists out her certain costs, rent and commute, the cost of the sofa, and puts in her estimate of the utilities bill. She calculates in F12 the balance of her income given her total expenses, calculated by the formula in F10.

Clara wants the sofa but doesn't want her total expenses to exceed her income. She is uncertain about her utilities bill. How can the spreadsheet help her decide what to do?

---

[3] To be clear, Clara is a fictional character, contrived to illustrate these technical solutions to representing uncertainty. Our interview study looked at how actual users deal with uncertainty in today's spreadsheets. It would be interesting in future to get the reactions of real people to the technical solutions presented in this section.

2.2 Managing Uncertainty with Uncertain Values

In this section, we turn to a technical approach to handling uncertainty in spreadsheets based on storing *uncertain values* in cells. Most spreadsheet systems allow a cell to hold only an individual value, such as a text or a number. Streit [36] proposed and implemented a spreadsheet where a cell can hold an value of an *uncertainty type*, such as a number explicitly tagged as an estimate, or a numeric interval like 7 ± 2, or a probability distribution such as a normal distribution with parameters for mean and standard deviation.

Other researchers have proposed aggregate values in cells, such as arrays [2], which could in principle represent uncertain values, but to the best of our knowledge Streit was the first to consider uncertain values explicitly.

Streit proposed to enrich the spreadsheet interface in several ways:

1. The user can input uncertainty information into cells. Input can be via a textual notation, such as a numeric range. Input could also be assisted by some interface that gathers parameters of a probability distribution, for instance.
2. Uncertainty information propagates through formula evaluation.
3. The presence of uncertainty information is indicated within cells. A most likely value might be displayed, for example, together with an indication that the value is uncertain.

Unlike Streit, in the following classification we do not consider visualization techniques. Visualization of uncertainty [37] is an important subject, but outside the scope of this paper. In terms of semantics, uncertain values and their propagation through formula evaluation are a kind of computational effect [1]. To the best of our knowledge, there has been no formal semantics for Streit's uncertain values. That too would be an interesting challenge for future research.

Next, we discuss three kinds of uncertain value—Qualitative, Possibilistic, and Probabilistic—that could be implemented in spreadsheet systems.

Qualitative. The simplest form of uncertain value is an *estimate*, a qualitative indication that the value is approximate.

- The function ESTIMATE(V) returns the value V tagged as uncertain.

Clara can indicate that she is uncertain about her utilities bill as follows:

E9 = "Utilities"; F9 = ESTIMATE(100)

(In practice, Clara would likely use some graphical interface to help enter the ESTIMATE tag, but we omit the details.)

Estimates propagate through calculation, so that the formulas depending on cell F9 also return estimates.[4]

F10 = SUM(F6:F9) // value: ESTIMATE(1985)
F12 = F4−F10 // value: ESTIMATE(15)

---

[4] Readers might notice a parallel to the literature on information flow and 'tainting'.

Clara's simple spreadsheet gains little from this qualitative tracking of uncertainty. Still, larger spreadsheets with many values, some uncertain, some not, may benefit more as there would be a clear indication of which results are certain, in spite of the presence of uncertain values.

The implementation of qualitative uncertainty as the function ESTIMATE(V) is merely one possibility. One might also imagine an alternative implementation, invoked through buttons/menus, where the 'tag' is applied as cell metadata, much like cell formatting. While easier to access for non-expert end-users, this implementation would also be less flexible; it would not be possible to mark values within larger expressions as being uncertain, for example.

Possibilistic. There are several possibilistic approaches to uncertainty, where an uncertain value represents a set of possible values. In one, the *multiple scenarios* approach, the uncertain value consists of a tuple of values, each corresponding to one of N scenarios. For example, $N = 3$ could represent best case, most-likely case, and worst case.

- The function SCENARIOS(V1,...,VN) returns an uncertain value representing N scenarios.

Suppose Clara considers the best case cost of her utilities to be 50, the most-likely 100, and the worst case to be 150. She can write this as follows:

E9 = "Utilities"; F9 = SCENARIOS(50,100,150)

These scenarios propagate through her calculations like this:

F10 = SUM(F6:F9) // value: SCENARIOS(1935,1985,2035)
F12 = F4−F10 // value: SCENARIOS(65,15,−35)

The spreadsheet tells Clara that, given her assumptions, her best case balance is 65, most-likely 15, and worst case -15.

Having a multiple-scenario value propagate through the sheet is equivalent to Clara trying out the three input values one at a time: the classic 'what-if' analysis afforded by spreadsheets. Still, the advantage over one at a time entry is that the spreadsheet can display the three possibilities side-by-side in the cell. Also, if the other costs are uncertain, Clara can enter those too as multiple-scenarios. One implementation of uncertainty propagation could automatically combine all the 1st cases, all the 2nd cases and all the 3rd cases (which in this simple example corresponds to best-case, most-likely case, and the worst-case), which is easier than for Clara to manually manage the correspondences.[5]

An alternative to these *dependent* scenarios, where an elementwise product of scenarios is taken, is *independent* scenarios (which we have not discussed), where the cartesian product of the individual scenarios would be taken.

---

[5] Some versions of Microsoft Excel provide a specialist tool, the Scenario Manager, that evaluates multiple copies of a template sheet, one copy for each scenario, and produces a report. The Scenario Manager is a wizard, outside the formula language. Its results are not subject to automatic recalculation. See https://www.youtube.com/watch?v=c0tdVlPvFZ4.

Another form of possibilistic uncertainty, considered by Streit and others, is a numeric interval, where propagation consists of interval arithmetic [17].

Probabilistic. A possibilistic uncertain value consists of a set of possible values (such as a finite set of scenarios or an infinite set of real numbers in a numeric interval). Additionally, a *probabilistic* uncertain value is a probability distribution: roughly a possibilistic value, a set, together with a weight for each value.

A simple way to introduce weighted estimates is the *triangular distribution*:

- The function DIST.TRIANG(a, b, c) constructs a value distributed according to a *triangular distribution*, with lower bound a, upper bound b, and mode c, where $a < b$ and $a \leq c \leq b$.

The triangular distribution can capture subjective judgments of probabilities, where c is someone such as Clara's best guess, and a and b are subjectively the intuitive minimum and maximum. The shape of the density of this continuous distribution is a triangle that peaks at c and falls to zero at a and b.

If Clara enters a probabilistic approximate value using this function it propagates as follows.

```
E9 = "Utilities"; F9 = DIST.TRIANG(50,150,100)
F10 = SUM(F6:F9)  // value: DIST.TRIANG(1935,2035,1985)
F12 = F4−F10  // value: DIST.TRIANG(−35,65,15)
```

In the case above, the distribution can be propagated exactly. Streit only considered propagation of parametric probability distributions (such as the families of normal distributions or triangular distributions) that can be calculated exactly. We cannot always compute propagated distributions exactly (for example, the sum of two triangular distributions is not a triangular distribution, and in general it may not be possible to derive a closed-form expression for a distribution generated through any arbitrary arithmetic on other probability distributions). Instead approximate representations based on Monte Carlo sampling can be used. A general but approximate representation of a probabilistic uncertain value is an array of samples from the distribution. Such an array is known as a *stochastic information packet* (SIP) [30].[6] A key advantage of SIPs is that it is easy to compute with them; for example, the sum of two triangular distributions represented as SIPs, can be computed (and thereby represented) as the elementwise addition of the individual SIPs.

How does the spreadsheet visualise a probabilistic uncertain value in a cell? A simple option is to show "∼15", that is, the mean 15 together with the sign "∼" to indicate the uncertainty. A more sophisticated representation is to visualize the density, perhaps as a histogram, although it takes practice for end users to make sense of a density.

---

[6] Use of SIPs to represent probabilistic values is equivalent to the mechanisms used by @Risk and Crystal Ball. Guesstimate is a spreadsheet-like system that uses SIPs to represent uncertain values. See https://www.getguesstimate.com/.

In Clara's case, she may be most interested in the risk that her balance goes below zero. If she writes the Boolean formula F12<0 to test this event, propagation of the triangular distribution in F12 yields the probability that the formula is true.

E13="Chance of overdraft"; F13=F12<0 // value: 25%

So with probabilistic assumptions, the spreadsheet can tell Clara the probability of modelled events such as her overdraft. In this case, she really wants that sofa, and will take the risk!

2.3   Managing Uncertainty with Sheet-Defined Functions

Observe that when considering multiple chosen scenarios in the possibilistic case, or when considering many randomly drawn scenarios in the probabilistic case, we are replaying the original spreadsheet model with different parameters.

A user can replay a calculation by copying part of a sheet, and changing parameter values. Still, this practice is prone to error as the model gets larger, or if it and its copies need to be updated.

Peyton Jones, Blackwell, and Burnett [26] proposed *sheet-defined functions* as an automatic general alternative to manual replication. It is simply the pervasive idea of procedural abstraction from programming languages, but applied to spreadsheets. A sheet-defined function f is specified by a *body*, a piece of grid, and has a number of input parameters identified by ranges in the body, and an output, also identified by a range in the body. A formula $f(V_1, \ldots, V_n)$ is computed by making a copy of the body, pasting the values $V_1, \ldots, V_n$ into the parameter ranges, evaluating the body, and then returning the value in the result range.

Sestoft's monograph [34] describes how to implement sheet-defined functions. He also describes example usages of sheet-defined functions, including idioms for both possibilistic and probabilistic uncertainty.

To explain these idioms, we first turn Clara's budget into a sheet-defined function as follows. The function Budget has a single parameter, the uncertain utilities bill, held in range F9, and it returns the balance together with a numeric indicator of whether the balance is overdrawn in the range F12:H12.

```
function BUDGET( F9 ) returns F12:G12 {
  E1 = "Budget for Jan 2018"
  E3 = "Income"; F4 = 2000
  E5 = "Expenses"
  E6 = "Rent"; F6 = 1100  // A certain cost
  E7 = "Commute"; F7 = 85 // Another certain cost
  E8 = "Sofa!"; F8 = 700  // Clara is deciding whether to buy this item
  E9 = "Utilities"; F9 = 100  // This cell is the parameter to the SDF
  E10 = "Total expenses"; F10 = SUM(F6:F9)
  E12 = "Balance"; F12 = F4−F10; G12 = IF(F12<0,1,0) // Result cells
}
```

(Here we rely on an extension [22] of Calculation View [29] for a textual notation for sheet-defined functions. Instead, an implementation of sheet-defined functions would provide a graphical interface for the user to specify the function name, parameters, and other metadata.)

Clara can calculate her three scenarios of possibilistic reasoning with the formulas below. She can write the formula first in C54, and then drag fill to C55 and C56.

```
B54=50;  C54 = BUDGET(B54)  // value: {65,0}
B55=100; C55 = BUDGET(B55)  // value: {15,0}
B56=150; C56 = BUDGET(B56)  // value: {35,1}
```

In some recent spreadsheet systems, an array held in a cell such as C54 *spills* for display into adjacent cells.[7] In this case, the formula in cell C54 returns the array {65, 0}, and therefore the cell C54 displays the number 65, and cell D54 displays the number 0 (that is, the second item in the array *spills* into the adjacent cell D54. Similarly, the arrays returned into C55 and C56 spill into the corresponding cells in column D.

Clara can make changes to her budget just once in the sheet-defined function, and they automatically propagate to the three scenarios.

Turning to probabilistic reasoning, we start with the observation that a SIP, the underlying representation for a probabilistic uncertain value, is simply an array. We can achieve the effect of Monte Carlo modelling by mapping the sheet-defined function for our budget over a SIP array. The array is drawn from the distribution modelling our probabilistic uncertainty about the parameter.

Here is how Clara can do this in her situation:

```
B58 = SIP.TRIANGULAR(B54,B56,B55,1000) // column vector of samples
C58 = VMAP(B58,Budget) // returns: SIPs with means {15.51, 25%}
```

The column vector in B58 has 1000 samples from a triangular distribution. The function call VMAP(B58,Budget) calculates $\{b_i, o_i\} = Budget(s_i)$ for each sample s_i with $i \in 1..1000$, and returns a $[1000 \times 2]$ array $\{b_i, o_i \mid i \in 1..1000\}$. In this example, there is only a single uncertain parameter. To handle N uncertain inputs, the sheet-defined function would take N arguments, and the input would be an N-column array of random samples.

We are simply using the advanced encapsulation afforded by sheet-defined functions to more robustly implement the Monte Carlo simulation idiom described earlier in Sect. 1.3! This is a more robust implementation as the logic for each trial is defined exactly once and can be easily modified.

Sheet-defined functions and arrays are general-purpose tools, and so there is no inbuilt interface for viewing the resulting array as a probability distribution. Still, it is possible to write formulas, or even sheet-defined functions, to calculate summary statistics of the resulting array.

---

[7] https://support.office.com/en-us/article/Dynamic-arrays-and-spilled-array-behavior-205c6b06-03ba-4151-89a1-87a7eb36e531.

3 End-User Behaviour with Uncertainty

Clara's fictitious example is useful to illustrate the differences between formalisms, but is clearly contrived. In order to ground our analysis in examples of real user behaviour, we designed and conducted an interview study.

We aimed to investigate the following questions:

1. What types of uncertainty do spreadsheet users deal with?
2. How do they manage these various types of uncertainty?

We conducted semi-structured interviews of 11 participants, who walked us through their spreadsheets. We analysed the audio transcripts of these interviews, identifying six qualitatively different types of uncertainty, as well as six categories of strategies participants used to cope with the uncertainty. A summary of our study protocol, and the results, follows. These results are presented in greater detail in our paper "Somewhere Around That Number"[3].

3.1 Interview Study

Participants. We interviewed 11 participants who worked in finance, construction, IT consulting, the oil and gas industry, business administration, and academic research. The size of participants' spreadsheets ranged from 40 rows to thousands of rows. Participants were recruited using convenience sampling via email invitation and were eligible to participate if they used spreadsheets that contained uncertain data. We did not filter participants based on whether they used spreadsheets for work or personal use, but all participants we recruited dealt with uncertainty in spreadsheets for work purposes. To elicit participation, the invitation gave several examples of spreadsheet tasks which could involve uncertainty, such as budgeting, planning, business forecasting, data collection and analysis, scientific modelling, and making predictions. Interviews lasted on average 60 min, and participants were reimbursed with a £30 voucher for an online store.

Procedure. We asked participants to bring one or more of their own spreadsheets which contained uncertain data to the interview session. They were instructed to remove any sensitive information that they did not want to share.

In the first part of the session, participants were asked to talk about their work, and how uncertainty and spreadsheets are a part of this work. In the second part of the session, we discussed if participants gain insights from uncertainty, what tools or strategies they use to gain this insight, and what challenges they perceive in doing so. In the final part, we asked participants to walk us through their spreadsheets, and explain how these spreadsheets were constructed, and what they were used for. The session was audio recorded, and participants' walk-through of their spreadsheets was screen recorded.

Data Analysis. The audio recordings were transcribed verbatim and analysed using iterative coding based on an inductive approach of thematic analysis [5]. There was no pre-existing coding scheme, but we did approach the data with a specific focus to uncover uncertainty types, and user strategies for managing uncertainty. Through a detailed analysis of the transcripts, we identified key features of the spreadsheets and work practices that related to participant concerns with uncertainty.

3.2 Findings

Types of Uncertainty. Based on participants' descriptions of their spreadsheets during the interviews, we identified six types of uncertainty in spreadsheets: *estimates, dynamic data, errors, missing data, unfindable data,* and *untraceable data.*

1. *Estimates* were the most common type of uncertainty among participants, and refer to approximated values of which the precise value is not known, such as the expected profit of a project: *'We're talking about the future. We don't know exactly what's going to happen. All we can do is make best estimates'* (P8).
2. *Dynamic data* refers to data of which the values changed dependent on time, for instance stock market information.
3. *Errors* were either (1) data that users believed to be incorrect based on their prior knowledge and expectations (these could be caused, for example, by measurement errors, transcription errors), or (2) a spreadsheet error value, such as those created by mis-typed formulas, or formulas receiving arguments of incorrect types, or broken links to external sources. Spreadsheet errors and their sources are an important area of research in their own right, with sophisticated existing taxonomies [19] that we shall not replicate here.
4. *Missing data* were values that were not recorded in the dataset, such as gaps in measured sensor data.
5. *Unfindable data* is information which in principle could be computed from data contained within the spreadsheet, but was hard to extract, and was thus experienced by the user as uncertain. If users were unable to find the information, they regularly used their own estimate instead. For example, P7 dealt with timesheets which gave an overview of hours that all employees of his department had worked per day. He wanted to know how many of these hours were worked on the weekend, but did not know the correct spreadsheet formula to retrieve this data from the timesheet: *'There's a second unknown, which is the weekends [...] Well for me it's difficult, I'm sure there's probably people that can extract it out there'* (P7). This type of uncertainty is particularly interesting, because its presence depends on the circumstances of the spreadsheet user; it is not merely a static property of the data itself.
6. *Untraceable data* refers to data for which the source could not be traced. For instance, participants described situations where it was unclear whether data they received from other people was derived from a computational model, or whether it was *'completely based on their [colleagues'] intuition'* (P11).

Our claim is that these six categories are useful distinctions between different kinds of unknowns arising in our sample of spreadsheets, and by inference in the whole population of spreadsheets. Estimates, Errors and Missing data are common types of uncertainty found in earlier work [4,33]. In addition to these uncertainty types, we found additional three types of uncertainty: Dynamic data, Unfindable data, and Untraceable data.

A further distinction from the literature on uncertainty is between *aleatoric uncertainty*, due to some random process, and *epistemic uncertainty*, due to lack of knowledge. Our classes of estimates and dynamic data are aleatoric uncertainties, while our classes of errors, missing data, unfindable data, and untraceable data are epistemic uncertainties.

Participants' Strategies to Cope with Uncertainty. Participants described various strategies to cope with uncertainty in their spreadsheets. Through our analysis we identified 35 different strategies, which were categorised into six high-level categories: *Minimise, Understand, Communicate, Ignore, Exploit,* and *Add* uncertainty. This is an extension of the categories observed by Boukhelifa et al. [4], with a specific focus on spreadsheets (the original categorisation was tool-agnostic).

1. *Minimising* uncertainty was the most common strategy. Examples of Minimise strategies were to acquire more data, or compare the data with historic data from past situations to try and come to more accurate estimates.
2. The second most common strategy was to *Understand* uncertainty. Participants discussed uncertain data with colleagues, read literature on the subject, researched why the dataset was uncertain, and compared a subset of alternative scenarios.
3. Participants also tried to *Communicate* uncertainty to both themselves and to others. This type of strategy did not necessarily improve people's understanding of why data was uncertain, but its aim was to highlight that there was uncertainty present in the data. To communicate uncertainty to themselves, participants gave spreadsheet cells different colours, or added comments. Participants communicated uncertainty to others through presentations, reports or by providing a verbal narrative.
4. Three participants said they also *Ignored* uncertainty at times. P6, P10 and P11 explained it could be difficult to conduct analysis on a dataset that contained errors or missing values, and these were removed during the analysis process.
5. Sometimes however, it was valuable to know there was uncertainty in a dataset. In this case, an *Exploit* strategy was used, and the uncertainty was extracted or quantified. For example, P11 exploited missing values by adding weights to them in model building, based on how often they occurred. The amount of missing values in a dataset could provide valuable information about what caused the uncertainty, and how important they were to consider.

6. Interestingly, one coping strategy to deal with uncertainty was to minimise one type of uncertainty, by *Adding* another type of uncertainty. For example, if participants dealt with unfindable data that they could not extract from a spreadsheet, they used an estimate instead. Uncertainty was also added by only considering a summary or a subset of the data. P11 dealt with datasets of measured sensor data, which could be tens of thousands of rows. To be able to view and easily digest this data, he would replace multiple data points with one data point, such as the average value of those data points.

Table 2. An overview of each of the six types of uncertainty, showing the definition of each, and an example quote where a participant described the type of uncertainty.

Type of uncertainty	Definition	Example quote
Estimates	Data approximated by the user of which the precise value is not known, such as the expected number of attendees to an event.	*'The final value will be somewhere around that figure. However, I haven't gotten to the point yet where I'm able to say: it's in this range, it's in the +/− 3% range'* (P8)
Dynamic data	Data which is not static and changes over time, such as stock market information.	*'When you go to open that [spreadsheet] next month, the information has changed'* (P9)
Errors	Data that contains errors, such as formula or transcription errors.	*'You also get a few very unusual values, I seem to get negative infinities quite a lot. Which clearly is not possible (...) So there's a whole load of error values in there'* (P11)
Missing data	Data that is missing from the data set, such as gaps in measured sensor data.	*'The weather file that's used to generate the data there in the other program is missing large chunks of wind data (...) That was one of those kind of reports that was fairly heavily caveated as being 'We've had to make quite a lot of numbers up here, to get any idea of what might happen with this'* (P6)
Unfindable data	Data which technically is contained within a spreadsheet, but which cannot easily be found by the user, such as the total amount of hours worked on the weekend in an employee timesheet. Usually, being unable to find it, the user uses an estimate instead.	*'There's a second unknown [in the spreadsheet], which is the weekends. (...) I think that kind of formula that we're trying Excel to do is probably difficult. Well for me it's difficult, I'm sure there's probably people that can extract it out there'* (P7)
Untraceable data	Data from which the user cannot trace its original source, and whether or how it is calculated, such as subjective estimates made by other people, or complex and inaccessible formulas.	*'Sometimes it would just be numbers in a spreadsheet, and you couldn't really identify what the hell was going on (...) And half the time, to actually check it, the easiest way is almost to make a new spreadsheet doing it your own way, and see if you get the same number'* (P6)

Table 3. An overview of the six categories of coping strategies, its definition and an example quote where a participant gave an example of the strategy.

Coping strategy	Definition	Example quote
Add	The user adds additional uncertainty to a spreadsheet. For example, the user is unable to find data in a sheet, and uses an estimate or interval instead. In this situation, data is contained within a spreadsheet, but cannot easily be found by the user, and uncertainty is added by using an approximated value	*'A lot of the time, there are those things which just aren't [worth the effort to figure out the right formula to extract data]. So they either get done quite slowly, the manual way, or they don't get done at all. And you get people putting like fudge figures (...) and say, 'Oh well it's anywhere between here and here. That's the best we can do'* (P6)
	Alternatively, a collection of data points is reduced to one data point, such as the average value, to make a large spreadsheet easier to view and digest	*'It gets very confusing when I start having dozen sheets, with 1,000 columns, and 10,000 rows in each. (...) I have a range of values, but the algorithms only give me an average, across say 100 points'* (P11)
Communicate	The user communicates uncertainty to others verbally, and through reports and presentations	*'Although we provide estimates, I do provide an estimate which is a hard figure. But what I tell them [clients] is: it's around that figure'* (P8)
	Users also communicate uncertainty to themselves, by highlighting cells in their spreadsheets that contain uncertain data	*'Quite a lot of colouring. Just to highlight particular aspects. So I'd do green for a particular area of what I think 'These are definites'. And light-blue or something for unknowns'* (P4)
Exploit	The user uses the amount of uncertainty in a spreadsheet as a valuable piece of information about the data	*'Sometimes it [uncertainty] contributes to the forecast, because you want to know sometimes if there's a specific reason for the missing data'* (P10)
Ignore	The user ignores uncertainty, by removing it from the spreadsheet or replacing it with other values	*'I use a filter on Excel to filter the values (...) So I try to identify them and then find and replace with NAs, most of the time'* (P11)
Minimise	The user minimises uncertainty, by acquiring more data, or discussing it with colleagues	*'We will liaise with our front office to say, 'Does this look correct? Have things like this happened in the past?' If it's chartered territory, the stuff that we've seen something like this before, we can get more of a better estimate'* (P3)
Understand	The user tries to understand uncertainty by reading literature, discussing it with colleagues, plotting data, evaluating the data source, and analysing a subset of possible scenarios	*'I would read as much around the literature as I possibly can, I will get in different views that people have (...). And then I'll work from all that to try and inform myself'* (P2)

Table 4. An overview of the six types of uncertainty (the counts and percentages indicate the frequency of each uncertainty type in our dataset of 352 observations) and the three formalisms. The table indicates how well we think a type of uncertainty could have been supported by a particular formalism, on the scale: none, weak support, some support, good support. We used the following rough guidelines to make this classification: None: this formalism will not help the users in this uncertainty type. Weak: this formalism will help with this uncertainty type, but has significant limitations in comparison to the others. Some: this formalism would help in some cases where users encounter this uncertainty type. Good: (features supported by) this formalism would help in many or most cases where users that encounter this uncertainty type. An asterisk (*) indicates that we observed at least one participant directly state that type of uncertainty would be supported by that formalism.

	Dynamic data (52, 15%)	Estimates (175, 50%)	Errors (47, 13%)	Missing data (44, 12%)	Unfindable data (24, 7%)	Untraceable data (10, 3%)
Qualitative	Some	Good*	Good*	Weak	None	None
Possibilistic	Some*	Good*	None	Some	None	None
Probabilistic	Good*	Weak	Some*	Some*	None	None

4   Design Implications of the Interview Study

So if the makers of spreadsheet software wished to implement user-facing tools for the representation and management of uncertainty,[8] what should they focus on? What features would solve the most common use cases we observed? In other words, what should we build to get the most bang for our buck? In this section we present some *implications for design* (notwithstanding the criticism of that phrase [7,35]).

In the previous section we have been introduced to a broad categorisation of uncertainty formalisms into Qualitative, Possibilistic, and Probabilistic, and how the formalism determines the family of features that can be implemented to support them – tags for Qualitative uncertainty, the scenario manager or intervals for Possibilistic uncertainty, and so on. At the same time, our research has identified six types of uncertainty in spreadsheets. We therefore examine how each formalism can support the management of each type of uncertainty, on a case by case basis, by considering how the user might use the various implementations of each formalism in those cases.

1. *Dynamic data* changes over time, either refreshing automatically or needing manual re-entry. A Qualitative tag may help to indicate the presence of uncertainty. A Possibilistic or Probabilistic representation may help to keep a record of historic values for a cell, with a Probabilistic representation capturing additional information, helpful for statistical inference.

---

[8] We began this paper with a set of premises leading to the hypothesis: introducing probabilistic programming to spreadsheets is a Good Idea. Through our user research, we refined our ideas from focusing on probabilistic programming, to ways of representing and managing uncertainty more generally.

Table 5. Participant quotes exemplifying how types of uncertainty could have been supported by a particular formalism.

Formalism - Type of uncertainty	How the formalism would be used	Example quote
Qualitative - Estimates	The user can input their own values of what they think an estimate is	'We have to convert that qualitative scenario into some quantitative numbers, to say 'OK, if X wins, it's likely that the market's going to react a little bit better, just because of her economic policies, therefore we would expect GBP to strengthen.' And then we would revise our forecast based on that, going forward. But then again, those are unknown numbers, we're just using, we would use estimates to try and say, 'This is where we'd end up' (P3)
Possibilistic - Dynamic data	The user can input intervals to indicate the range that dynamic data can fluctuate in, and/or can compare a subset of outcomes of dynamic data	'You'd have a day rate, and multiply it in terms of days. And then you add a certain percentage on afterwards, just so you have some space' (P4)
Possibilistic - Estimates	The user can compare a subset of possible estimates	'You get a huge extra benefit from looking at one policy to looking at two policies. It's a bit more time-consuming, it takes twice as long to run, but it's worth doing. But to go from two to three policies, it's not so clear that you get enough extra benefit from that, from the extra complexity' (P2)
Probabilistic - Dynamic data	Based on historic values, the user can quantify the likelihood of future values	'Once you get the data, then it becomes a modelling exercise. Like what is my chance of selling something next month? Well that depends on how much of that thing has been selling this month' (P9)
Probabilistic - Errors	If the data errors in a spreadsheet are quantifiable, probabilistic measures can be used to track and quantify uncertainty as it propagates through the calculations of a spreadsheet	'At every stage, a lot of the time we have: if we know we've got however many unknown values in spreadsheet 1, by the time we get to spreadsheet 10, there's a debate about whether the data's of any use at all, because the errors propagate, and multiply. Now I don't have a measure of how that affects it. I'd love to be able to measure that, but I have no idea how to do that within the tools' (P11)
Probabilistic - Missing data	The user can make estimates regarding missing data by analysing the distributions of the data that does exist in the spreadsheet	'If we get certain data, and we want to make sure: does that follow a certain distribution, or something like that? Or is that data based on a certain distribution? There are tests that give you a p value, very likely that this data point's origins are the same as the other data points. (...) So that gives you some confidence data in how well you can rely on this model, on that model, and stuff like that' (P10)

2. *Estimates* can be supported by Qualitative tags, which can propagate through the calculation dependency chain so that it is made apparent which calculations depend on estimated data. In entering estimates, users may also be able to provide upper and lower bounds, or a set of possible values, which are well-supported by the Possibilistic formalism. Finally, if possible, it would be ideal to elicit a formal probability distribution, but our experience with users suggests that non-expert end-users find it challenging to understand and confidently assign parameters to a probability distribution. Consider the difference between a non-expert having to produce "upper and lower bounds" for a cost in a budget, and having to produce a "mean and standard deviation for a Gaussian distribution". This makes the Probabilistic scenario a weak overall fit for estimates, although we recognise that it might fit the needs of some experts very well.

3. *Errors* benefit most from Qualitative tagging, as a way of annotating their presence and communicating them to others. It is unclear whether Possibilistic or Probabilistic uncertainty can help in error cases, although some cases of data entry errors can be detected using the insight that an erroneous data value is often very unlikely given the distribution of the rest of the data.

4. *Missing data* can be supported by Qualitative tagging, if calculations are done on ranges with missing values. Missing values can be inferred and represented either in a Possibilistic manner (for instance, if the missing value is known to be a member of a finite set) or in a Probabilistic manner (for instance, if the missing value is part of a dataset with a known or calculable distribution).

5. *Unfindable data* – that is, answers that are prohibitively difficult to extract from the data in the spreadsheet, even if it is possible in principle. This type of uncertainty cannot be solved by implementing better tools for representing and manipulating uncertainty; the solutions here depend more on enabling users to store and retrieve data in different ways, from enabling the pivoting and refactoring of existing spreadsheet layouts, to better assistance in formula authoring.

6. *Untraceable data* might be supported using a Qualitative tag, to indicate and communicate its presence. However, better tracking of data provenance would help deal with the issue more holistically. Users should be able to inspect the history of edits to a spreadsheet cell, and answer the questions: who edited this cell and when? Was the data manually typed in, copied in from a document, or written by a macro or other script? The implementation of some of these ideas (such as tracking and linking to documents from which data is copied) might go beyond the scope of the spreadsheet software, and require the participation of other tools and the operating system to track provenance.

In summary, solutions focused around enabling users to formally *represent* uncertainty offer varying levels of support dynamic data, errors, estimates and missing data, but are unlikely to help with unfindable data and untraceable data. Table 4 shows a matrix of the three categories and the six types of uncertainty, summarising the discussion above. For many pairings where we claim that the formalism would support the user, we provide a participant quote in Table 5 – these entries are indicated with an asterisk (*).

It should be noted that the user behaviour we observed was shaped and constrained by the affordances of spreadsheets today; users may appear to make limited use of uncertainty in spreadsheets simply because of the lack of native support for uncertainty in spreadsheets. There may be a larger latent demand for such features, which is currently unobservable due to this 'technological determinism'. Or, there may be no such demand and the observed behaviour is a true reflection of society's needs. It is impossible to tell using our interview methodology alone.

Bearing this in mind, let us return to the question of what designers of spreadsheet systems can take away from this. One clear conclusion is that no single formalism can solve most user problems. Of Qualitative, Possibilistic, and Probabilistic formalisms, Qualitative tagging appears to be the most widely applicable, but what the user can do with it is limited. Our recommendation is therefore to offer features from multiple formalisms. Furthermore, if we were to pick one to exclude, it would be Possibilistic, as the technique of managing different scenarios across different columns (or rows, or cells, or sheets, or even whole workbooks), as well as tinkering with values in cells, are both already effective coping strategies in many use cases for Possibilistic data. This recommendation comes despite the fact that Possibilistic data is likely to be easier to input and reason about than Probabilistic data, for most end-users. Finally, solutions focused around *representing* uncertainty are unlikely to help with unfindable and untraceable data—for those cases, we need tools offering education, assistance with authoring, layout and refactoring, and provenance tracking.

5   Conclusions

In this paper, we formally acknowledge the human activity of *end-user probabilistic programming*, in which a computer user, likely not a statistical expert, creates models to support their own decision making. While probabilistic programming is a powerful tool, its widespread adoption is limited by the high statistical and programming expertise required. Spreadsheets, on the other hand, enjoy the status of being the *premier* end-user programming tool, and are arguably the venue of much of the world's data and decision making. What if we were to bring probabilistic programming to spreadsheets?

As a stepping stone, we consider the simpler question of bringing native support for *uncertain* values to spreadsheets. The choice of whether we support qualitative (such as simple tags), possibilistic (such as sets or intervals) or probabilistic (such as distributions) uncertainty will have profound implications for the user experience, which we illustrated through the fictitious case study of Clara's budget. To study these implications empirically, we interviewed 11 spreadsheet users, focusing on their current use cases and coping strategies for uncertainty.

We produced a taxonomy of uncertain data as experienced by end-users, and analysed how each uncertainty formalism could support the different categories, finding the following:

- No single formalism emerges the clear winner, and though Possibilistic uncertainty is well-understood by end-users, the data suggests that it would add the least benefits, since end-users already have several effective strategies for coping with it.
- We also identified that users experience types of uncertainty, such as unfindable and untraceable data, that cannot be solved by representations of uncertainty alone.
- Estimates were the most common type of uncertainty experienced by end-users in our sample, with 157 occurrences (nearly 50% of all observations). While we should be careful with drawing quantitative inferences from small samples such as ours, this is still an indication that Estimates are an important category of uncertainty cases.

As well as considering special-purpose uncertain values, we showed idioms for representing probabilistic and possibilistic uncertainty using the general-purpose concept of sheet-defined functions.

The idea of probabilistic programming in spreadsheets is certainly alluring. But for full effect, we must ensure that the idea works for the tens (or even hundreds) of millions of spreadsheet users facing uncertainty in their decision making. At present, neither uncertain values nor sheet-defined functions are available in mainstream spreadsheets. Will one or both revolutionise probabilistic programming in spreadsheets? Time will tell.

Acknowledgements. Thanks to Breck Baldwin and Matthijs Vákár for information regarding Stan. We are grateful to Alan Blackwell, Eunice Jun, Tom Minka, Simon Peyton Jones for their helpful comments on a draft of this paper.

References

1. Benton, N., Hughes, J., Moggi, E.: Monads and effects. In: Barthe, G., Dybjer, P., Pinto, L., Saraiva, J. (eds.) APPSEM 2000. LNCS, vol. 2395, pp. 42–122. Springer, Heidelberg (2002). https://doi.org/10.1007/3-540-45699-6_2
2. Blackwell, A.F., Burnett, M.M., Peyton Jones, S.L.: Champagne prototyping: A research technique for early evaluation of complex end-user programming systems. In: VL/HCC, pp. 47–54. IEEE Computer Society (2004)
3. Borghouts, J., Gordon, A.D., Sarkar, A., O'Hara, K.P., Toronto, N.: Somewhere around that number: An interview study of how spreadsheet users manage uncertainty. arXiv preprint arXiv:1905.13072 (2019)
4. Boukhelifa, N., Perrin, M.E., Huron, S., Eagan, J.: How data workers cope with uncertainty: a task characterisation study. In: Proceedings of the 2017 CHI Conference on Human Factors in Computing Systems, pp. 3645–3656. ACM (2017)
5. Braun, V., Clarke, V.: Using thematic analysis in psychology. Qual. Res. Psychol. **3**(2), 77–101 (2006)

6. Carpenter, B., Gelman, A., Homan, M.D., Lee, D., Goodrich, B., Betancourt, M., Brubaker, M., Guo, J., Li, P., Riddell, A.: Stan: a probabilistic programming language. J. Stat. Softw. **76**(1), 1–32 (2017)
7. Dourish, P.: Implications for design. In: Proceedings of the SIGCHI Conference on Human Factors in Computing Systems, pp. 541–550. ACM (2006)
8. Gelman, A., Hill, J.: Data Analysis Using Regression and Multilevel/Hierarchical Models. Cambridge University Press, Cambridge (2007)
9. Gilks, W.R., Thomas, A., Spiegelhalter, D.J.: A language and program for complex Bayesian modelling. The Statistician **43**, 169–178 (1994)
10. Goodman, N., Mansinghka, V.K., Roy, D.M., Bonawitz, K., Tenenbaum, J.B.: Church: a language for generative models. In: Uncertainty in Artificial Intelligence (UAI 2008), pp. 220–229. AUAI Press (2008)
11. Gordon, A.D., Graepel, T., Rolland, N., Russo, C.V., Borgström, J., Guiver, J.: Tabular: a schema-driven probabilistic programming language. In: Principles of Programming Languages (POPL 2014) (2014)
12. Gordon, A.D., Henzinger, T.A., Nori, A.V., Rajamani, S.K.: Probabilistic programming. In: Herbsleb, J.D., Dwyer, M.B. (eds.) Proceedings of the on Future of Software Engineering, FOSE 2014, Hyderabad, India, 31 May–7 June 2014, pp. 167–181. ACM (2014)
13. Gorinova, M.I., Sarkar, A., Blackwell, A.F., Syme, D.: A live, multiple-representation probabilistic programming environment for novices. In: Proceedings of the 2016 CHI Conference on Human Factors in Computing Systems, pp. 2533–2537. ACM (2016)
14. Goudie, R., Thomas, A.: MultiBUGS: A parallel implementation of the bugs modelling framework for faster Bayesian inference (2019). talk at workshop on Advances and challenges in Machine Learning Languages. https://people.ds.cam.ac.uk/rg447/2019-05-21-goudie-acmll-slides.pdf
15. Grad, B.: The creation and the demise of VisiCalc. IEEE Ann. Hist. Comput. **29**(3), 20–31 (2007)
16. Hermans, F., Jansen, B., Roy, S., Aivaloglou, E., Swidan, A., Hoepelman, D.: Spreadsheets are code: an overview of software engineering approaches applied to spreadsheets. In: FOSE@SANER, pp. 56–65. IEEE Computer Society (2016)
17. Hyvönen, E., De Pascale, S.: A new basis for spreadsheet computing: interval solver for microsoft excel. AI Magazine **21**(4), 83–92 (2000)
18. Ko, A.J., et al.: The state of the art in end-user software engineering. ACM Comput. Surv. (CSUR) **43**(3), 21 (2011)
19. Kulesz, D., Wagner, S.: Asheetoxy: a taxonomy for classifying negative spreadsheet-related phenomena. arXiv preprint arXiv:1808.10231 (2018)
20. Lunn, D., Jackson, C., Best, N., Thomas, A., Spiegelhalter, D.: The BUGS Book. CRC Press, Florida (2013)
21. Martin, O.: Bayesian Analysis with Python: Introduction to statistical modeling and probabilistic programming using PyMC3 and ArviZ, 2nd edn. (2018)
22. McCutchen, M., Borghouts, J., Gordon, A.D., Jones, S.P., Sarkar, A.: Elastic sheet-defined functions: Generalising spreadsheet functions to variable-size input arrays (2019). (In submission)
23. Minka, T., Winn, J., Guiver, J., Zaykov, Y., Fabian, D., Bronskill, J.: Infer.NET 0.3 (2018), Microsoft Research Cambridge https://dotnet.github.io/infer

24. Nardi, B.A., Miller, J.R.: The spreadsheet interface: a basis for end user programming. In: Diaper, D., Gilmore, D.J., Cockton, G., Shackel, B. (eds.) Proceedings of the IFIP TC13 Third Interantional Conference on Human-Computer Interaction, INTERACT 1990, Cambridge, UK, 27–31 August, 1990, pp. 977–983. North-Holland (1990)
25. Ntzoufras, I.: Bayesian Modeling Using WinBUGS. Wiley, Hoboken (2009)
26. Peyton Jones, S.L., Blackwell, A.F., Burnett, M.M.: A user-centred approach to functions in excel. In: International Conference on Functional Programming, pp. 165–176 (2003)
27. Pfeffer, A.: Figaro: an object-oriented probabilistic programming language. Technical report, Charles River Analytics (2009)
28. Powell, S.G., Baker, K.R.: The Art of Modeling with Spreadsheets. John Wiley & Sons, Inc., New York (2003)
29. Sarkar, A., Gordon, A.D., Peyton Jones, S., Toronto, N.: Calculation view: multiple-representation editing in spreadsheets. In: VL/HCC, pp. 85–93. IEEE Computer Society (2018)
30. Savage, S., Scholtes, S., Zweidler, D.: Probability Management. OR/MS Today (2006)
31. Savage, S.L.: The Flaw of Averages. Wiley, Hoboke (2009)
32. Scaffidi, C., Shaw, M., Myers, B.: Estimating the numbers of end users and end user programmers. In: 2005 IEEE Symposium on Visual Languages and Human-Centric Computing (VL/HCC 2005), pp. 207–214. IEEE (2005)
33. Schunn, C.D., Trafton, J.G.: The psychology of uncertainty in scientific data analysis. In: Handbook of the Psychology of Science. Springer (2012)
34. Sestoft, P.: Spreadsheet Implementation Technology: Basics and Extensions. The MIT Press, Cambridge (2014)
35. Stolterman, E.: The nature of design practice and implications for interaction design research. Int. J. Des. 2(1), 55–65 (2008)
36. Streit, A.: Encapsulation and abstraction for modeling and visualizing information uncertainty. Ph.D. thesis, Queensland University of Technology (2008)
37. Streit, A., Pham, B., Brown, R.: A spreadsheet approach to facilitate visualization of uncertainty in information. IEEE Trans. Vis. Comput. Graph. 14(1), 61–72 (2008). https://doi.org/10.1109/TVCG.2007.70426
38. Tolpin, D., van de Meent, J.-W., Wood, F.: Probabilistic programming in Anglican. In: Bifet, A., et al. (eds.) ECML PKDD 2015. LNCS (LNAI), vol. 9286, pp. 308–311. Springer, Cham (2015). https://doi.org/10.1007/978-3-319-23461-8_36
39. Winston, W.L.: Microsoft Excel 2019 Data Analysis and Business Modeling, 6th edn. Microsoft Press, USA (2019)
40. Wu, M., Perov, Y.N., Wood, F.D., Yang, H.: Spreadsheet probabilistic programming. CoRR abs/1606.04216 (2016), (see also the Scenarios tool at. invrea.com)

The Logical Path to Autonomous Cyber-Physical Systems

(Invited Paper)

André Platzer$^{(\boxtimes)}$ (iD)

Computer Science Department, Carnegie Mellon University, Pittsburgh, USA
aplatzer@cs.cmu.edu

Abstract. Autonomous cyber-physical systems are systems that combine the physics of motion with advanced cyber algorithms to act on their own without close human supervision. The present consensus is that reasonable levels of autonomy, such as for self-driving cars or autonomous drones, can only be reached with the help of artificial intelligence and machine learning algorithms that cope with the uncertainties of the real world. That makes safety assurance even more challenging than it already is in cyber-physical systems (CPSs) with classically programmed control, precisely because AI techniques are lauded for their flexibility in handling unpredictable situations, but are themselves harder to predict. This paper identifies the logical path toward autonomous cyber-physical systems in multiple steps. First, differential dynamic logic (dL) provides a logical foundation for developing cyber-physical system models with the mathematical rigor that their safety-critical nature demands. Then, its ModelPlex technique provides a logically correct way to tame the subtle relationship of CPS models to CPS implementations. Finally, the resulting logical monitor conditions can then be exploited to safeguard the decisions of learning agents, guide the optimization of learning processes, and resolve the nondeterminism frequently found in verification models. Overall, logic leads the way in combining the best of both worlds: the strong predictions that formal verification techniques provide alongside the strong flexibility that the use of AI provides.

Keywords: Autonomous cyber-physical systems · Safe AI ·
Hybrid systems · Differential dynamic logic · Formal verification ·
Runtime verification

1 Introduction

Autonomous cyber-physical systems (autonomous CPS) use sophisticated software to control the physics of motion. They plan their own goals and actions in

This material is based upon work supported by the Alexander von Humboldt Foundation, National Science Foundation under NSF CAREER Award CNS-1054246 and CNS-1446712, and US Air Force and DARPA under Contract No. FA8750-18-C-0092.

© Springer Nature Switzerland AG 2019
D. Parker and V. Wolf (Eds.): QEST 2019, LNCS 11785, pp. 25–33, 2019.
https://doi.org/10.1007/978-3-030-30281-8_2

pursuit of them. And if things go wrong, they react to situation changes in order to prevent problems on their own without close human supervision. Autonomous cyber-physical systems are a technological dream come true. Or are they?

Well, for one thing, cyber-physical systems have found frequent use, but are not yet operated very autonomously. Certainly, the prospects that autonomous cyber-physical systems promise are very appealing, but it is precisely their goal of autonomy and lack of human supervision that also makes them fairly challenging to build just right. Granted, it is also challenging to design an ordinary CPS with human supervision because humans need sufficiently early warning to gain situational awareness and react, which, in turn, requires ample foresight in the CPS design. But the desire for autonomy changes the state of affairs considerably.

From a performance perspective, the biggest difference compared to ordinary CPS is that autonomous CPS do not need to be monitored all the time, but "do the right thing" on their own. The biggest difference from a safety perspective is that it's not clear what the right thing is and humans cannot save the day if the autonomous CPS goes awry, because the whole point is that they are not supervised closely. Autonomy benefits from the help of artificial intelligence and machine learning algorithms that cope with the uncertainties of the real world [27]. Of course, the added flexibility in handling unpredictable situations makes the safety impact of the addition of AI themselves harder to predict.

Formal methods provide ways of establishing safety properties for ordinary CPS [2,17,19,26,28,34,36,40], and AI provides ways of giving autonomy to CPS. This calls for a combination of formal methods and artificial intelligence [1,9,13, 14,16], just not by a friendly ignorance of one another. Instead, the trick is to combine both in a way that each field actually retains its benefits for the CPS in the end. This paper surveys an approach for Safe AI in CPS in which logic leads the way in combining the best of both worlds.

2   Challenge

Cyber-Physical Systems combine cyber capabilities such as communication, computation and control with physical capabilities such as the motion of robots, cars, or aircraft. Mathematical models for such CPS are based on hybrid systems, which combine discrete dynamical systems with continuous dynamical systems, e.g., because discrete change one step at a time fits well to computation, while continuous dynamics along differential equations fits well to their motion.

Formal Verification uses the descriptive models of hybrid systems for predicting, with the help of model checking [8,11] or proof [31,36], whether all their behavior satisfies safety properties of interest, such as collision freedom. Especially in the case of logical proofs, formal methods enable very strong guarantees about *all* behavior of the mathematical models from a small reasoning basis [35]. In order to overcome complexity challenges, it is often important to work with models that use simplifying abstractions, because models that include literally all implementation detail quickly become prohibitively expensive to analyze.

Machine Learning forgoes the principle of explicitly programming all behavior of a system and, instead, uses learning algorithms that generalize responses from static data (e.g., a set of labeled images to classify) or from dynamic experience (e.g., responses to trial and error). *Reinforcement learning* (RL) [39], for example, repeatedly tries out an action, observes what the overall outcome of a sequence of actions was, and then increases the probability with which its policy decides for actions that have had large fractions of good outcomes so far. The big advantage of reinforcement learning is that it can be used with very minimal assumptions on the system to be controlled. All it takes is a black-box way of executing actions and reliably observing the outcome, e.g., in a simulator. In practice, learning systems are also lauded for their flexibility in responding to situations that were not directly programmed into the system design. Learning is, thus, presently considered crucial to reach reasonable levels of autonomy.

Of course, guarantees are harder to come by. At the very least, one has to assume that the outcomes observed for the individual actions in the individual states are strongly correlated (in fact, Markovian) with outcomes at other times. Under suitable assumptions, finite-state cases provide elegant theoretical guarantees [39]. But the infinite-state case of CPSs is significantly more complex, because even the luxury of an arbitrary countable amount of experiments is not enough to try all actions in all states. Indeed, black-box uses require fairly strong additional assumptions to enable any correct predictions at all [6,37,41], and many of those assumptions need to be provided as explicit inputs into safety analysis algorithms for soundness. In particular, a white-box model is required to obtain guarantees even if only an executable model is needed during learning.

Safety for Autonomous CPS requires direct attention to the interplay of learning systems with hybrid system models. Even if the combination of learning algorithms with the CPS dynamics formally are hybrid systems again, they cannot be considered quite as naïvely due to the resulting scale. Without summarizing symbolic abstractions, it would have been completely infeasible, for example, to verify the hybrid systems model of the next-generation Airborne Collision Avoidance System ACAS X [18] defined by interpolation of a Markov Decision Process policy on its half a trillion different discretized state regions. Instead, the computational complexities call for approaches that establish safety from simpler models that do not include full detail on the learning while still benefiting from the flexibility advantages of learning without risking unsafety.

3 Approach

As a foundation for the safe design of autonomous CPS, this approach uses *differential dynamic logic* dL [30,31,33,36] which provides modalities $[\alpha]$ and $\langle\alpha\rangle$ for every hybrid system model α such that the dL formula $[\alpha]\phi$ is true in a state whenever the postcondition ϕ is true after all runs of α (safety) and the dL formula $\langle\alpha\rangle\phi$ is true in a state whenever ϕ is true after at least one run of α (liveness). Besides serving as a flexible specification language, dL also comes with axiomatizations [30,32,33,35] that enable its use for verification purposes.

CPS Modeling is the first step and culminates in a hybrid system α describing all possible behavior of the CPS. For both complexity control reasons and flexibility reasons, it is best *not* to describe completely accurately under which exact circumstance the learning system decides upon which exact control action. Instead, the hybrid system α describes all actions that are possible as well as the continuous dynamics of the system.

Elaborate modeling advice can be found elsewhere [36,38], but nondeterminism is frequently used for this purpose. For example, a hybrid systems model

$$((\beta \cup \gamma); x' = f(x))^* \tag{1}$$

expresses that the CPS can nondeterministically choose (by operator $\cup$) to either run control action β or control action γ and will then (after the ; operator) follow the continuous dynamics of the differential equation $x' = f(x)$ for a certain period of time, before repeating (by operator * for repetition) the sequence of discrete and continuous actions any number of times. For example, β could be the action of accelerating while γ could be braking (additional actions such as turning left add more $\cup$ operators, accordingly). A model of this shape is fairly noncommittal, because its use of nondeterminism in action choices, differential equation durations, and repetition counts deliberately leaves open how exactly it is run, giving the learning CPS a lot of flexibility in filling in these choices at its leisure later without requiring any change in the model.

KeYmaera X: Hybrid Systems Model Safety can be established by proving in the tool KeYmaera X [12] a dL safety property of the form

$$\phi \to [\alpha]\psi \tag{2}$$

which, if proved, implies that, if the system starts in any initial state satisfying formula ϕ, then all states reached after all runs of the hybrid systems model α satisfy formula ψ. Formal proofs of dL formulas such as (2) are highly trustworthy, not just because of the clever design of KeYmaera X that reduces its soundness-critical core to less than 2000 lines of code [12] but also because of the cross-verification of the soundness of dL in both Isabelle/HOL and Coq [3].

A formal proof of (2) justifies that all behavior of α satisfies the safety property. The most valuable takeaway lesson besides the formal proof itself are the additional requirements inevitably found during the proof, which characterize when it is even safe to use the various control actions in the model α. For example, in the initial model (1), actions β and γ were unconstrained, but it may not always be safe to accelerate without first checking a condition C that, e.g., relates the obstacle distance to the present velocities and braking capabilities:

$$(((?C; \beta) \cup \gamma); x' = f(x))^* \tag{3}$$

This refined hybrid system (3) includes an additional test (written ?C) that needs to pass since C holds true before running action β. If C is true in the present state, then both β and γ can be run by a nondeterministic choice ($\cup$), otherwise only γ is available, because the condition $?C$ would fail. *All* such additional constraints that are required for safety will be discovered during the proof of (2), because a sound proof could not otherwise succeed, and dL is sound [30,32,35].

ModelPlex: Model Safety Transfer provides the correctness bridge between a verified hybrid systems model and its implementation by synthesizing correct-by-construction runtime monitors. A dL proof of formula (2) in KeYmaera X is a great achievement, but, due to its (desirable) modeling simplifications, does not provide an answer for the full complexities of a learning CPS. Usually, there is a discrepancy between the implementation detail of the autonomous CPS and the simplified descriptions that were chosen to be included in the verified model. Fortunately, the ModelPlex procedure [24] can overcome such discrepancies. Given a verified dL model, ModelPlex synthesizes a monitor along with a dL correctness proof for it, saying that the real implementation is safe as long as it satisfies that runtime monitor (and will always remain safe when continuing the model).

The same relationship between verified model and runtime monitor also is the cornerstone to safeguard the decisions of learning agents [14], which is crucial to obtain safety after deployment unless ModelPlex has already been used during learning to guide the learning process toward safe answers (which speeds up convergence). The logical monitor conditions obtained from a ModelPlex proof can be directly exploited as a safety signal for learning. Since it is challenging to implement learning algorithms in a provably correct way, the continued use of ModelPlex monitors after deployment is advisable even if ModelPlex monitor outputs were used to steer learning toward safe answers during training.

VeriPhy: Executable Proof Transfer synthesizes executable machine code binaries (e.g., for x64 or ARM) that inherit the safety theorems such as (2) by a chain of formal proofs in theorem provers [4]. The resulting executables are not just formally verified to be safe for the CPS, but also accept control input from unverified controllers that will be checked against ModelPlex monitors for safety before execution and are vetoed otherwise. This input decides how to resolve the nondeterminism in the hybrid systems model, e.g., whether to run β or γ in (3). But the verified controller sandbox generated by VeriPhy only accepts β if the condition C was true that was required for the safety proof. While the need to test C when deciding on β was evident from the way model (3) was written, other conditions are more difficult to read off, and the key is to find them all and then prove safety of the control sandbox, which VeriPhy does automatically.

Safe Learning in CPSs is made possible by the combination of a hybrid systems model verified in KeYmaera X [12], whose safety-critical monitor conditions were extracted along with a proof of correctness by ModelPlex [24], and whose verified controller sandbox was synthesized along with a chain of correctness proofs by VeriPhy [4]. This combination enables any reinforcement learning algorithm to be run as a black box [14]. The VeriPhy output provides a verified CPS sandbox within which the reinforcement learning can experiment safely. The reinforcement learning algorithm can focus on identifying the most optimal decisions, which is usually replaced by nondeterminism in verification for the sake of simplicity. Convergence of the learning algorithm is improved, because the ModelPlex monitors give immediate feedback about which individual action might cause an unsafe future in which state. This is faster than having to wait until an entire sequence of actions has been chosen that, say, lead to a collision, and then facing the nontrivial task of retroactively identifying to what extent which action contributed to this collision and back-propagate generalizable knowledge.

If the physical behavior was modeled adequately, then this approach leads to a provably safe policy [14]. Otherwise, quantitative ModelPlex, which gives a real-valued (instead of boolean-valued) degree of compliance, is experimentally shown to guide the optimization of reinforcement learning (RL) off model to a graceful recovery using the ability of boolean ModelPlex to reliably spot when the real behavior is outside the verified model. The question is what could then prove safety regardless, not just observe recovery. Clearly, if all model assumptions are completely wrong, then no amount of analysis will make the system safe but magic is needed instead. Yet, if there merely is uncertainty about which one of a whole pile of models is the right one, yet they are not all wrong, then not only is safety preserved, but learning can also optimize the system by actively experimenting to find out which model accurately reflects the present reality [15]. Conjunctions of the ModelPlex monitors for all plausible models keep the learning AI safe. Solving for distinct monitor predictions makes it possible to plan differentiating experiments to converge a.s. to the true model, if possible. When the verified models are given together with a tactic that proves them, then safety proofs can be reified, such that both the model and its safety proof can be adapted to better fit observations with *verification-preserving model updates.*

4   Summary and Outlook

Overall, logic leads the way in combining the best of both worlds: the strong predictions that formal verification techniques provide for CPS alongside the strong flexibility that the use of AI provides. Table 1 summarizes the logical technologies that enable the respective combinations, their uses in CPS design, and their corresponding counterpart in AI. Each element of the safety transition stack

Table 1. Logical triumvirate of technologies for transitioning trustworthiness to autonomous cyber-physical systems

dL proof technology	RL learning technology
KeYmaera X: identify safe actions in CPS model	RL optimizes action choice
ModelPlex: safe model to safe implementation	Safe reward signal for RL
VeriPhy: monitored sandbox to safe executables	CPS sandbox for RL

fulfills a different purpose and integrates the benefits of learning and proving in different ways. They all share differential dynamic logic dL as a common logical foundation and reinforcement learning RL as a learning foundation.

While logic paints a particularly clear picture of how to safely navigate the path to autonomous CPSs, and while its efficacy has been demonstrated throughout on small scale [5], numerous interesting challenges remain that go beyond the ones of interest already for ordinary CPS [34]. The guarantees, even in the presence of learning, are strong on the controls side of CPS. The safety-relevant control error is provably reduced to zero thanks to the logical safety stack, but only under the assumption of bounded deviations in sensing [24].

The picture is not so rosy on the sensing side of CPS. And I argue that this is not a coincidence. Of course, no amount of reasoning can bypass the sensory illusions of the Cartesian Demon that fooled all but René Descartes' existence of thoughts [7]. If literally *all* sensors and actuators of a CPS could be arbitrarily wrong, then no connection can be made between the suspected and real state of the system. But even if sensors are almost always a little wrong, they are not usually all that wrong, which enables a logical angle of attack [21,23,25] for guarantees despite bounded sensor errors. Now, how can concrete bounds be substantiated for sensor errors with as little doubt as possible? An answer to this question is the true challenge beyond recent progress in verified perception [10,29].

References

1. Alshiekh, M., Bloem, R., Ehlers, R., Könighofer, B., Niekum, S., Topcu, U.: Safe reinforcement learning via shielding. In: McIlraith, Weinberger [22]
2. Alur, R.: Formal verification of hybrid systems. In: Chakraborty, S., Jerraya, A., Baruah, S.K., Fischmeister, S. (eds.) EMSOFT, pp. 273–278. ACM, New York (2011). https://doi.org/10.1145/2038642.2038685
3. Bohrer, B., Rahli, V., Vukotic, I., Völp, M., Platzer, A.: Formally verified differential dynamic logic. In: Bertot, Y., Vafeiadis, V. (eds.) Certified Programs and Proofs - 6th ACM SIGPLAN Conference, CPP 2017, Paris, France, 16–17 January 2017, pp. 208–221. ACM, New York (2017). https://doi.org/10.1145/3018610.3018616
4. Bohrer, B., Tan, Y.K., Mitsch, S., Myreen, M.O., Platzer, A.: VeriPhy: verified controller executables from verified cyber-physical system models. In: Grossman, D. (ed.) Proceedings of the 39th ACM SIGPLAN Conference on Programming Language Design and Implementation, PLDI 2018, pp. 617–630. ACM (2018). https://doi.org/10.1145/3192366.3192406
5. Bohrer, B., Tan, Y.K., Mitsch, S., Sogokon, A., Platzer, A.: A formal safety net for waypoint following in ground robots. IEEE Robot. Autom. Lett. **4**(3), 2910–2917 (2019). https://doi.org/10.1109/LRA.2019.2923099
6. Collins, P.: Optimal semicomputable approximations to reachable and invariant sets. Theory Comput. Syst. **41**(1), 33–48 (2007). https://doi.org/10.1007/s00224-006-1338-3
7. Descartes, R.: Meditationes de prima philosophia, in qua Dei existentia et animae immortalitas demonstratur (1641)

8. Doyen, L., Frehse, G., Pappas, G.J., Platzer, A.: Verification of hybrid systems. Handbook of Model Checking, pp. 1047–1110. Springer, Cham (2018). https://doi.org/10.1007/978-3-319-10575-8_30

9. Dreossi, T., Donzé, A., Seshia, S.A.: Compositional falsification of cyber-physical systems with machine learning components. In: Barrett, C., Davies, M., Kahsai, T. (eds.) NFM 2017. LNCS, vol. 10227, pp. 357–372. Springer, Cham (2017). https://doi.org/10.1007/978-3-319-57288-8_26

10. Dvijotham, K., et al.: Training verified learners with learned verifiers. CoRR abs/1805.10265 (2018)

11. Frehse, G., et al.: SpaceEx: scalable verification of hybrid systems. In: Gopalakrishnan, G., Qadeer, S. (eds.) CAV 2011. LNCS, vol. 6806, pp. 379–395. Springer, Heidelberg (2011). https://doi.org/10.1007/978-3-642-22110-1_30

12. Fulton, N., Mitsch, S., Quesel, J.-D., Völp, M., Platzer, A.: KeYmaera X: an axiomatic tactical theorem prover for hybrid systems. In: Felty, A.P., Middeldorp, A. (eds.) CADE 2015. LNCS (LNAI), vol. 9195, pp. 527–538. Springer, Cham (2015). https://doi.org/10.1007/978-3-319-21401-6_36

13. Fulton, N., Platzer, A.: Safe AI for CPS. In: IEEE International Test Conference, ITC 2018, Phoenix, AZ, USA, October 29–November 1 2018, pp. 1–7. IEEE (2018). https://doi.org/10.1109/TEST.2018.8624774

14. Fulton, N., Platzer, A.: Safe reinforcement learning via formal methods: toward safe control through proof and learning. In: McIlraith, Weinberger [22], pp. 6485–6492. https://www.aaai.org/ocs/index.php/AAAI/AAAI18/paper/view/17376

15. Fulton, N., Platzer, A.: Verifiably safe off-model reinforcement learning. In: Vojnar, T., Zhang, L. (eds.) TACAS 2019. LNCS, vol. 11427, pp. 413–430. Springer, Cham (2019). https://doi.org/10.1007/978-3-030-17462-0_28

16. Gillula, J.H., Tomlin, C.J.: Guaranteed safe online learning via reachability: tracking a ground target using a quadrotor. In: IEEE International Conference on Robotics and Automation, ICRA 2012, St. Paul, Minnesota, USA, 14–18 May 2012, pp. 2723–2730. IEEE (2012). https://doi.org/10.1109/ICRA.2012.6225136

17. Henzinger, T.A., Sifakis, J.: The discipline of embedded systems design. Computer 40(10), 32–40 (2007). https://doi.org/10.1109/MC.2007.364

18. Jeannin, J., et al.: A formally verified hybrid system for safe advisories in the next-generation airborne collision avoidance system. STTT 19(6), 717–741 (2017). https://doi.org/10.1007/s10009-016-0434-1

19. Larsen, K.G.: Verification and performance analysis for embedded systems. In: Chin, W., Qin, S. (eds.) TASE 2009, Third IEEE International Symposium on Theoretical Aspects of Software Engineering, Tianjin, China, 29–31 July 2009, pp. 3–4. IEEE Computer Society (2009). https://doi.org/10.1109/TASE.2009.66

20. 2012 27th Annual IEEE Symposium on Logic in Computer Science (LICS). IEEE, Los Alamitos (2012)

21. Martins, J., Platzer, A., Leite, J.: Dynamic doxastic differential dynamic logic for belief-aware cyber-physical systems. In: Cerrito, S., Popescu, A. (eds.) TABLEAUX. LNCS, vol. 11714. Springer, Cham (2019). https://doi.org/10.1007/978-3-030-29026-9_24

22. McIlraith, S.A., Weinberger, K.Q. (eds.): Proceedings of the Thirty-Second AAAI Conference on Artificial Intelligence, New Orleans, Louisiana, USA, 2–7 February 2018. AAAI Press (2018)

23. Mitsch, S., Ghorbal, K., Vogelbacher, D., Platzer, A.: Formal verification of obstacle avoidance and navigation of ground robots. Int. J. Robot. Res. 36(12), 1312–1340 (2017). https://doi.org/10.1177/0278364917733549

24. Mitsch, S., Platzer, A.: ModelPlex: verified runtime validation of verified cyber-physical system models. Form. Methods Syst. Des. **49**(1–2), 33–74 (2016). https://doi.org/10.1007/s10703-016-0241-z
25. Mitsch, S., Platzer, A.: Verified runtime validation for partially observable hybrid systems. CoRR abs/1811.06502 (2018). http://arxiv.org/abs/1811.06502
26. Nerode, A.: Logic and control. In: Cooper, S.B., Löwe, B., Sorbi, A. (eds.) CiE 2007. LNCS, vol. 4497, pp. 585–597. Springer, Heidelberg (2007). https://doi.org/10.1007/978-3-540-73001-9_61
27. Paden, B., Cáp, M., Yong, S.Z., Yershov, D.S., Frazzoli, E.: A survey of motion planning and control techniques for self-driving urban vehicles. IEEE Trans. Intell. Veh. **1**(1), 33–55 (2016). https://doi.org/10.1109/TIV.2016.2578706
28. Pappas, G.J.: Wireless control networks: modeling, synthesis, robustness, security. In: Caccamo, M., Frazzoli, E., Grosu, R. (eds.) HSCC, pp. 1–2. ACM, New York (2011). https://doi.org/10.1145/1967701.1967703
29. Pei, K., Cao, Y., Yang, J., Jana, S.: Towards practical verification of machine learning: the case of computer vision systems. CoRR abs/1712.01785 (2017)
30. Platzer, A.: Differential dynamic logic for hybrid systems. J. Autom. Reasoning **41**(2), 143–189 (2008). https://doi.org/10.1007/s10817-008-9103-8
31. Platzer, A.: Logical Analysis of Hybrid Systems: Proving Theorems for Complex Dynamics, vol. 1. Springer, Heidelberg (2010). https://doi.org/10.1007/978-3-642-14509-4
32. Platzer, A.: The complete proof theory of hybrid systems. In: LICS [20], pp. 541–550. https://doi.org/10.1109/LICS.2012.64
33. Platzer, A.: Logics of dynamical systems. In: LICS [20], pp. 13–24. https://doi.org/10.1109/LICS.2012.13
34. Platzer, A.: Logic & proofs for cyber-physical systems. In: Olivetti, N., Tiwari, A. (eds.) IJCAR. LNCS, vol. 9706, pp. 15–21. Springer, Cham (2016). https://doi.org/10.1007/978-3-319-40229-1_3
35. Platzer, A.: A complete uniform substitution calculus for dierential dynamic logic. J. Autom. Reasoning **59**(2), 219–265 (2017). https://doi.org/10.1007/s10817-016-9385-1
36. Platzer, A.: Logical Foundations of Cyber-Physical Systems. Springer, Cham (2018). https://doi.org/10.1007/978-3-319-63588-0
37. Platzer, A., Clarke, E.M.: The image computation problem in hybrid systems model checking. In: Bemporad, A., Bicchi, A., Buttazzo, G. (eds.) HSCC 2007. LNCS, vol. 4416, pp. 473–486. Springer, Heidelberg (2007). https://doi.org/10.1007/978-3-540-71493-4_37
38. Quesel, J.D., Mitsch, S., Loos, S., Aréchiga, N., Platzer, A.: How to model and prove hybrid systems with KeYmaera: a tutorial on safety. STTT **18**(1), 67–91 (2016). https://doi.org/10.1007/s10009-015-0367-0
39. Sutton, R.S., Barto, A.G.: Reinforcement Learning. The MIT Press, Cambridge (1998)
40. Tiwari, A.: Logic in software, dynamical and biological systems. In: LICS, pp. 9–10. IEEE Computer Society (2011). https://doi.org/10.1109/LICS.2011.20
41. Zuliani, P., Platzer, A., Clarke, E.M.: Bayesian statistical model checking with application to simulink/stateflow verification. Form. Methods Syst. Des. **43**(2), 338–367 (2013). https://doi.org/10.1007/s10703-013-0195-3

Probabilistic Verification

Model Checking Constrained Markov Reward Models with Uncertainties

Giovanni Bacci$^{(\boxtimes)}$, Mikkel Hansen, and Kim Guldstrand Larsen

Department of Computer Science, Aalborg University, Aalborg, Denmark
{giovbacci,mhan,kgl}@cs.aau.dk

Abstract. We study the problem of analysing Markov reward models (MRMs) in the presence of imprecise or uncertain rewards. Properties of interests for their analysis are (i) probabilistic bisimilarity, and (ii) specifications expressed as probabilistic reward CTL formulae.

We consider two extensions of the notion of MRM, namely (a) *constrained Markov reward models*, i.e., MRMs with rewards parametric on a set variables subject to some constraints, and (b) *stochastic Markov reward models*, i.e., MRMs with rewards modelled as real-valued random variables as opposed to precise rewards. Our approach is based on quantifier elimination for *linear* real arithmetic. Differently from existing solutions for parametric Markov chains, we avoid the manipulation of rational functions in favour of a symbolic representation of the set or parameter valuations satisfying a given property in the form of a quantifier-free first-order formula in the linear theory of the reals.

Our work finds applications in model repair, where parameters need to be tuned so as to satisfy the desired specification, as well as in robustness analysis in the presence of stochastic perturbations.

Keywords: Parameter synthesis · Markov chains · Markov reward models · Model checking

1 Introduction

The wide-spread diffusion of cyber-physical systems (CPSs) poses the challenge of handling their growing complexity, while meeting requirements on correctness, predictability, performance without compromising time- and cost-to-market. Their analysis requires one to address a number of non-functional properties related to the quantitative aspects that are typical of such systems.

Finite-state Markovian models are popular modelling formalisms for the quantitative analysis of systems with probabilistic behaviours. Among these, Markov reward models (MRMs) were proposed as a natural extension of the usual notion of discrete-time Markov chain with (real-valued) state rewards.

Interesting properties of MRMs may be expressed by means of quantitative extensions of CTL, such as Probabilistic Reward CTL (PRCTL), *cf.* e.g., [1].

Work supported by the Advanced ERC Grant nr. 867096 (LASSO) and the Innovation Fund Denmark Center DiCyPS (nr. 864701).

D. Parker and V. Wolf (Eds.): QEST 2019, LNCS 11785, pp. 37–51, 2019.
https://doi.org/10.1007/978-3-030-30281-8_3

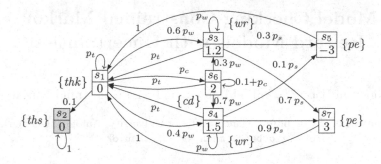

Fig. 1. Model of a Ph.D. student in CS.

As a running example, consider the MRM in Fig. 1 modelling the stress level of a Ph.D. student in computer science. It has seven states, namely $s_1, \ldots, s_7$, annotated with the propositions *thk* (thinking), *ths* (thesis), *wr* (writing paper), *tc* (tool coding), and *pe* (paper evaluation). The level of stress of the student is modelled by associating with each state a reward that represents the stress that the student accumulates in the state at each time unit. After spending some time thinking, the student starts developing a tool with probability $p_c = 0.2$, or writing a paper with probability $p_w = 0.5$ (40% of the times for a journal and 60% for a conference), otherwise the student submits the thesis with probability 0.1. Once the tool is mature enough, the student starts writing a paper about it with probability p_w (70% of which for a journal, otherwise for a conference). When a paper has been completed it is submitted for evaluation with probability $p_s = 0.3$. The paper may be rejected or accepted (moving resp. to s_7 or s_5) according to some acceptance rate. At any moment before the thesis is completed, the student may move back to s_1 with probability $p_t = 0.2$ to think on how to proceed with his/her Ph.D.

We can use PRCTL formulae for specifying properties such as "the average level of stress accumulated until the thesis is submitted doesn't exceed 10", "the student completes the thesis without passing through a state with stress level higher than 2 with probability greater than or equal to 0.9", or "the probability of eventually start coding and, subsequently, submitting the thesis with expected accumulated stress within $[x, y]$ is less than or equal to z".

In the above MRM we fixed specific reward values. This is an unrealistic simplification since the level of stress at each state may vary depending on different factors, possibly external to the student. The same argument applies to CPSs, that typically rely on sensor measurements which are inherently imprecise, but also to other systems, e.g., resource-management protocols that depends on stochastic assumptions on the future workload.

Typically, there are two ways for dealing with uncertain measurements: (i) determine the precision of the instrument and associate an error ϵ with each measurement, or (ii) perform estimation statistics (e.g., by recursive Bayesian estimation [16]) and use random variables to model each measurement.

In this paper we address the problem of analysing MRMs in presence of imprecise rewards. We present two extensions of the notion of MRM: (i) *constrained Markov reward models* (CMRMs), i.e., MRMs with rewards and transition probabilities parametric on a set variables subject to some constraints expressed as a first-order formula in the linear theory of the reals; (ii) *stochastic Markov reward models* (SMRMs), i.e., MRMs with rewards modelled as real-valued random variables as opposed to exact rewards.

Intuitively, a CMRM models a family of MRMs arising by plugging in concrete values for the parameters provided that they satisfy its constraints. Analogously, an SMRM models a probability distribution over a family of MRMs.

We are interested in the analysis of these models with respect to (i) probabilistic bisimilarity, and (ii) specifications expressed as PRCTL formulae.

On the one hand, the analysis of CMRMs is done by inferring constraints over its parameters characterising the valuations satisfying the property then, verify the robustness of the model within the given precision. On the other hand, analysing an SMRM consists in measuring the likelihood that an MRM obtained by plugging in real-valued outcomes of its random variables satisfies a given property.

Our Contribution: In relation with the analysis of CMRMs our contribution is twofold. First, we show that the computation of the set of parameter valuations ensuring that some states are probabilistic bisimilar with each other can be done using quantifier elimination in the linear fragment of the theory of the reals. Secondly, we demonstrate that for CMRMs having parameters only on state rewards, also the PRCTL model checking problem can be solved within the linear fragment of the theory of the reals. This allows one to employ SMT solvers and quantifier elimination procedures specialised on linear constraints [14], avoiding to use more inefficient and less scalable solvers for bilinear inequalities [17] as suggested e.g., in the case of interval Markov chains [15].

As for the analysis of SMRMs, we describe how one can apply the results for CMRMs described above for estimating the probability of satisfying a given property by employing Mote Carlo simulation techniques.

Related Work: The analysis of CMRMs falls into the research area of parameter synthesis for parametric Markov chains (pMCs) [4,6–10,12] and Interval Markov chains (IMC) [2,3,15]. Most of these works [4,5,7,8,10] consider pMCs without state rewards and focus mainly on computing closed form solutions for reachability probabilities in pMCs as rational functions. The same approach was recently extended to the computation of closed form solutions for expected rewards and long-run average rewards [6,9].

In the context of model checking pMCs and IMCs, quantifier elimination for the first-order theory of the reals has been mainly applied as a theoretical tool for proving complexity upper-bounds of the model checking problem for IMCs [2,3,15] and pMCs [9]. Notably, the model checker PROPhESY [5] relies on SMT solving via the existential theory of the reals to determine approximate covering of the parameter space.

2   Preliminaries and Notation

We denote by $\mathbb{R}$, $\mathbb{Q}$, and $\mathbb{N}$ respectively the sets of real, rational, and natural numbers. We denote by Σ^n, Σ^* and, Σ^ω respectively the set of words of length $n \in \mathbb{N}$, finite length, and infinite length, built over the finite alphabet Σ.

The dot product of two vectors $\mathbf{a} = (a_1, \ldots, a_n)$ and $\mathbf{b} = (b_1, \ldots, b_n)$ in $\mathbb{R}^n$ is defined as $\mathbf{a} \cdot \mathbf{b} = \sum_{i=1}^n a_i b_i$. For $c \in \mathbb{R}$, we use $\mathbf{c} \in \mathbb{R}^n$ to denote the constant vector $(c, \ldots, c)$. For a function $f : I \to \mathbb{R}$ with finite domain with cardinality $|I| = n$ we may use f to denote also the I-indexed vector $(f(i))_{i \in I} = (f(i_1), \ldots, f(i_n))$.

Measure Theory. Let Ω be a set. A family $\Sigma \subseteq 2^\Omega$ is called a *σ-algebra* if it contains the empty set and is closed under complement and countable unions, in this case (Ω, Σ) is called a *measurable space* and elements of Σ *measurable sets*. When $\Sigma = 2^\Omega$, the measure space (Ω, Σ) is *discrete*.

A *measure* on (Ω, Σ) is a σ-additive function $\mu : \Sigma \to \mathbb{R}$, i.e. a map satisfying $\mu(\bigcup_{i \in I} E_i) = \sum_{i \in I} \mu(E_i)$ for any countable family of pairwise disjoint measurable sets $(E_i)_{i \in I}$, in this case (Ω, Σ, μ) is called a *measure space*. If additionally μ satisfies $\mu(\Omega) = 1$, it is called a *probability measure* and (Ω, Σ, μ) a *probability space*. We denote by $\mathcal{D}(\Omega)$ the set of discrete probability distributions on Ω. For $x \in \Omega$, the *Dirac distribution* concentrated at x is the distribution $1_x \in \mathcal{D}(\Omega)$ defined, for arbitrary $y \in \Omega$, as $1_x(y) = 1$ if $x = y$, 0 otherwise.

For measurable spaces (Ω, Σ) and (Γ, Θ), a map $f : \Omega \to \Gamma$ is *measurable* if for all $E \in \Theta$, $f^{-1}(E) = \{x \mid f(x) \in E\} \subseteq \Sigma$. Given a measurable map $f : \Omega \to \Gamma$ and a measure μ on (Ω, Σ) we define the measure $\mu[f]$ on (Γ, Θ), called the *push forward of μ under f*, as $\mu[f](E) = \mu(f^{-1}(E))$ for $E \in \Theta$.

A real-valued *random variable* $X : \Omega \to \mathbb{R}$ is a measurable function from a probability space (Ω, Σ, P) to the Borel space $(\mathbb{R}, \mathcal{B}(\mathbb{R}))$. Intuitively, X can be seen as the outcome of some experiment, e.g. measuring some sensor value. Given a "test" $A \in \mathcal{B}(\mathbb{R})$, we write $P[X \in A]$ for the probability that X has a value in A, i.e. $P[X \in A] = P[X](A)$. A random variable is associated with its *cumulative distribution function* (CDF) $F_X : \mathbb{R} \to [0, 1]$ defined as $F_X(x) = P[X \in (-\infty, x]]$; and a *probability density function* (PDF) f_X, a non-negative Lebesgue-integrable function satisfying $P[X \in [a, b]] = \int_a^b f_X(x) \mathrm{d}x$. The *expected value* of X, written $E[X]$, is intuitively understood as the long-run average of repetitions of the experiment X, formalised by the Lebesgue integral $\int_\Omega X \mathrm{d}P$ (corresponding to $\int_\mathbb{R} f_X(x) \mathrm{d}x$ when X admits density function f_X).

3   Markov Reward Models

In this section we recall the definitions of Markov reward models (MRMs), *probabilistic bisimulation*, and *probabilistic reward CTL* (PRCTL).

In what follows we fix a finite set of atomic propositions $\mathcal{AP}$.

Definition 1. *A Markov chain is a tuple $\mathcal{M} = (S, \tau, \ell)$ consisting of a finite nonempty set of states S, a transition probability function $\tau : S \to \mathcal{D}(S)$, and a labelling function $\ell : S \to 2^{\mathcal{AP}}$ mapping states to atomic propositions.*

Intuitively, if $\mathcal{M}$ is in state s it moves to state s' with probability $\tau(s)(s')$. In this sense, $\mathcal{M}$ can be thought as a state-machine that generates paths in S^ω. A *path* is an infinite sequence of states $\pi = s_1 s_2 s_3 \cdots \in S^\omega$; for $i \geq 1$ we denote by $\pi[i]$ the i-th state of π, i.e., $\pi[i] = s_i$, and $\pi|^i$ the prefix of length i of π, i.e., $\pi|^i = s_1 \cdots s_i$ and $\pi|^0 = \varepsilon$.

We denote by $G_\mathcal{M} = (S, \rightarrow)$ the underlying graph of $\mathcal{M}$, where $s, s' \in S$ are connected by a directed edge, written $s \rightarrow s'$, if and only if $\tau(s)(s') > 0$. We indicate by $\rightarrow^*$, the transitive and reflexive closure of $\rightarrow$.

In order to associate probabilities to events, we adopt the classical cylinder set construction (*cf.* [1, Ch10]). For $w \in S^*$, the cylinder set of w is the set of all paths having prefix w, i.e., $cyl(w) = wS^\omega$. Given an *initial probability distribution* $\iota \in \mathcal{D}(S)$, we define the probability space $(S^\omega, \Sigma_\mathcal{M}, Pr_\iota^\mathcal{M})$, where $\Sigma_\mathcal{M} = \sigma(\{cyl(w) \mid w \in S^*\})$ is the smallest σ-algebra that contains all the cylinder sets, and $Pr_\iota^\mathcal{M}$ is the unique probability measure such that, for all $w = s_1 \cdots s_n \in S^*$, $Pr_\iota^\mathcal{M}(cyl(w)) = \iota(s_1) \cdot \prod_{0<i<n} \tau(s_i)(s_{i+1})$.

When $\iota = 1_s$ we write $Pr_s^\mathcal{M}$, or just Pr_s when $\mathcal{M}$ is clear from the context.

Definition 2. *A Markov reward model is a tuple* $\mathcal{R} = (S, \tau, \ell, \rho)$ *where* (S, τ, ℓ) *is a MC, and* $\rho : S \rightarrow \mathbb{R}$ *is the* reward function *assigning a reward to each state.*

A Markov reward model generates paths in S^ω according to its underlying Markov chain; in addition, whenever a transition is performed, say from s to s', the system is rewarded by $\rho(s)$. It is worth noting that the reward is given after leaving the current state.

In the following, it may be convenient to represent the reward function ρ and the probability transition distributions $\tau(s)$ (for $s \in S$) as S-indexed vectors.

Example 1. Consider the MRM depicted in Fig. 1. Its reward function ρ and the distribution $\tau(s_1)$ are respectively represented in vector notation as

$$\rho = (0, 0, 1.2, 1.5, -3, 2, 3), \quad \text{and} \quad \tau(s_1) = (p_t, 0.1, 0.6 p_w, 0.4 p_w, 0, p_c, 0).$$

The next definition extends the classic notion of probabilistic bisimulation for Markov chains by Larsen and Skou [13] to the case of MRMs.

Definition 3 (Bisimulation). *Let* $\mathcal{R} = (S, \tau, \ell, \rho)$ *be an MRM. An equivalence relation* $R \subseteq S \times S$ *is a* probabilistic bisimulation *if whenever* $(s, s') \in R$ *then,* $\ell(s) = \ell(s')$, $\rho(s) = \rho(s')$, *and for all* R-*equivalence class* $C \in S/_R$, $\tau(s)(C) = \tau(s')(C)$.

Two states $s, s' \in S$ *are* probabilistic bisimilar, *written* $s \sim_\mathcal{R} s'$, *if they are related by some probabilistic bisimulation.*

Example 2. Consider the MRM depicted in Fig. 1. As it is presented, none of its states are bisimilar with each other because each pair of states differs on the rewards or the labels. If we consider instead an MRM with underlying chain as in Fig. 1 and rewards $\rho = (0, 0, 1, 1, 3, 2, 3)$, now we have that $s_5 \sim s_7$ and, consequently, $s_3 \sim s_4$ because $\tau(s_3)(\{s_5, s_7\}) = p_s = \tau(s_4)(\{s_5, s_7\})$. □

Before presenting probabilistic reward CTL, we need to introduce the concept of expected cumulative reward for reaching a set of states $B \subseteq S$.

Let $\hat{\rho}_B \colon S^\omega \to \mathbb{R}$ be the random variable representing the reward accumulated along a prefix of the path belonging to $(S \setminus B)^* B$. This is formalised as $\hat{\rho}_B(\pi) = \sum_{i=1}^{n-1} \rho(\pi[i])$ if $\pi|^n \in (S \setminus B)^* B$ for some $n \in \mathbb{N}$, otherwise 0.

We denote by $E[\hat{\rho}_B|s]$ the expected value of $\hat{\rho}_B$ with respect to the probability distribution Pr_s. Following [1, Ch10], $E[\hat{\rho}_B|s]$ can be computed as the (unique) solution r_s of the following system of linear equations

$$
r_s = \begin{cases} 0 & \text{if } s \in B \text{ or } s \not\rightarrow^* B \\ \rho(s) + \left(\sum_{t \in S} \tau(s)(t) \cdot r_t \right) & \text{otherwise.} \end{cases}
\tag{1}
$$

We are now ready to present Probabilistic Reward CTL. PRCTL allows for *state formulae* describing properties about states in a MRM, and *path formulae* describing properties about paths in a MRM. State formulae Φ, Ψ and path formulae φ are constructed over the following abstract syntax:

$$\Phi, \Psi ::= t\!t \mid a \mid \rho \bowtie r \mid \neg \Phi \mid \Phi \wedge \Psi \mid \mathbb{P}_J(\varphi) \mid \mathbb{E}_R(\Phi), \quad \text{(STATE FORMULAE)}$$

$$\varphi ::= \mathcal{X}\Phi \mid \Phi \mathcal{U} \Psi \mid \Phi \mathcal{U}_R^n \Psi \quad \text{(PATH FORMULAE)}$$

where $a \in \mathcal{AP}$, $r \in \mathbb{Q}$, $n \in \mathbb{N}$, $\bowtie \in \{=, <, >\}$, and $J \subseteq [0,1]$ and $R \subseteq \mathbb{R}$ are intervals with rational bounds.

Given a MRM $\mathcal{R} = (S, \tau, \ell, \rho)$, a state $s \in S$, and a path $\pi \in S^\omega$, we denote by $\mathcal{R}, s \models \Phi$ (resp. $\mathcal{R}, \pi \models \varphi$) the fact that the state s satisfies the state formula Φ (resp. the path π satisfies the path formula φ). $Sat(\Phi)$ denotes the set of all states satisfying the property Φ, i.e. $Sat(\Phi) = \{s \in S \mid \mathcal{R}, s \models \Phi\}$. Formally, the *satisfiability relation* $\models$ is inductively defined as:

$$\mathcal{R}, s \models t\!t \qquad \text{always holds}$$

$$\mathcal{R}, s \models a \qquad \text{iff} \qquad a \in \ell(s)$$

$$\mathcal{R}, s \models \rho \bowtie r \qquad \text{iff} \qquad \rho(s) \bowtie r$$

$$\mathcal{R}, s \models \neg \Phi \qquad \text{iff} \qquad \mathcal{R}, s \not\models \Phi$$

$$\mathcal{R}, s \models \Phi \wedge \Psi \qquad \text{iff} \qquad \mathcal{R}, s \models \Phi \text{ and } \mathcal{R}, s \models \Psi$$

$$\mathcal{R}, s \models \mathbb{P}_J(\varphi) \qquad \text{iff} \qquad Pr_s(\{\pi \in S^\omega \mid \mathcal{R}, \pi \models \varphi\}) \in J$$

$$\mathcal{R}, s \models \mathbb{E}_R(\Phi) \qquad \text{iff} \qquad E[\hat{\rho}_{Sat(\Phi)}|s] \in R$$

$$\mathcal{R}, \pi \models \mathcal{X}\Phi \qquad \text{iff} \qquad \mathcal{R}, \pi[2] \models \Phi$$

$$\mathcal{R}, \pi \models \Phi \mathcal{U} \Psi \qquad \text{iff} \qquad \text{there exists } j \geq 1 \text{ such that } \pi[j] \models \Psi, \text{and}$$
$$\mathcal{R}, \pi[j'] \models \Phi \text{ for all } 1 \leq j' < j$$

$$\mathcal{R}, \pi \models \Phi \mathcal{U}_R^n \Psi \qquad \text{iff} \qquad \text{there exists } 1 \leq j \leq n \text{ such that } \pi[j] \models \Psi \text{ and}$$
$$\text{for all } 1 \leq k \leq j, \ \sum_{i=1}^{k-1} \rho(\pi[i]) \in R \text{ and,}$$
$$\text{for all } 1 \leq h < j, \ \mathcal{R}, \pi[h] \models \Phi.$$

As usual, we can derive the logical operators $f\!f$, $\lor$, and $\rightarrow$ as follows: $f\!f \stackrel{\text{def}}{=} \neg t\!t$, $\Phi \lor \Psi \stackrel{\text{def}}{=} \neg(\neg\Phi \land \neg\Psi)$, and $\Phi \rightarrow \Psi \stackrel{\text{def}}{=} \neg\Phi \lor \Psi$. Similarly, we can derive the temporal operators $\Diamond$ and $\Box$ as: $\Diamond\Phi \stackrel{\text{def}}{=} t\!t\,\mathcal{U}\,\Phi$ and $\Box\Phi \stackrel{\text{def}}{=} \neg\Diamond(\neg\Phi)$.

Example 3. Consider the MRM $\mathcal{R}$ depicted in Fig. 1. We can verify that "the average level of stress accumulated by the student until the thesis is submitted doesn't exceed 23" by proving that $\mathcal{R}, s_1 \models \mathbb{E}_{\leq 23}(ths)$ holds true. Analogously, we can check that the property "the student completes the thesis without passing through a state with stress level higher than 2 with probability greater than or equal to 0.5" does not hold true by proving $\mathcal{R}, s_1 \models \neg\mathbb{P}_{\geq 0.5}((\rho \leq 2)\,\mathcal{U}\,ths)$. $\Box$

4 Constrained Markov Reward Models

Constrained Markov reward models (CMRMs) model families of MRMs where both transition probabilities and state rewards are parametric on a set of real-valued variables subject to constraints expressed as a first-order formula in the linear theory of the reals.

Let $\mathbf{x} = (x_1, \ldots, x_k)$ be a vector of real-valued parameters. We denote by $\mathcal{E}$ the set of affine maps $f : \mathbb{R}^k \rightarrow \mathbb{R}$ of the form $f(\mathbf{x}) = \mathbf{a} \cdot \mathbf{x} + b$ with $\mathbf{a} = (a_1, \ldots, a_k) \in \mathbb{Q}^k$ and $b \in \mathbb{Q}$, i.e. $f(x_1, \ldots, x_k) = \left(\sum_{i=1}^{k} a_i x_i\right) + b$.

Definition 4. *A constrained Markov reward model is a tuple $\mathcal{F} = (S, \tau, \ell, \rho, F)$ where S and ℓ are defined as for MCs, $\tau : S \rightarrow (S \rightarrow \mathcal{E})$ is a parametric transition function, $\rho : S \rightarrow \mathcal{E}$ is a parametric reward function, and $F(\mathbf{x})$ is a linear first-order formula s.t., for all $s \in S$, $F(\mathbf{x})$ implies $\tau(s) \geq \mathbf{0} \land \mathbf{1} \cdot \tau(s) = 1$.*

Intuitively, a CMRM $\mathcal{F} = (S, \tau, \ell, \rho, F)$ defines a family of MRMs arising by plugging in concrete values for the parameters. A parameter valuation $\mathbf{v} \in \mathbb{R}^k$ is *admissible* (or *feasible*) if $F(\mathbf{v})$ holds true. By abuse of notation, we may write F to indicate the set of admissible valuations. Given $\mathbf{v} \in F$ we denote by $\mathcal{F}(\mathbf{v})$ the MRM associated with $\mathbf{v}$. In this respect, it will be convenient to think of $\mathcal{F}$ as a function $\mathcal{F} : F \rightarrow$ MRM. The semantics of $\mathcal{F}$, written $[\mathcal{F}]$, is defined as the image of $\mathcal{F}$, i.e. $[\mathcal{F}] = \{\mathcal{F}(\mathbf{v}) \mid \mathbf{v} \in F\}$.

Example 4. As already mentioned in the introduction, the stress level of a Ph.D. student may be influenced by several factors. For instance, we can define the following CMRM $\mathcal{F} = (S, \tau, \ell, \bar{\rho}, F)$ having as underlying Markov chain that of Fig. 1, parametric vector of rewards $\rho = (0, 0, h+c, h+j, r, 0, a)$, and constraints $F = (-5 \leq h \leq 0) \land (1 \leq c \leq j \leq 5) \land (-3 \leq a+r \leq 3) \land a \leq r$. Here, the parameter h models the help of the supervisor, whereas c and j represent the stress that accumulates while writing a conference paper c, or a journal paper j. The stress caused by having a paper accepted or rejected in modelled by the parameters a and r respectively.

It is easy to note that the MRM in Fig. 1 is an instance of $\mathcal{F}$. $\Box$

Remark 1. Our notion of CMRM trivially extends that of IMC w.r.t. the so-called uncertain Markov chain interpretation [15]. The condition that transitions probabilities lie within some intervals is trivially expressed as a conjunction of liner inequalities. We shall also recall, that our definition is similar to that of *augmented* pMCs given by Hutschenreiter et al. [9]. Nevertheless, we do not require all admissible valuations to induce the same underlying graph; instead, we require linear constraints as opposed to polynomial constraints.   □

Given a CMRM $\mathcal{F}$ we are interested in finding suitable symbolic representations of the set of parameter valuations that ensure the corresponding MRMs to enjoy some property. Properties of interests for this paper are bisimulation and PRCTL formulae. Specifically, given states $s, s' \in S$, and a PRCTL state formula Φ, we are interested in the set of parameter valuations for which s becomes bisimilar to s' or such that Φ holds at s. These are formalised as follows.

$$[\![s \sim_{\mathcal{F}} s']\!] \stackrel{\text{def}}{=} \left\{ \mathbf{v} \in F \mid s \sim_{\mathcal{F}(\mathbf{v})} s' \right\}, \quad [\![\mathcal{F}, s \models \Phi]\!] \stackrel{\text{def}}{=} \left\{ \mathbf{v} \in F \mid \mathcal{F}(\mathbf{v}), s \models \Phi \right\}.$$

Example 5. Consider the CMRM $\mathcal{F}$ defined in Example 4 and the PCTL formula $\Phi = \mathbb{E}_{[0,10]}(ths)$. The set of feasible parameter valuations satisfying Φ, i.e., $[\![\mathcal{F}, s \models \Phi]\!]$, can be symbolically represented by the following formula

$$\phi = 0 \leq 57a + 575c + 750h + 175j + 168r \leq \frac{1750}{3} \wedge F. \tag{2}$$

The above formula can be used for instance to determine how much effort the supervisor needs to provide to ensure that the formula Φ is satisfied. In this case, this can be done by maximising h under the constraints ϕ.   □

5   Analysing Bisimilarity on Constrained MRMs

In this section we address the problem of finding symbolic representations of the set of parameter valuations ensuring that two given states of a CMRM are probabilistic bisimilar.

Before diving in this problem, it will be convenient to present an alternative characterisation for probabilistic bisimilarity on MRMs. The characterisation is a variant of that of Jonsson and Larsen [11] based on the notion of *coupling*.

Given $\mu, \nu \in \mathcal{D}(X)$, we denote by $\Gamma(\mu, \nu)$ the set of *couplings* for (μ, ν), i.e., probability distributions $\gamma \in \mathcal{D}(X \times X)$ such that, for all $x, y \in X$

$$\sum_{x' \in X} \gamma(x', y) = \nu(y) \quad \text{and} \quad \sum_{y' \in X} \gamma(x, y') = \mu(x). \tag{3}$$

The probability measures μ and ν are respectively called left and right marginals of γ. A coupling can be seen as a redistribution of the "probability mass" from the right marginal to the left and *vice versa*.

Lemma 1. $s \sim s'$, *if and only if there exists a relation* $R \subseteq S \times S$, *such that* $s \, R \, s'$ *and, whenever* $m \, R \, n$ *the following conditions hold*

(i) $\ell(m) = \ell(n)$, $\rho(m) = \rho(n)$, and
(ii) there exists $\gamma \in \Gamma(\tau(m), \tau(n))$ such that $\{(u, v) \in S \times S \mid \gamma(u, v) > 0\} \subseteq R$.

The above lemma allows us to encode $[\![s \sim_{\mathcal{F}} s']\!]$ as the following first-order formula in the linear theory of the reals.

$$(\! | s \sim_{\mathcal{F}} s' | \!) \stackrel{\text{def}}{=} F \wedge \exists \mathbf{b}. \; b_{ss'} = 1 \wedge \bigwedge_{m,n \in S} \beta(m, n) \wedge \gamma(m, n) \wedge \sigma(m, n) \qquad (4)$$

where the sub-formulae β, γ, and σ are defined as follows:

$$\beta(m, n) \stackrel{\text{def}}{=} \bigwedge_{m,n \in S} (b_{mn} = 0 \vee b_{mn} = 1) \qquad (5)$$

$$\gamma(m, n) \stackrel{\text{def}}{=} (b_{mn} = 1) \rightarrow \exists \mathbf{c} \geq \mathbf{0}. \; \left(\begin{array}{l} \bigwedge_{u \in S} \sum_{v \in S} c_{uv} = \tau(m)(u) \wedge \\ \bigwedge_{v \in S} \sum_{u \in S} c_{uv} = \tau(n)(v) \wedge \\ \bigwedge_{u,v \in S} c_{uv} \leq b_{uv} \end{array} \right) \qquad (6)$$

$$\sigma(m, n) \stackrel{\text{def}}{=} (b_{mn} = 1) \rightarrow (\ell(m) = \ell(n) \wedge \rho(m) = \rho(n)). \qquad (7)$$

In the formula $(\! | s \sim_{\mathcal{F}} s' | \!)$, $\mathbf{b} = (b_{mn})_{m,n \in S}$ represents a selection of a relation $R \in S \times S$ by means of binary variables (cf. (5)) such that $b_{m,n} = 1$ iff $(m, n) \in R$. According to Lemma 1, the selection needs to include the pair (s, s') and satisfy the conditions (i) and (ii). This is modelled by imposing $b_{ss'} = 1$, $\sigma(m, n)$ and $\gamma(m, n)$ for all $m, n \in S$. In particular, in (6), the variables $\mathbf{c} = (c_{uv})_{u,v \in S}$ model coupling for $(\tau(m), \tau(n))$ the condition (ii) is enforced by requiring $c_{u,v} \leq b_{u,v}$ for all $u, v \in S$.

By Lemma 1, the elimination of the existential quantifiers in $(\! | s \sim_{\mathcal{F}} s' | \!)$ yields a Boolean formula with linear predicates on $\mathbf{x}$ representing the set of valuations $[\![s \sim_{\mathcal{F}} s']\!]$. Notably, the formula $(\! | s \sim_{\mathcal{F}} s' | \!)$ is a first-order formula in the existential theory of the reals which involves only linear predicates. Therefore, quantifier elimination can be performed by using tools specialised for the linear theory of real-arithmetic, such as MJOLLNIR [14].

6   Model Checking Constrained MRMs

In this section we consider the model checking problem of CMRMs against PRCTL formulae for the class of constrained MRMs having parameters only on state rewards. For this class, we show that the set of parameter valuations satisfying given PRCTL formula can, again, be encoded as a first-order formula in the linear theory of real-arithmetic.

In this section assume that in the CMRM $\mathcal{P} = (S, \tau, \ell, \rho, F)$, parameters occur only in the state rewards, that is, $\tau(s)(u) \in [0, 1]$ for all $s, u \in S$.

For a state $s \in S$, and a PRCTL state formula Φ, we characterise $[\![\mathcal{P}, s \models \Phi]\!]$ by means of the set of valuations satisfying $F \wedge (\! | s, \Phi | \!)$. The formula $(\! | s, \Phi | \!)$ encodes the satisfiability of $\mathcal{P}$ in s up-to the constraints F, and it is defined by induction on the structure of Φ.

For the Boolean fragment of PRCTL, the reduction is as one may expect.

$$(\!|s, t\!|) = t\,, \qquad (\!|s, a\!|) = a \in \ell(s)\,, \qquad (\!|s, \rho \bowtie r\!|) = \rho(s) \bowtie r\,,$$
$$(\!|s, \neg\Phi\!|) = \neg(\!|s, \Phi\!|)\,, \qquad (\!|s, \Phi \wedge \Psi\!|) = (\!|s, \Phi\!|) \wedge (\!|s, \Psi\!|)\,.$$

For all other formulae, it is convenient to use $(\!|u, \Diamond\Phi\!|)$ as short for $\bigvee_{u \to^* v}(\!|v, \Phi\!|)$.

$$(\!|s, \mathbb{P}_J(\mathcal{X}\Phi)\!|) = \exists \mathbf{p}.\, \mathbf{1} \cdot \mathbf{p} \in J \wedge \bigwedge_{u \in S} ((p_u = \tau(s)(u) \wedge (\!|u, \Phi\!|)) \vee (p_u = 0 \wedge \neg(\!|u, \Phi\!|)))$$

$$(\!|s, \mathbb{P}_J(\Phi\mathcal{U}\Psi)\!|) = \exists \mathbf{p}.\, p_s \in J \wedge \bigwedge_{u \in S} \left(\begin{array}{c} (p_u = 1 \wedge (\!|u, \Psi\!|)) \vee (p_u = 0 \wedge \neg(\!|u, \Diamond\Psi\!|)) \vee \\ (p_u = 0 \wedge \neg(\!|u, \Phi\!|) \wedge \neg(\!|u, \Psi\!|)) \vee \\ (p_u = \tau(u) \cdot \mathbf{p} \wedge (\!|u, \Phi\!|) \wedge \neg(\!|u, \Psi\!|) \wedge (\!|u, \Diamond\Psi\!|)) \end{array} \right)$$

$$(\!|s, \mathbb{P}_J(\Phi\mathcal{U}_R^n\Psi)\!|) = \exists \mathbf{q}.\, \mathbf{1} \cdot \mathbf{q} \in J \wedge \bigwedge_{w \in S^n} \left(\begin{array}{c} (q_w = Pr_s(cyl(w)) \wedge \beta(w)) \vee \\ (q_w = 0 \wedge \neg\beta(w)) \end{array} \right)$$

$$(\!|s, \mathbb{E}_R(\Phi)\!|) = \exists \mathbf{r}.\, r_s \in R \wedge \bigwedge_{u \in S} ((r_u = 0 \wedge \alpha_u) \vee (r_u = \rho(u) + \tau(u) \cdot \mathbf{r} \wedge \neg\alpha_u))$$

where $\mathbf{p} = (p_s)_{s \in S}$, $\mathbf{r} = (r_s)_{s \in S}$, $\mathbf{q} = (q_w)_{w \in S^n}$, $\alpha_u = (\!|u, \Phi\!|) \vee \neg(\!|u, \Diamond\Phi\!|)$, and

$$\beta(s_1 \ldots s_n) = \bigvee_{1 \leq j \leq n} \left((\!|s_j, \Psi\!|) \wedge \bigwedge_{1 \leq k \leq j} \sum_{i=1}^{k-1} \rho(s_i) \in R \wedge \bigwedge_{1 \leq h < j}(\!|s_h, \Phi\!|) \right).$$

In $(\!|s, \mathbb{P}_J(\mathcal{X}\Phi)\!|)$, the variable p_u represents the probability of moving from s to u, where u satisfies the property Φ, therefore $\mathbf{1} \cdot \mathbf{p} = \sum_{u \in S}$ is the probability of moving in one step from s to a state satisfying Φ. Analogously, in $(\!|s, \mathbb{P}_J(\Phi\mathcal{U}\Psi)\!|)$, the variable p_u models the probability of u satisfying the property $\Phi\mathcal{U}\Psi$: if u satisfies the property Φ then p_u is 1; if u satisfies neither Φ nor Ψ or cannot reach a state satisfying Ψ, then p_u is 0; otherwise, p_u amounts to the probability of moving in one step to some other state u' and, from there, satisfying $\Phi\mathcal{U}\Psi$.

In $(\!|s, \mathbb{P}_J(\Phi\mathcal{U}_R^n\Psi)\!|)$, q_w is the probability of executing the trace w in exactly n steps starting from s where $\beta(w)$ models that fact that w satisfies $\Phi\mathcal{U}_R^n\Psi$. Therefore, $\mathbf{1} \cdot \mathbf{q}$ amounts to the probability of executing from s a path that satisfies $\Phi\mathcal{U}_R^n\Psi$.

Finally, in $(\!|s, \mathbb{E}_R(\Phi)\!|)$, one can readily see that $\mathbf{r}$ models the unique solution of the system of linear equations (1), where formula α_u capture the fact that u either satisfies the property Φ or cannot reach any state that does.

The following result states the correctness of the above characterisation.

Theorem 1. $\mathbf{v} \models F \wedge (\!|s, \Phi\!|)$ iff $\mathbf{v} \in [\![\mathcal{P}, s \models \Phi]\!]$.

Example 6. Consider the CMRM $\mathcal{F}$ and the formula Φ from Example 5. The encoding of the satisfiability of s_1 with respect to Φ, namely $(\!|s, \Phi\!|)$, boils down to the following first order linear formula.

$$\exists \mathbf{x}.\ x_1 \geq 0 \wedge x_1 \leq 10 \wedge x_1 = \frac{x_1}{5} + \frac{x_2}{10} + \frac{3x_3}{10} + \frac{x_4}{5} + \frac{x_6}{5} \wedge$$

$$x_2 = 0 \wedge x_3 = c + h + \frac{x_1}{5} + \frac{x_3}{2} + \frac{9x_5}{100} + \frac{21x_7}{100} \wedge$$

$$x_4 = h + j + \frac{x_1}{5} + \frac{x_3}{2} + \frac{3x_5}{100} + \frac{27x_7}{100} \wedge$$

$$x_5 = a + x_1 \wedge x_6 = \frac{x_1}{5} + \frac{3x_3}{20} + \frac{7x_4}{20} + \frac{3x_6}{10} \wedge x_7 = r + x_1$$

By using the tool MJOLLNIR [14] we are able to eliminate the existential quantifiers on $\mathbf{x} = (x_1, \ldots, x_7)$ obtaining the formula ϕ of Eq. (2).   □

7 Markov Models with Stochastic Rewards

It's common practice to model experimental measurements by means of real-valued random variables distributed according to well studied families of distributions (e.g., normal or student's T).

In this section we introduce the notion of *stochastic Markov reward models* (SMRMs) where state rewards are real-valued random variables. Then, we present a PRCTL model checking framework for SMRMs.

From here on we fix the probability space (Ω, Σ, P) representing the environment where the experiments are performed, and we use $\mathbb{Y}$ to denote the set of real-valued random variables of the form $Y : \Omega \to \mathbb{R}$.

Definition 5. *A stochastic Markov reward model is a tuple $\mathcal{J} = (S, \tau, \ell, \rho)$ where (S, τ, ℓ) is a Markov chain, and $\rho : S \to \mathbb{Y}$ is a reward function assigning a real-valued random variable to each state.*

An SMRM $\mathcal{J} = (S, \tau, \ell, \rho)$ can be intuitively interpreted as a measurable function $\mathcal{J} : \Omega \to \mathrm{MRM}_{\mathcal{J}}$, where $\mathcal{J}(\omega)$ is the MRM having (S, τ, ℓ) as underlying Markov chain and $(\rho(s)(\omega))_{s \in S}$ as vector of rewards. Such interpretation justifies the intuition of $\mathcal{J}$ being an experiment whose outcomes is a MRM.

The above intuition is formalised as follows. We denote by $\mathrm{MRM}_{\mathcal{J}}$ the set of all MRMs having the same underlying Markov chain as $\mathcal{J}$. We construct the σ-algebra $\Sigma_{\mathcal{J}}$ as the family of sets $A \subseteq \mathrm{MRM}_{\mathcal{J}}$ whose corresponding set of rewards vectors is Borel measurable in $\mathbb{R}^m$ ($m = |S|$). Formally,

$$A \in \Sigma_{\mathcal{J}} \quad \text{iff} \quad A \subseteq \mathrm{MRM}_{\mathcal{J}} \text{ and } \{\rho(\mathcal{R}) \mid \mathcal{R} \in A\} \in \mathcal{B}(\mathbb{R}^m).$$

Accordingly, the semantics of $\mathcal{J}$ is the probability space $(\mathrm{MRM}_{\mathcal{J}}, \Sigma_{\mathcal{J}}, P[\mathcal{J}])$.

Given a SMRM $\mathcal{J}$, a state $s \in S$, and a PRCTL formula Φ, it comes natural to ask how likely is that a concrete instance of $\mathcal{J}$ satisfies Φ at s, denoted by $P[\mathcal{J}, s \models \Phi]$. This model checking problem is formalised as follows

$$P[\mathcal{J}, s \models \Phi] \stackrel{\text{def}}{=} P[\mathcal{J}](\{\mathcal{R} \in \mathrm{MRM}_{\mathcal{J}} \mid \mathcal{R}, s \models \Phi\}). \tag{8}$$

We study the above model checking problem for a subclass of SMRMs having random variables $(Y \colon \Omega \to \mathbb{R}) \in \mathcal{E}_{\mathbf{X}}$ of the form $Y(\omega) = \mathbf{a} \cdot \mathbf{X}(\omega) + b$, with $\mathbf{a} \in \mathbb{Q}^k$, $b \in \mathbb{Q}$ and, where $\mathbf{X} = (X_1, \ldots, X_k)$ is a vector of pairwise *independent* real-valued random variables[1]. Observe that, elements in $\mathcal{E}_{\mathbf{X}}$ may not be independent from each other.

Hereafter we consider the SMRM $\mathcal{J} = (S, \tau, \ell, \rho)$ with $\rho \colon S \to \mathcal{E}_{\mathbf{X}}$, and we use $\mathcal{P}$ to refer to the CMRM obtained by replacing the random variables X_i in $\mathcal{J}$ with the parameters x_i $(i = 1, \ldots, k)$.

For Eq. (8) to be well-defined the set $\{\mathcal{R} \in \mathrm{MRM}_{\mathcal{J}} \mid \mathcal{R}, s \models \varPhi\}$ needs to be a measurable event in $\Sigma_{\mathcal{J}}$. The following result ensures that.

Lemma 2. $\{\mathcal{R} \in MRM_{\mathcal{J}} \mid \mathcal{R}, s \models \varPhi\} \in \Sigma_{\mathcal{J}}$.

Proof. By definition of $\mathcal{P}$ we have that

$$\{\mathcal{R} \in \mathrm{MRM}_{\mathcal{J}} \mid \mathcal{R}, s \models \varPhi\} = \{\mathcal{P}(\mathbf{v}) \mid \mathbf{v} \in [\![\mathcal{P}, s \models \varPhi]\!]\})$$

Therefore, by def. of $\Sigma_{\mathcal{J}}$ and measurability of affine transformations we have that the claim holds iff $[\![\mathcal{P}, s \models \varPhi]\!] \in \mathcal{B}(\mathbb{R}^k)$. By Theorem 1, $[\![\mathcal{P}, s \models \varPhi]\!]$ can be described by means of a Boolean formula with linear predicates. Since σ-algebras are closes under complement (i.e., negation), countable unions (i.e., disjunctions) and countable intersections (i.e., conjunctions) and, affine transformations are measurable, we conclude that $\{\mathcal{R} \in \mathrm{MRM}_{\mathcal{J}} \mid \mathcal{R}, s \models \varPhi\} \in \Sigma_{\mathcal{J}}$. ∎

The following theorem characterises the model checking problem for the SMRM $\mathcal{J}$ in terms of the model checking problem for the CMRM $\mathcal{P}$.

Theorem 2. $P[\mathcal{J}, s \models \varPhi] = P[\mathbf{X} \in [\![\mathcal{P}, s \models \varPhi]\!]]$.

Proof. The claim holds true according to the following equalities.

$$\begin{aligned}
P[\mathcal{J}, s \models \varPhi] &= P[\mathcal{J}](\{\mathcal{R} \in \mathrm{MRM}_{\mathcal{J}} \mid \mathcal{R}, s \models \varPhi\}) && \text{(by Eq. (8))} \\
&= P[\mathcal{P} \circ \mathbf{X}](\{\mathcal{R} \in \mathrm{MRM}_{\mathcal{J}} \mid \mathcal{R}, s \models \varPhi\}) && (\mathcal{J} = \mathcal{P} \circ \mathbf{X}) \\
&= P[\mathcal{P} \circ \mathbf{X}](\mathcal{P}([\![\mathcal{P}, s \models \varPhi]\!])) && \text{(def. } \mathcal{P} \text{ and def. } [\![\mathcal{P}, s \models \varPhi]\!]) \\
&= P((\mathcal{P} \circ \mathbf{X})^{-1}(\mathcal{P}([\![\mathcal{P}, s \models \varPhi]\!]))) && \text{(def. push-forward)} \\
&= P(\mathbf{X}^{-1}([\![\mathcal{P}, s \models \varPhi]\!])) && ((\mathcal{P} \circ \mathbf{X})^{-1} = \mathbf{X}^{-1} \circ \mathcal{P}^{-1}) \\
&= P[\mathbf{X} \in [\![\mathcal{P}, s \models \varPhi]\!]]. && \text{(def. push-forward)}
\end{aligned}$$

∎

By Theorem 2 we can estimate the value p of $P[\mathcal{J}, s \models \varPhi]$ by applying Monte Carlo simulation techniques. For this, we sample n independent repetitions of $\mathbf{X}$, associating with each repetition a Bernoulli random variable B_i. A realisation b_i of B_i is 1 if the corresponding sampled value of $\mathbf{X}$ lays in $[\![\mathcal{P}, s \models \varPhi]\!]$, and

---

[1] In fact, the vector $\mathbf{X}$ is a multivariate random variable $\mathbf{X} \colon \Omega \to \mathbb{R}^n$ with marginals $X_i \colon \Omega \to \mathbb{R}$ $(i = 1, \ldots, n)$.

0 otherwise. Finally, we estimate p by means of the observed relative success rate $\tilde{p} = (\sum_{i=1}^{n} b_i)/n$. The absolute error ε of the estimation can be bound with a certain degree of confidence $\delta \in (0, 1]$ by tuning the number of required simulations based on the Hoeffding's inequality $P(|\tilde{p} - p| \geq \varepsilon) \leq 2e^{-2n\varepsilon^2}$. Thus, we can determine the number of samples required to estimate p with absolute error ε and confidence δ by imposing $2e^{-2n\varepsilon^2} = 1 - \delta$, from which we obtain

$$n = \left\lceil -\frac{\ln(\delta/2)}{2\varepsilon^2} \right\rceil. \tag{9}$$

For example, we can estimate p with an error $\varepsilon = 0.01$ with a confidence of 95% (i.e., $\delta = 0.95$) by drawing $n = 18445$ samples.

Example 7. Let $\mathbf{X} = (H, C, J, A, R)$ be a vector of random variables respectively distributed as $H \sim \text{unif}(-5, 0)$, $C \sim \text{unif}(1, 2)$, $J \sim \mathcal{N}(2, 0.1)$, $A \sim \mathcal{N}(-3, 0.5)$, and $R \sim \mathcal{N}(-3, 0.5)$. We define the SMRM $\mathcal{J} = (S, \tau, \ell, \rho)$ having as underlying Markov chain that of Fig. 1, and the following random vector of rewards

$$\rho = (0, 0, H + C, H + J, R, 0, A),$$

which shall be understood as a stochastic version of the one presented in Example 4. By taking the formula $\Phi = \mathbb{E}_{[0,10]}(ths)$ from Example 5, we can estimate $P[\mathcal{J}, s_1 \models \Phi] \cong 0.156061$ with an error $\varepsilon = 0.005$ and confidence of 99.99% (i.e., $\delta = 0.0001$) by generating $n = 198070$ samples.   □

8   Conclusion

We described a framework for the analysis of Markov reward models in presence of uncertain rewards. To this end we propose two extensions of the notion of MRM: (a) *constrained Markov reward models*, having state rewards parametric on a set variables subject to some constraints, and (b) *stochastic Markov reward models*, having rewards modelled as random variables.

We demonstrated that the analysis CMRMs with respect to probabilistic bisimilarity and PRCTL formulae, can be reduced to perform quantifier elimination on fist-order formulae in the linear fragment of the theory of the reals. Our reduction does not lead to an improvement on the theoretical complexity of the model checking problem for (augmented) parametric Markov chains (*cf.* [9, Theorem 4]). However, we believe that our reduction is important from the perspective of implementation in practice, because it allows one to employ SMT solvers or quantifier elimination procedures specialised on linear constraints such as MJOLLNIR [14]. It is worth noting that our reduction can be also applied with PRCTL formulas with parametric bounds, extending even further the applicability of our approach.

Finally, we provided a characterisation of the model checking problem for SMRMs in terms of the model checking problem for CMRMs. As we have shown, this result allows one to estimate the probability that a given SMRM satisfies a given specification by employing Monte Carlo simulation techniques.

All the calculations presented in the examples have been done by using a prototype implementation of the above described algorithms[2] coded in Mathematica [18]. For the quantifier elimination we employed MJOLLNIR instead of the built-in solution offered by Mathematica.

Our work finds applications in model repair, where parameters need to be tuned so as to satisfy a desired specification, as well as in robustness analysis in presence of stochastic perturbations.

References

1. Baier, C., Katoen, J.: Principles of Model Checking. MIT Press, Cambridge (2008)
2. Benedikt, M., Lenhardt, R., Worrell, J.: LTL model checking of interval Markov chains. In: Piterman, N., Smolka, S.A. (eds.) TACAS 2013. LNCS, vol. 7795, pp. 32–46. Springer, Heidelberg (2013). https://doi.org/10.1007/978-3-642-36742-7_3
3. Chatterjee, K., Sen, K., Henzinger, T.A.: Model-checking ω-regular properties of interval Markov chains. In: Amadio, R. (ed.) FoSSaCS 2008. LNCS, vol. 4962, pp. 302–317. Springer, Heidelberg (2008). https://doi.org/10.1007/978-3-540-78499-9_22
4. Daws, C.: Symbolic and parametric model checking of discrete-time Markov chains. In: Liu, Z., Araki, K. (eds.) ICTAC 2004. LNCS, vol. 3407, pp. 280–294. Springer, Heidelberg (2005). https://doi.org/10.1007/978-3-540-31862-0_21
5. Dehnert, C., Junges, S., Jansen, N., Corzilius, F., Volk, M., Bruintjes, H., Katoen, J.-P., Ábrahám, E.: PROPhESY: A PRObabilistic ParamEter SYnthesis Tool. In: Kroening, D., Păsăreanu, C.S. (eds.) CAV 2015. LNCS, vol. 9206, pp. 214–231. Springer, Cham (2015). https://doi.org/10.1007/978-3-319-21690-4_13
6. Gainer, P., Hahn, E.M., Schewe, S.: Accelerated model checking of parametric Markov chains. In: Lahiri, S.K., Wang, C. (eds.) ATVA 2018. LNCS, vol. 11138, pp. 300–316. Springer, Cham (2018). https://doi.org/10.1007/978-3-030-01090-4_18
7. Hahn, E.M., Hermanns, H., Wachter, B., Zhang, L.: PARAM: a model checker for parametric Markov models. In: Touili, T., Cook, B., Jackson, P. (eds.) CAV 2010. LNCS, vol. 6174, pp. 660–664. Springer, Heidelberg (2010). https://doi.org/10.1007/978-3-642-14295-6_56
8. Hahn, E.M., Hermanns, H., Zhang, L.: Probabilistic reachability for parametric markov models. STTT **13**(1), 3–19 (2011). https://doi.org/10.1007/s10009-010-0146-x
9. Hutschenreiter, L., Baier, C., Klein, J.: Parametric Markov chains: PCTL complexity and fraction-free Gaussian elimination. In: Bouyer, P., Orlandini, A., Pietro, P.S. (eds.) Proceedings 8th International Symposium on Games, Automata, Logics and Formal Verification, GandALF 2017. EPTCS, vol. 256, pp. 16–30 (2017). https://doi.org/10.4204/EPTCS.256.2
10. Jansen, N., Corzilius, F., Volk, M., Wimmer, R., Ábrahám, E., Katoen, J.-P., Becker, B.: Accelerating parametric probabilistic verification. In: Norman, G., Sanders, W. (eds.) QEST 2014. LNCS, vol. 8657, pp. 404–420. Springer, Cham (2014). https://doi.org/10.1007/978-3-319-10696-0_31
11. Jonsson, B., Larsen, K.G.: Specification and refinement of probabilistic processes. In: Proceedings of the Sixth Annual Symposium on Logic in Computer Science (LICS 1991), Amsterdam, The Netherlands, 15–18 July 1991, pp. 266–277. IEEE Computer Society (1991). https://doi.org/10.1109/LICS.1991.151651

---

[2] Available at http://people.cs.aau.dk/~giovbacci/tools.html.

12. Lanotte, R., Maggiolo-Schettini, A., Troina, A.: Parametric probabilistic transition systems for system design and analysis. Formal Asp. Comput. **19**(1), 93–109 (2007). https://doi.org/10.1007/s00165-006-0015-2
13. Larsen, K.G., Skou, A.: Bisimulation through probabilistic testing. Inf. Comput. **94**(1), 1–28 (1991)
14. Monniaux, D.: Quantifier elimination by lazy model enumeration. In: Touili, T., Cook, B., Jackson, P. (eds.) CAV 2010. LNCS, vol. 6174, pp. 585–599. Springer, Heidelberg (2010). https://doi.org/10.1007/978-3-642-14295-6_51
15. Sen, K., Viswanathan, M., Agha, G.: Model-checking Markov chains in the presence of uncertainties. In: Hermanns, H., Palsberg, J. (eds.) TACAS 2006. LNCS, vol. 3920, pp. 394–410. Springer, Heidelberg (2006). https://doi.org/10.1007/11691372_26
16. Srkk, S.: Bayesian Filtering and Smoothing. Cambridge University Press, New York (2013)
17. Tran, D.Q., Gumussoy, S., Michiels, W., Diehl, M.: Combining convex-concave decompositions and linearization approaches for solving BMIs, with application to static output feedback. IEEE Trans. Automat. Contr. **57**(6), 1377–1390 (2012). https://doi.org/10.1109/TAC.2011.2176154
18. Wolfram Research Inc.: Mathematica, Version 11.2, Champaign, IL (2017)

A Modest Approach to Modelling and Checking Markov Automata

Yuliya Butkova[1] , Arnd Hartmanns[2]([⊠]) , and Holger Hermanns[1,3]

[1] Saarland University, Saarland Informatics Campus, Saarbrücken, Germany
[2] University of Twente, Enschede, The Netherlands
a.hartmanns@utwente.nl
[3] Institute of Intelligent Software, Guangzhou, China

Abstract. Markov automata are a compositional modelling formalism with continuous stochastic time, discrete probabilities, and nondeterministic choices. In this paper, we present extensions to the MODEST language and the mcsta model checker to describe and analyse Markov automata models. MODEST is an expressive high-level language with roots in process algebra that allows large models to be specified in a succinct, modular way. We explain its use for Markov automata and illustrate the advantages over alternative languages. The verification of Markov automata models requires dedicated algorithms for time-bounded probabilistic reachability and long-run average rewards. We describe several recently developed such algorithms as implemented in mcsta and evaluate them on a comprehensive set of benchmarks. Our evaluation shows that mcsta improves the performance and scalability of Markov automata model checking compared to earlier and alternative tools.

1 Introduction

Studying dependability and performance aspects of critical designs or implementations [4] requires a formal mathematical model that captures the core quantitative aspects of such systems. In particular, we need *stochastic continuous time* to model delays of which we only know averages, e.g. the mean time to failure, *discrete probabilistic choices* to describe instantaneous uncertain decisions, as in e.g. randomised algorithms, and *nondeterminism* to be able to deal with underspecification, abstraction, unquantified uncertainty, and concurrency. Markov automata (MA, [18,20]) extend the classical formalisms of continuous-time Markov chains and discrete-time Markov decision processes (MDP) to encompass all three of these aspects. In contrast to continuous-time MDP (CTMDP), they are compositional: there is a natural parallel composition operator for networks of MA that provides for both interleaved and synchronising transitions without the need for ad-hoc operations to combine transition rates.

MA are the semantic basis for generalised stochastic Petri nets [19] and dynamic extensions of fault trees [6,31]. Several publications studied algorithmic

---

Authors are listed alphabetically. This work has received financial support by DFG grant 389792660 as part of TRR 248 (see perspicuous-computing.science) by ERC Advanced Grant 69561 (POWVER), and by NWO VENI grant 639.021.754.

D. Parker and V. Wolf (Eds.): QEST 2019, LNCS 11785, pp. 52–69, 2019.
https://doi.org/10.1007/978-3-030-30281-8_4

problems related to the efficient analysis of MA [2,12–15,23,24,30]. In this light, it is disappointing that tool support for MA is thus far rather brittle. The one dedicated tool for compositional modelling with MA, SCOOP [38], is unmaintained, as is the corresponding lower-level MA model checker IMCA [22]. The one other actively developed tool with comprehensive MA support is STORM [17], which however lacks built-in support for high-level compositional modelling.

Using the mathematical formalism of MA directly to build complex models is cumbersome. For their use to be practical, we need a higher-level *modelling language*. Aside from a parallel composition operator, such languages typically provide variables over finite domains that can be used in expressions to e.g. enable or disable transitions. Their semantics is then an MA whose states are the valuations of the variables, allowing to compactly describe very large MA. In this paper, we present recent extensions to MODEST [27], a high-level modelling language for stochastic hybrid systems, that add support for expressing MA models. Rooted in process algebra, MODEST provides various composition operators that allow large models to be assembled from small, easy-to-understand components. In Sect. 3, we illustrate the use of MODEST for MA, and we compare its succinctness, expressivity, and readability with alternative languages.

We build MA models to compute quantitative properties of systems such as safety (the probability to reach an unsafe state), reliability (doing so within a time bound), or throughput (the long-run average amount of work completed per time unit). Probabilistic model checking techniques [3] can be applied to MA to effectively compute or approximate such values. While the computation of *unbounded* reachability probabilities and expected accumulated rewards can be reduced to checking the MA's embedded MDP, *time-bounded* probabilities and *long-run average* rewards require dedicated algorithms. We summarise the currently available algorithms, their particular characteristics, and notable implementation considerations, in Sect. 4. To complement our extension of the MODEST language with suitable analysis facilities, we have implemented the most promising of these algorithms in the mcsta model checker of the MODEST TOOLSET [28]. We use the MA models of the Quantitative Verification Benchmark Set [29] to evaluate the performance of our implementation and of the different algorithms in Sect. 5. We compare the results with IMCA and STORM.

2   Markov Automata

The mathematical formalism of Markov automata provides nondeterministic choices as in labelled transition systems (LTS, or Kripke structures or finite automata), discrete probabilistic decisions as in discrete-time Markov chains (DTMC), and states with exponentially distributed residence times as in continuous-time Markov chains (CTMC). The relationships between these formalisms are visualised in Fig. 1. We now define MA formally and describe their semantics.

Preliminaries. We write $\{x_1 \mapsto y_1, \dots\}$ to denote the function that maps all x_i to y_i, and if necessary in the respective context, implicitly maps to 0 all x

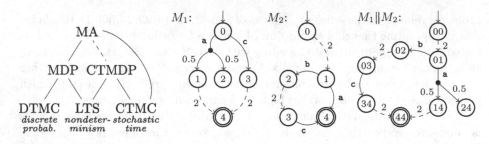

Fig. 1. The MA family tree   **Fig. 2.** Example Markov automata

for which no explicit mapping is specified. Given a set S, its powerset is 2^S. A (discrete) probability distribution over S is a function $\mu \in S \to [0,1]$ such that $spt(\mu) \stackrel{\text{def}}{=} \{\, s \in S \mid \mu(s) > 0 \,\}$ is countable and $\sum_{s \in spt(\mu)} \mu(s) = 1$. $Dist(S)$ is the set of all probability distributions over S, and $\mu_1 \otimes \mu_2$ is the product distribution of μ_1 and μ_2 defined by $(\mu_1 \otimes \mu_2)(s) = \mu_1(s) \cdot \mu_s(s)$. We refer to discrete random choices as *probabilistic* and to continuous ones as *stochastic*.

Definition 1. *A* Markov automaton *(MA) is a tuple*

$$M = \langle S, s_0, A, P, Q, rr, br \rangle$$

where
- *S is a finite set of* states, *with $s_0 \in S$ being the* initial state,
- *A is a finite set of* actions,
- *$P \in S \to 2^{A \times Dist(S)}$ is the* probabilistic transition *function,*
- *$Q \in S \to 2^{\mathbb{Q} \times S}$ the* Markovian transition *function,*
- *$rr \in S \to [0, \infty)$ is the* rate reward *function, and*
- *$br \in S \times Tr(M) \times S \to [0, \infty)$ is the* branch reward *function*

with $Tr(M) \stackrel{\text{def}}{=} \bigcup_{s \in S} P(s) \cup \bigcup_{s \in S} Q(s)$. $P(s)$ and $Q(s)$ must be finite sets for all $s \in S$. We define the exit rate *of $s \in S$ as $E(s) = \sum_{\langle \lambda, s' \rangle \in Q(s)} \lambda$.*

Example 1. Fig. 2 shows two MA M_1 and M_2 without rewards. We draw probabilistic transitions as solid, Markovian ones as dashed lines. If a transition leads to a single target state, we omit the intermediate probabilistic branching node.

The semantics of an MA is that, in state s, (1) the probability to take Markovian transition $\langle \lambda, s' \rangle \in Q(s)$ and move to state s' within t time units is

$$\lambda/E(s) \cdot (1 - e^{-E(s) \cdot t}), \tag{1}$$

i.e. the residence time follows the exponential distribution with rate $E(s)$ and the choice of transition is weighted by their rates; and (2) at any point in time, a probabilistic edge $\langle a, \mu \rangle \in P(s)$ can be taken with the successor state being chosen according to μ. MA thus separate interaction from timing: the former is represented by the action-labelled probabilistic transitions, and the latter is governed by the rates of the Markovian transitions. This is the key difference to CTMDP, which have one kind of transitions with both actions and rates. It enables parallel composition operators with action synchronisation for MA without any need to prescribe an ad-hoc operation for combining rates.

Definition 2. *Given two MA $M_i = \langle S_i, s_{0_i}, A_i, P_i, Q_i \rangle$, $i \in \{1, 2\}$, their parallel composition is $M_1 \parallel M_2 \stackrel{\text{def}}{=} \langle S_1 \times S_2, \langle s_{0_1}, s_{0_2} \rangle, A_1 \cup A_2, P, Q \rangle$ with P the smallest function such that*

$$(\langle a, \mu \rangle \in P_1(s_1) \wedge a \notin A_2 \Rightarrow \langle a, \mu \otimes \{ s_2 \mapsto 1 \} \rangle \in P(\langle s_1, s_2 \rangle))$$
$$\wedge (\langle a, \mu \rangle \in P_2(s_2) \wedge a \notin A_1 \Rightarrow \langle a, \{ s_1 \mapsto 1 \} \otimes \mu \rangle \in P(\langle s_1, s_2 \rangle))$$
$$\wedge (\langle a, \mu_1 \rangle \in P_1(s_1) \wedge \langle a, \mu_2 \rangle \in P_2(s_2) \wedge a \in A_1 \cap A_2 \Rightarrow \langle a, \mu_1 \otimes \mu_2 \rangle \in P(\langle s_1, s_2 \rangle))$$

and Q is the smallest function s.t. $(\langle \lambda, s_1' \rangle \in Q_1(s_1) \Rightarrow \langle \lambda, \langle s_1', s_2 \rangle \rangle \in Q(\langle s_1, s_2 \rangle))$ and vice-versa for Q_2.

The operator above uses multi-way synchronisation on the shared alphabet of the two automata; similar operators could be defined for other synchronisation mechanisms, e.g. to define input-output MA. Fig. 2 includes the parallel composition of the example M_1 and M_2, where we write nm for state $\langle n, m \rangle$. The two automata synchronise on the shared actions a and c.

We defined MA as *open* systems [8]: probabilistic transitions can interact with, wait for, and be blocked by other MA in parallel composition. For verification, we make the usual *closed system* and *maximal progress* assumptions, i.e. we assume that probabilistic transitions face no further interference and take place without delay. If multiple probabilistic transitions are available in a state, however, the choice between them remains nondeterministic. Since the probability that a Markovian transition is taken in zero time is 0, the maximal progress assumption allows us to remove all Markovian transitions from states that also have a probabilistic transition. In such *closed MA*, we can thus distinguish between Markovian states (where $P(s) = \varnothing$) and probabilistic states (where $Q(s) = \varnothing$). The behaviour of a closed, deadlock-free MA M is defined via its paths:

Definition 3. *A path $\pi \in \Pi(M)$ is an infinite sequence*
$$\pi = s_0 \, t_0 \, tr_0 \, s_1 \ldots \in (S \times [0, \infty) \times Tr(M))^\omega$$

such that $Q(s_i) = \varnothing \Rightarrow t_i = 0$ and $tr_i \in P(s_i) \cup Q(s_i)$. We write $\Pi_f(M)$ for the set of all path prefixes π_f ending in a state. Let $\pi_{\leq i} \stackrel{\text{def}}{=} s_0 t_0 \ldots s_i$. The duration $\text{dur}(\pi_f)$ of a path prefix is the sum of its residence times t_i. A path's reward is
$$\text{rew}(\pi) = \sum_{i=0}^{\infty} t_i \cdot rr(s_i) + br(s_i, tr_i, s_{i+1})$$
and is analogously defined for prefixes.

A path prescribes a resolution of all nondeterministic, probabilistic, and stochastic choices. To define a probability measure, we resolve nondeterminism only:

Definition 4. *Given an MA M as above, a scheduler in $\mathfrak{S}(M)$ is a function $\sigma \in \Pi_f(M) \to Tr(M)$ s.t. $\forall s \in S \colon \sigma(s) = tr \Rightarrow tr \in P(s) \cup Q(s)$. A time-dependent scheduler is in $S \times [0, \infty) \to Tr(M)$; a memoryless one in $S \to Tr(M)$.*

We define deterministic schedulers only since randomised schedulers are in practice only needed for multi-objective problems [34]. We note that CTMDP with early schedulers [36] can be encoded as closed MA. A scheduler induces a probability measure over sets of measurable paths in the usual way [30]. For all of the following types of properties, we are interested in the maximum (supremum) and minimum (infimum) values when ranging over all schedulers $\sigma \in \mathfrak{S}(M)$:

Reachability probabilities: Given a set of goal states $G \subseteq S$, compute the probability of the set of paths that include a state in G. Memoryless schedulers suffice to achieve optimal results (i.e. the max. and min. probabilities).

Time-bounded reachability: Additionally restrict to paths where the sum of delays up to reaching the first state in G is below a bound $b \in [0, \infty)$. Here, time-dependent schedulers with input $b - \text{dur}(\pi_f)$ suffice.

Expected accumulated rewards: For $G \subseteq S$, compute the expected value of the random variable[1] that assigns to path π the value $\text{rew}(\pi_f)$ where π_f is the shortest prefix of π with a state in G. Memoryless schedulers suffice.

Long-run average rewards: Compute the expected value of the random variable that assigns to path π the value $\lim_{i \to \infty} \text{rew}(\pi_{\leq i}) / \text{dur}(\pi_{\leq i})$. Memoryless schedulers suffice.

Example 2. Consider MA $M_1 \parallel M_2$ of Fig. 2 and the probability to reach state $\langle 4, 4 \rangle$ within 1 time unit. In state $\langle 0, 1 \rangle$, we have to decide whether to choose action a or b. The optimal decision depends on the amount of time t that has passed in state $\langle 0, 0 \rangle$. In the plot on the right, we show the probability of reaching state $\langle 4, 4 \rangle$ (y-axis) depending on $1 - t$ (x-axis). The blue line represents the reachability probability for the memoryless scheduler that always chooses a and the red one is for

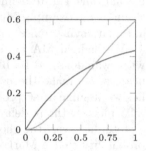

the scheduler that always takes action b. A time-dependent scheduler can make better decisions than either of these two by determining the values of t for which a results in a higher probability than b and vice-versa. The optimal scheduler thus chooses a if and only if $1 - t \leq 0.63$ approximately.

3 Modelling

Tools for the automated analysis of MA need a syntax in which the model and the properties of interest are specified. As noted in Sect. 1, such a modelling language needs to provide a parallel composition operator such that large MA can be built from small specifications, and will typically support modelling with variables that can be used in guards and assignments. In the context of such symbolic formalisms, we have *locations* and *edges* that each induce (many) states and transitions, respectively, in the formalism's plain-MA semantics.

3.1 MODEST for Markov Automata

As part of implementing the JANI [10] model exchange format, we recently introduced support for MA into the syntax and semantics of the MODEST modelling language [27]. MODEST previously supported MDP and more complex continuous-time formalisms such as stochastic hybrid automata, but did not

---

[1] This is well-defined if the maximum (minimum) probability to reach G is 1; otherwise, we define the minimum (maximum) expected accumulated reward to be ∞.

```
const real B;
int(0..2) succ = 0;
action a, b, c;
property P_Min = Pmin(<>[T<=B] (succ == 2));
property P_Max = Pmax(<>[T<=B] (succ == 2));
process M1()
{
  bool fail = false;
  alt {
  :: a palt {
    :1: {==}
    :1: {= fail = true =}
    }
  :: c
  };
  when(!fail) rate(2) {= succ++ =}
}
process M2()
{
  rate(2) tau;
  alt {
  :: a {= succ++ =}
  :: b; rate(2) tau; c {= succ++ =}
  }
}
par {
:: M1()
:: M2()
}
```

Fig. 3. MODEST for MA

```
global succ:{0..2} = 0
DONE = done.DONE[]
M1  = a.psum(0.5 -> M1a[] ++ 0.5 -> DONE[])
      ++ c.M1a[]
M1a = <2>.setGlobal(succ, succ + 1)
      .DONE[]
M2  = <2>.(a.M2a[] ++ b.<2>.c.M2a[])
M2a = setGlobal(succ, succ + 1).DONE[]
init M1[] || M2[]
comm (a, a, a), (c, c, c)
reachCondition (succ = 2)
```

Fig. 4. MAPA process algebra

```
ma
const double B
module M1
  s1: [0..4];
  [a] s1=0 -> 0.5:(s1'=1) + 0.5:(s1'=2);
  [c] s1=0 -> 1:(s1'=3);
  <>  s1=1 | s1=3 -> 2:(s1'=4);
endmodule
module M2
  s2: [0..4];
  <>  s2=0 -> 2:(s2'=1);
  [a] s2=1 -> 1:(s2'=4);
  [b] s2=1 -> 1:(s2'=2);
  <>  s2=2 -> 2:(s2'=3);
  [c] s2=3 -> 1:(s2'=4);
endmodule
"P_Min": Pmin=? [F<=B (s1=4 & s2=4)];
"P_Max": Pmax=? [F<=B (s1=4 & s2=4)];
```

Fig. 5. PRISM dialect supporting MA

```
#INITIALS
s00
#GOALS
s44
#TRANSITIONS
s00 !
* s01 2
s01 a
* s02 1
s01 b
* s14 0.5
* s24 0.5
s14 !
* s44 2
s02 !
* s03 2
s03 c
* s34 1
s34 !
* s44 2
```

Fig. 6. IMCA state space format

have provisions for succinctly annotating edges with rates. We added the rate(e) construct for this purpose, which behaves analogously to the existing when(e) construct for specifying the enabling condition of an edge. MODEST enforces the separation of probabilistic and Markovian transitions by requiring edges for which a rate is specified to have the predefined and non-synchronising τ action label. If this restriction is not met, the model is recognised as a CTMDP.

At its core, MODEST is a process algebra: it provides various operations such as parallel composition (par), sequential composition (;), parameterised process definitions, process calls, and guards (when) to flexibly construct complex models out of small and reusable components. Its syntax however borrows heavily from commonly used programming languages, and it provides high-level conveniences such as do loops and a full-fledged mechanism for throwing (throw) and handling (try-catch) exceptions. As such, MODEST tends to be more verbose than classic process algebras, but also more readable and beginner-friendly. To specify complex behaviour in a succinct manner, MODEST also provides variables of standard basic types (e.g. bool, int, or bounded int), arrays, and user-defined recursive datatypes akin to functional programming languages. Its syntax for expressions again is aligned with C-like programming languages for ease of use.

In Fig. 3, we show a MODEST representation of the parallel composition of MA M_1 and M_2 of Fig. 2. M_1 has been slightly optimised by merging states 1 and 3 into the last line of process M1; this actually came naturally when modelling due to the ease in which behaviours can be combined and shared in MODEST. The model also includes the declaration of two properties of interest for verification, P_Min and P_Max, which ask for the probability to reach state $\langle 4, 4 \rangle$—made observable via the global variable succ—within time B akin to Example 2. B is an open parameter for which values can be specified at verification time. There are many features of MODEST not used in this small model; the interested reader may find more complex MODEST MA models, in particular with arrays and rewards, in the Quantitative Verification Benchmark Set [29] at qcomp.org.

Tool Support. The MODEST TOOLSET [28] is a comprehensive suite of tools for quantitative modelling and verification. Its primary input languages are MODEST and JANI. MA are supported in its mosta, moconv[2], mcsta, and modes tools. mosta visualises the symbolic semantics of models and is useful for model debugging. moconv transforms models between modelling languages (it can e.g. convert MODEST to JANI) and performs syntactic rewriting and optimisations. mcsta is an explicit-state model checker; we present and evaluate its MA-specific algorithms in Sects. 4 and 5. modes [9] is a statistical model checker with automated rare event simulation capabilities. It implements the lightweight scheduler sampling approach [32] for nondeterministic models, including MA [16]. The MODEST TOOLSET is written in C#, works cross-platform on Linux, Mac OS, and Windows, and is freely available at modestchecker.net. All its tools share a common infrastructure for parsing and syntactic transformations. mcsta and modes additionally build on the same state space exploration engine that compiles models to bytecode at runtime for memory efficiency and performance.

3.2 Alternative Modelling Languages

MODEST is not the only modelling language for MA. These are the alternatives:

State Space Files for IMCA. The first MA-specific algorithms were implemented in the IMCA tool [22]. Its only input language is a text-based explicit state space format as illustrated for our example of $M_1 \parallel M_2$ in Fig. 6. This is clearly not a useful modelling language, but a format to be automatically generated by tools.

---

[2] moconv can also export CTMDP to JANI, but due to their lack of a natural parallel composition operator, the analysis of CTMDP is not supported in the other tools.

Guarded Commands with STORM. Of the alternative tools for MA, STORM [17] is the only one that is actively maintained. It provides many input languages, with MA being supported through a state space format similar to IMCA's, via JANI, as the semantics of generalised stochastic Petri nets [19] in GREATSPN format [1], and through an extension of the PRISM guarded command language. We show our example in the latter in Fig. 5. It is a very simple, small language that is easy to learn, however it completely lacks higher-level constructs to structure and compose models aside from the implicit parallel composition of its *modules*.

Process Algebra with SCOOP. MAPA [38] is a dedicated process algebra for MA. It is supported by SCOOP [38], which can linearise, reduce, and finally export MAPA models to IMCA for verification. We show the example of M_1 and M_2 in MAPA in Fig. 4. As a classic concise process algebra, MAPA tends to be very succinct, but also difficult to read. MAPA models can be much more flexibly composed than PRISM models, yet there is less syntactic structure than in MODEST—although the languages conceptually share many operators. MAPA notably has a predefined *queue* datatype, and users can specify custom non-recursive datatypes.

JANI Model Interchange. JANI [10] is a model interchange format designed to ease tool development and interoperation. It is JSON-based and thus human-debuggable, but not intended as human-writable. It represents networks of automata with variables symbolically. Since both the MODEST TOOLSET and STORM support JANI, it is possible to e.g. build MA models in the MODEST language, export them to JANI with moconv, and then verify them with STORM. Likewise in the other direction, we can e.g. create a Petri net with GREATSPN, convert to JANI with STORM, and analyse it with mcsta or modes. In this way, the most appropriate modelling language can be combined with the best analysis method and tool for every specific scenario.

4 Algorithms

While the values for some classes of properties can be computed by checking the embedded MDP of an MA, most need dedicated MA-specific algorithms. We briefly describe the algorithms implemented for MA in mcsta, STORM and IMCA.

4.1 Untimed and Expected-Reward Properties

Like for CTMC, properties that do not refer to time, or that only refer to expected times, can be computed on the embedded MDP of the Markov automaton. These properties include unbounded as well as branch reward-bounded reachability probabilities and expected accumulated rewards. For simplicity, we will refer to all of these as "unbounded properties". The available algorithms include all the standard exhaustive model checking algorithms for MDP [33], in particular using linear programming (LP), policy iteration, value iteration, interval iteration [5, 25], and sound value iteration [35]. Standard "unsound" value iteration and typical LP solvers do not provide any guarantees (such as ϵ-closeness to the true probability or value) on their results, while interval iteration and sound value iteration do. To combat the state space explosion problem of the exhaustive methods, the BRTDP learning-based approach [7] can be used for probabilities. It attempts to explore only a small part of the state space that is sufficient to provide a lower and an upper bound on the result that are close enough. Its efficiency both in terms of runtime and in terms of memory reduction highly depend on the structure of the model, though.

Tool Support. mcsta implements value iteration, LP, and interval iteration for expected rewards and unbounded reachability probabilities. It is being extended to support sound value iteration. It also provides BRTDP as in [2] where simulations with the uniform probabilistic scheduler are used to explore a part of the state space. After every batch of simulation runs, interval iteration is used to compute bounds. STORM implements value and policy iteration, LP, interval iteration, sound value iteration, and a variant of BRTDP. It also provides algorithms to compute exact (rational) solutions using exact arithmetic, but they are currently limited to small models. IMCA supports value iteration only.

4.2   Time-Bounded Reachability

Time-bounded properties pose one of the most challenging problems in MA model checking. Several algorithms with rather different characteristics are currently available for approximating time-bounded reachability probabilities: The **discretisation** approach [23] discretises the time horizon into small intervals, such that the MA will likely perform at most one Markovian transition within each interval. **Unif+** was first presented for CTMDP [13] and later extended to MA [21] in the straightforward way. It is based on an approximation of the optimal time-bounded reachability probability over timed schedulers with that same value but ranging over untimed schedulers. The **switch-step** algorithm [12] attempts to compute *switching points*: the points at which the optimal scheduler changes the action for at least one state, as illustrated in Example 2. Finally, the **BRTDP** idea for time-bounded reachability properties on CTMDP [2] can be extended to MA straightforwardly: the simulation phase performs CTMC-style simulation for Markovian states and MDP-style simulation over probabilistic states. Time progresses only over Markovian states and the simulation stops whenever the time bound expires or a target state is reached. Resolution of nondeterminism is performed via the randomised scheduler that samples the next action uniformly at random from the enabled actions. The analysis phase can be performed by any of the other algorithms for time-bounded analysis on MA.

Tool Support. mcsta implements Unif+ and switch-step while STORM supports Unif+ and the discretisation approach. Both provide sound implementations of these algorithms (i.e. they guarantee ϵ-correct results). IMCA implements only discretisation and uses unsound techniques for certain subproblems.

4.3   Long-Run Average Rewards

There exist two approaches for computing long-run average rewards: one based on a reduction to a linear program [24], and a value iteration-based algorithm [14] that approximates the reward up to a user-specified (and guaranteed) precision. In both cases, first the long-run average reward is determined for each maximal end component, then the end components are collapsed, and the overall result is computed as an expected reward value on the collapsed state space.

Tool Support. mcsta and STORM implement both of the algorithms while IMCA implements only the linear programming-based approach.

4.4   Other Verification Problems

We now briefly summarise other MA verification problems, name the corresponding available algorithms, and mention where they are implemented.

Time-bounded expected rewards extend the time-bounded reachability problem to rewards. The property represents the expected accumulated reward until a time bound is reached. Algorithmic support for this property is limited to the discretisation-based approach of [24], which is implemented in IMCA.

Resource-bounded rewards generalise both time-bounded reachability and time-bounded expected rewards. A resource-bounded reward property represents the expected accumulated reward within a finite resource budget. The resource is formally represented by a second type of (branch or rate) reward in the model. The only algorithm available to date is presented in [30], with no tool support.

Discounted Rewards. Expected discounted reward properties ask for the expected total reward where rewards collected at a certain time point are discounted with a value, depending on this time point. For example, when dealing with income, discounted rewards allow to take inflation into account. Iterative algorithms for computing and approximating the value exist, such as policy and value iteration [15]. There is however no tool support so far.

Multi-objective Tradeoffs. Multi-objective MA model checking allows finding a scheduler that is optimal for several objectives, rather than only one. The only algorithm available to date and implemented in STORM is presented in [34]. It does not support the full range of properties, in particular excluding long-run average and discounted rewards. For the underlying time-bounded analysis, it resorts to discretisation, which tends to not scale well (see Sect. 5 below).

5   Experiments

The Quantitative Verification Benchmark Set (QVBS, [29]) currently contains 18 MA models, specified in MODEST, STORM's extension of the PRISM language for

MA (cf. Sect. 3.2), as GREATSPN Petri nets, and as fault trees in the GALILEO format [37]. For every model, there is also a JANI version. All models have open parameters (like B in our MODEST example of Fig. 3) to be scaled up from small to huge state spaces. We use most of these models, selecting parameters that make for challenging, but not impossible, state space sizes (up to a few millions of states), to compare the performance and scalability of the algorithms implemented in mcsta with IMCA and STORM. The models include variations of queueing systems, dependability models, scheduling problems, and security case studies. We excluded those models that only have spurious non-determinism (i.e. they are equivalent to a CTMC), and those that can be fully checked in just a few seconds for the given parameter valuations. Due to the absence of long-run average reward properties in most MA models of the benchmark set, we added sensible long-run average properties to most of the MODEST models (which are easy to modify by hand, in contrast to JANI) in order to be able to do a meaningful performance comparison. Those are mainly steady-state probabilities (i.e. the special case of a rate reward of 1 in some states and of 0 in all others), or properties describing long-run average costs of running the modelled system.

All experiments were conducted on two servers with Intel Core i7-4790 processors and 16 resp. 32 GB of RAM running 64-bit Ubuntu Linux 18.04. We keep the default values for all the command line arguments of the tools, unless we explicitly mention specific parameters being used. When we request a certain precision for results (with sound methods), we request absolute, not relative, precision. We show all results as scatter plots like the one below, with log-log axes. Every benchmark *instance*—a model, a valuation for its parameters, and a property to check—results in one point in these plots. A point $\langle x, y \rangle$ states that the runtime of the tool noted on the x-axis on one instance was x seconds while the runtime of the tool noted on the y-axis was y seconds. Thus points above the solid diagonal line indicate instances where the x-tool was the fastest; it was more than ten times faster (slower) on points above (below) the dotted line. We set the timeout to 30 min; a timeout is denoted by an "x" dot in the plots.

5.1   mcsta and IMCA

The plot on the right compares the runtime of mcsta and IMCA on time-bounded ("tbr"), long-run average ("lra"), and unbounded properties ("unb"). The input of IMCA is an explicit representation of a state space (cf. Sect. 3.2). Thus, before a model can be analysed with IMCA, the state space has to be fully explored, transformed into this format, and saved to disk. This takes additional time and memory. Models of a few kB in MODEST lead to IMCA files of several GB. We use mcsta to perform this transformation, which took up to 200 s on each of the benchmarks we selected for our

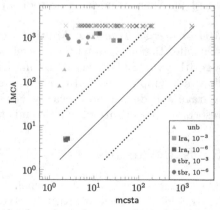

experiments. The runtime presented for IMCA does not include the time to generate input models, but only the time it takes to load them into memory and analyse them. For mcsta, we do include the time for state space exploration (from MODEST or JANI input). For all experiments, we chose the best runtime among all algorithms provided in each tool. For time-bounded properties we set the precision to 10^{-3} and 10^{-6}. The same holds for long-run properties for mcsta, but not for IMCA since its command-line interface does not support setting the precision for these properties. For unbounded properties we use the default parameters of both tools, including precision, since this once again cannot be changed for IMCA.

We see that IMCA performs far worse than mcsta. This is despite the fact that the considered runtime does not include time for model generation and that its only algorithm for time-bounded properties is unsound (with unsound methods tending to be faster than sound ones [35]), while the one of mcsta is sound. The performance gap is likely due to IMCA only implementing the discretisation-based approach, which is known to be inefficient [12,13], and not providing the most recent model checking algorithms for any of the property types.

5.2   mcsta and STORM

STORM, like mcsta, implements multiple and current algorithms. We thus present the results of this comparison in more detail. The runtimes for both tools include the time for state space exploration and for the numeric computations.

Time-Bounded Properties. Fig. 7 summarises the comparison of time-bounded solvers in mcsta and STORM. Once again we run experiments with precision values 10^{-3} and 10^{-6} and configure the tools to produce sound results. In the top-left plot we compare the best runtime for each tool among the algorithms that it implements; in the bottom-left plot, we compare mcsta's and STORM's implementations of Unif+. In both comparisons, mcsta achieves better runtimes than STORM. In particular, mcsta has no timeouts in the best-algorithm comparison. In the Unif+ comparison, mcsta and STORM both time out in some cases, yet whenever mcsta times out on a model, STORM does so, too (the "x" dot on the 45° line is actually a superposition of several such dots here). The two plots on the right compare the runtime of the switch-step implementation in mcsta with Unif+ in both tools. We do not compare to the discretisation algorithm for time-bounded properties implemented in STORM due to the consistent reports [12,13] of its inefficiency (which we confirmed in Sect. 5.1 with IMCA). We observe that neither Unif+ nor switch-step dominates the other, no matter which tool is used. This is because none of the two algorithms is strictly better than the other. Consider the top-right plot: it compares switch-step and Unif+ in mcsta and confirms the results presented in [12] that the algorithms are good in complementary scenarios. There are cases where one of them times out while the other finishes quite fast, and vice-versa. In particular, Unif+ performs somewhat better when a lower precision is required. Overall, the individual algorithms for

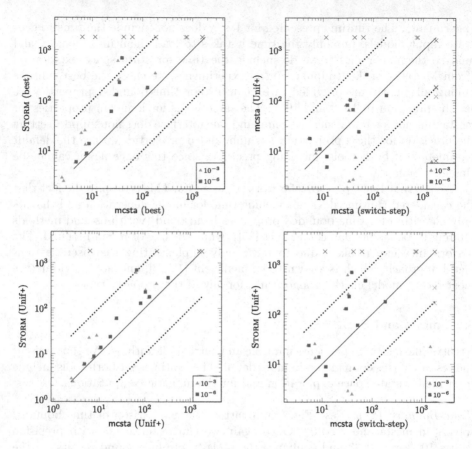

Fig. 7. Runtime of mcsta and STORM on time-bounded properties

time-bounded reachability in mcsta perform competitively, and especially when combined in a portfolio approach (i.e. using the best for each model, which could practically be done by running both concurrently on a multi-core system), offer noticeably better performance and scalability than STORM overall.

Long-Run Average Properties. Fig. 8 summarises the comparison of algorithms for model checking long-run average properties in mcsta and STORM. For value iteration-based algorithms ("VI"), we run experiments on precision values 10^{-3} and 10^{-6}, and use only its sound variations. For the linear programming-based approaches ("LP"), we set mcsta and STORM to use linear programming at *all* steps of the algorithm. The LP-based algorithms run with default parameters in both tools. For the top-left plot, we again chose the best runtime over the two algorithms for each tool. The LP-based approaches are not competitive: this can be seen from the three other plots. Here, the bottom-right plot shows that the LP-based algorithms in both mcsta and STORM run out of time on most of the benchmarks. In contrast, the VI-based solutions in both tools finish the

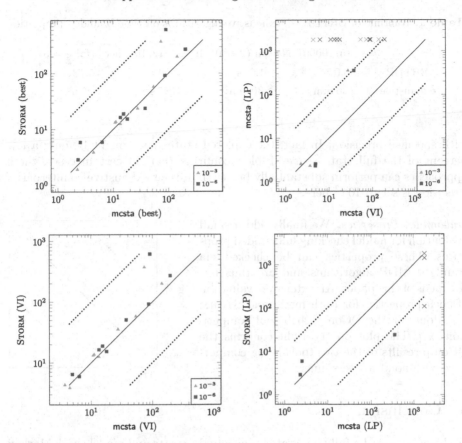

Fig. 8. Runtime of mcsta and STORM on long-run average reward properties

computations on the same benchmarks within the given time bound, as can be seen from the bottom-left and top-right plots. The exact reason for this is hard to extract. It may be possible that, when dealing with long-run properties, the LP-based approach itself is not as efficient as the one using VI, at least on existing benchmarks. Alternatively, it may be that the underlying LP algorithms or their implementations are not efficient. Overall, mcsta and STORM are roughly on par, albeit with mcsta having a few instances where it is significantly faster. The overall similarity is likely due to the set of implemented algorithms being exactly the same. We do notice, though, that specifically STORM's LP method appears to work better than mcsta's.

BRTDP. We compared exhaustive algorithms, i.e. those that perform computations on the full state space, with their BRTDP extensions in mcsta on a few benchmarks for time-bounded and unbounded properties. Table 1 summarises the results. BRTDP is useful in cases where the property under consideration does not require the full state space to be explored in order to achieve results

Table 1. Runtime of BRTDP vs. exhaustive algorithms on time-bounded properties

	vgs $(5, 10000)$	stream (20000)	ftwc $(512, 10)$	hecs $(false, 4, 3)$
BRTDP	$6.07\,s$	$1.91\,s$	$6.69\,s$	$5.78\,s$
exhaustive	$>30\,min$	$>30\,min$	$1077\,s$	$>30\,min$

with specified precision. In fact, the explored state space might be only a few percent of the full state space. Table 1 confirms that, in certain cases, these approaches can perform substantially better than their exhaustive counterparts. Precision is set to 10^{-3} here.

Unbounded Properties. We finally add a small evaluation for model checking unbounded properties. These properties can be checked via standard MDP algorithms and are thus not the focus of this paper. An extensive evaluation of such properties for both mcsta and STORM was done for the QComp 2019 tool competition [26]. The plot on the right confirms the QComp results of the two tools being competitive with no absolute winner.

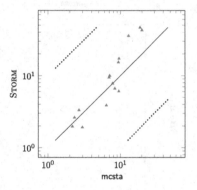

6   Conclusion

We have presented a fully integrated toolchain to create and model check Markov automata models based on the high-level compositional modelling language MODEST and the mcsta model checker of the MODEST TOOLSET. Other tools of the MODEST TOOLSET complement the approach, such as the modes simulator that helps deal with models too large for traditional model checking, or the moconv tool that can export MODEST models to JANI. We have compared the performance of the dedicated MA model checking algorithms in mcsta with IMCA and STORM. We found mcsta to significantly outperform IMCA, and to be faster than STORM in many cases. The JANI support in both the MODEST TOOLSET and STORM allows the user to choose the most appropriate tool in every instance, thus mcsta and STORM ought to be seen as complementary tools for a common goal. Overall, Markov automata now have user-friendly modelling and efficient verification support in tools that are actively maintained.

Data Availability. The data generated in our experimental evaluation as well as instructions to replicate the experiments are archived and available at DOI 10.4121/uuid:98d571be-cdd4-4e5a-a589-7c5b1320e569 [11].

References

1. Amparore, E.G., Balbo, G., Beccuti, M., Donatelli, S., Franceschinis, G.: 30 years of GreatSPN. In: Fiondella, L., Puliafito, A. (eds.) Principles of Performance and Reliability Modeling and Evaluation. SSRE, pp. 227–254. Springer, Cham (2016). https://doi.org/10.1007/978-3-319-30599-8_9
2. Ashok, P., Butkova, Y., Hermanns, H., Křetínský, J.: Continuous-time Markov decisions based on partial exploration. In: Lahiri, S.K., Wang, C. (eds.) ATVA 2018. LNCS, vol. 11138, pp. 317–334. Springer, Cham (2018). https://doi.org/10.1007/978-3-030-01090-4_19
3. Baier, C., de Alfaro, L., Forejt, V., Kwiatkowska, M.: Model checking probabilistic systems. In: Clarke, E., Henzinger, T., Veith, H., Bloem, R. (eds.) Handbook of Model Checking, pp. 963–999. Springer, Cham (2018). https://doi.org/10.1007/978-3-319-10575-8_28
4. Baier, C., Haverkort, B.R., Hermanns, H., Katoen, J.P.: Performance evaluation and model checking join forces. Commun. ACM **53**(9), 76–85 (2010)
5. Baier, C., Klein, J., Leuschner, L., Parker, D., Wunderlich, S.: Ensuring the reliability of your model checker: interval iteration for Markov decision processes. In: Majumdar, R., Kunčak, V. (eds.) CAV 2017. LNCS, vol. 10426, pp. 160–180. Springer, Cham (2017). https://doi.org/10.1007/978-3-319-63387-9_8
6. Boudali, H., Crouzen, P., Stoelinga, M.: A rigorous, compositional, and extensible framework for dynamic fault tree analysis. IEEE Trans. Dependable Sec. Comput. **7**(2), 128–143 (2010)
7. Brázdil, T., et al.: Verification of Markov decision processes using learning algorithms. In: Cassez, F., Raskin, J.-F. (eds.) ATVA 2014. LNCS, vol. 8837, pp. 98–114. Springer, Cham (2014). https://doi.org/10.1007/978-3-319-11936-6_8
8. Brázdil, T., Hermanns, H., Krcál, J., Kretínský, J., Rehák, V.: Verification of open interactive Markov chains. In: FSTTCS. LIPIcs, vol. 18, pp. 474–485. Schloss Dagstuhl - Leibniz-Zentrum fuer Informatik (2012)
9. Budde, C.E., D'Argenio, P.R., Hartmanns, A., Sedwards, S.: A statistical model checker for nondeterminism and rare events. In: Beyer, D., Huisman, M. (eds.) TACAS 2018. LNCS, vol. 10806, pp. 340–358. Springer, Cham (2018). https://doi.org/10.1007/978-3-319-89963-3_20
10. Budde, C.E., Dehnert, C., Hahn, E.M., Hartmanns, A., Junges, S., Turrini, A.: JANI: quantitative model and tool interaction. In: Legay, A., Margaria, T. (eds.) TACAS 2017. LNCS, vol. 10206, pp. 151–168. Springer, Heidelberg (2017). https://doi.org/10.1007/978-3-662-54580-5_9
11. Butkova, Y.: A Modest approach to modelling and checking Markov automata (artifact). 4TU.Centre for Research Data (2019). https://doi.org/10.4121/uuid:98d571be-cdd4-4e5a-a589-7c5b1320e569
12. Butkova, Y., Fox, G.: Optimal time-bounded reachability analysis for concurrent systems. In: Vojnar, T., Zhang, L. (eds.) TACAS 2019. LNCS, vol. 11428, pp. 191–208. Springer, Cham (2019). https://doi.org/10.1007/978-3-030-17465-1_11
13. Butkova, Y., Hatefi, H., Hermanns, H., Krčál, J.: Optimal continuous time Markov decisions. In: Finkbeiner, B., Pu, G., Zhang, L. (eds.) ATVA 2015. LNCS, vol. 9364, pp. 166–182. Springer, Cham (2015). https://doi.org/10.1007/978-3-319-24953-7_12
14. Butkova, Y., Wimmer, R., Hermanns, H.: Long-run rewards for Markov automata. In: Legay, A., Margaria, T. (eds.) TACAS 2017. LNCS, vol. 10206, pp. 188–203. Springer, Heidelberg (2017). https://doi.org/10.1007/978-3-662-54580-5_11

15. Butkova, Y., Wimmer, R., Hermanns, H.: Markov automata on discount! In: German, R., Hielscher, K.-S., Krieger, U.R. (eds.) MMB 2018. LNCS, vol. 10740, pp. 19–34. Springer, Cham (2018). https://doi.org/10.1007/978-3-319-74947-1_2
16. D'Argenio, P.R., Hartmanns, A., Sedwards, S.: Lightweight statistical model checking in nondeterministic continuous time. In: Margaria, T., Steffen, B. (eds.) ISoLA 2018. LNCS, vol. 11245, pp. 336–353. Springer, Cham (2018). https://doi.org/10.1007/978-3-030-03421-4_22
17. Dehnert, C., Junges, S., Katoen, J.-P., Volk, M.: A STORM is coming: a modern probabilistic model checker. In: Majumdar, R., Kunčak, V. (eds.) CAV 2017. LNCS, vol. 10427, pp. 592–600. Springer, Cham (2017). https://doi.org/10.1007/978-3-319-63390-9_31
18. Eisentraut, C.: Principles of Markov automata. Ph.D. thesis, Saarland University, Germany (2017)
19. Eisentraut, C., Hermanns, H., Katoen, J.-P., Zhang, L.: A semantics for every GSPN. In: Colom, J.-M., Desel, J. (eds.) PETRI NETS 2013. LNCS, vol. 7927, pp. 90–109. Springer, Heidelberg (2013). https://doi.org/10.1007/978-3-642-38697-8_6
20. Eisentraut, C., Hermanns, H., Zhang, L.: On probabilistic automata in continuous time. In: LICS, pp. 342–351. IEEE Computer Society (2010)
21. Gros, T.P.: Markov automata taken by Storm. Master's thesis, Saarland University, Germany (2018)
22. Guck, D., Han, T., Katoen, J.-P., Neuhäußer, M.R.: Quantitative timed analysis of interactive Markov chains. In: Goodloe, A.E., Person, S. (eds.) NFM 2012. LNCS, vol. 7226, pp. 8–23. Springer, Heidelberg (2012). https://doi.org/10.1007/978-3-642-28891-3_4
23. Guck, D., Hatefi, H., Hermanns, H., Katoen, J.P., Timmer, M.: Analysis of timed and long-run objectives for Markov automata. Logical Methods Comput. Sci. **10**(3), 1–29 (2014). https://doi.org/10.2168/LMCS-10(3:17)2014
24. Guck, D., Timmer, M., Hatefi, H., Ruijters, E., Stoelinga, M.: Modelling and analysis of Markov reward automata. In: Cassez, F., Raskin, J.-F. (eds.) ATVA 2014. LNCS, vol. 8837, pp. 168–184. Springer, Cham (2014). https://doi.org/10.1007/978-3-319-11936-6_13
25. Haddad, S., Monmege, B.: Interval iteration algorithm for MDPs and IMDPs. Theor. Comput. Sci. **735**, 111–131 (2018)
26. Hahn, E.M., et al.: The 2019 comparison of tools for the analysis of quantitative formal models. In: Beyer, D., Huisman, M., Kordon, F., Steffen, B. (eds.) TACAS 2019. LNCS, vol. 11429, pp. 69–92. Springer, Cham (2019). https://doi.org/10.1007/978-3-030-17502-3_5
27. Hahn, E.M., Hartmanns, A., Hermanns, H., Katoen, J.P.: A compositional modelling and analysis framework for stochastic hybrid systems. Formal Methods Syst. Des. **43**(2), 191–232 (2013)
28. Hartmanns, A., Hermanns, H.: The Modest Toolset: an integrated environment for quantitative modelling and verification. In: Ábrahám, E., Havelund, K. (eds.) TACAS 2014. LNCS, vol. 8413, pp. 593–598. Springer, Heidelberg (2014). https://doi.org/10.1007/978-3-642-54862-8_51
29. Hartmanns, A., Klauck, M., Parker, D., Quatmann, T., Ruijters, E.: The quantitative verification benchmark set. In: Vojnar, T., Zhang, L. (eds.) TACAS 2019. LNCS, vol. 11427, pp. 344–350. Springer, Cham (2019). https://doi.org/10.1007/978-3-030-17462-0_20
30. Hatefi, H.: Finite horizon analysis of Markov automata. Ph.D. thesis, Saarland University, Germany (2017). scidok.sulb.uni-saarland.de/volltexte/2017/6743/

31. Krcál, J., Krcál, P.: Scalable analysis of fault trees with dynamic features. In: DSN, pp. 89–100. IEEE Computer Society (2015)
32. Legay, A., Sedwards, S., Traonouez, L.-M.: Scalable verification of Markov decision processes. In: Canal, C., Idani, A. (eds.) SEFM 2014. LNCS, vol. 8938, pp. 350–362. Springer, Cham (2015). https://doi.org/10.1007/978-3-319-15201-1_23
33. Puterman, M.L.: Markov Decision Processes: Discrete Stochastic Dynamic Programming. Wiley, New York (1994)
34. Quatmann, T., Junges, S., Katoen, J.-P.: Markov automata with multiple objectives. In: Majumdar, R., Kunčak, V. (eds.) CAV 2017. LNCS, vol. 10426, pp. 140–159. Springer, Cham (2017). https://doi.org/10.1007/978-3-319-63387-9_7
35. Quatmann, T., Katoen, J.-P.: Sound value iteration. In: Chockler, H., Weissenbacher, G. (eds.) CAV 2018. LNCS, vol. 10981, pp. 643–661. Springer, Cham (2018). https://doi.org/10.1007/978-3-319-96145-3_37
36. Rabe, M.N., Schewe, S.: Finite optimal control for time-bounded reachability in CTMDPs and continuous-time Markov games. Acta Inf. **48**(5–6), 291–315 (2011)
37. Sullivan, K.J., Dugan, J.B., Coppit, D.: The Galileo fault tree analysis tool. In: FTCS-29, pp. 232–235. IEEE Computer Society (1999)
38. Timmer, M., Katoen, J.-P., van de Pol, J., Stoelinga, M.I.A.: Efficient modelling and generation of Markov automata. In: Koutny, M., Ulidowski, I. (eds.) CONCUR 2012. LNCS, vol. 7454, pp. 364–379. Springer, Heidelberg (2012). https://doi.org/10.1007/978-3-642-32940-1_26

Finite Approximation of LMPs for Exact Verification of Reachability Properties

Gildas Kouko, Josée Desharnais$^{(\boxtimes)}$, and François Laviolette

Department of Computer Science, Université Laval, Québec, Canada
gildas.kouko.1@ulaval.ca,
{josee.desharnais,francois.laviolette}@ift.ulaval.ca

Abstract. We give a discretization technique that allows one to check reachability properties in a family of continuous-state processes. We consider a sub-family of labelled Markov processes (LMP), whose transitions can be defined by uniform distributions, and simple reachability formulas.

The key of the discretization is the use of the mean-value theorem to construct, for a family of LMPs and reachability properties, a (finite) Markov decision process (MDP) equivalent to the initial (potentially infinite) LMP with respect to the formula. On the MDP obtained, we can apply known algorithms and tools for probabilistic systems with finite or countable state space. The MDP is constructed in such a way that the LMP satisfies the reachability property if and only if the MDP also satisfies it. Theoretically, our approach gives a precise final result. In practice, this is not the case, of course, but we bound the error on the formula with respect to the errors that can be introduced in the computation of the MDP. We also establish a bisimulation relation between the latter and the theoretical MDP.

1 Introduction

Systems verification is a major task and for critical systems, formal techniques are important. We focus on model-checking of continuous state-space systems with discrete time, called labelled Markov processes (LMPs) [7]. We consider a sub-family whose transition functions can be defined by uniform distributions.

In model-checking one first abstracts the behaviour of a system in the form of a transitions system, the model. Then, a property of interest is written in a temporal logic, and a software, called model-checker, is used to verify automatically if the model satisfies the property. There are models of continuous state space systems [1,12,16], and of continuous time processes [5], and combination of both, as well as continuous dynamics, in Stochastic Hybrid Systems [3]. Model-checking these systems is challenging, because of the continuous nature of the state spaces, and it usually involves approximating them.

In this paper, we develop a technique to construct a finite system from an LMP in such a way that a family of reachability properties will be checked exactly. We consider a sub-family of labelled Markov processes (LMP), whose

D. Parker and V. Wolf (Eds.): QEST 2019, LNCS 11785, pp. 70–87, 2019.
https://doi.org/10.1007/978-3-030-30281-8_5

transition function can be defined by uniform distributions, and simple reachability formulas (without nested reachability operator). More precisely, we use the *mean-value theorem* to prove that, for a family of LMPs and for a finite set of simple reachability properties, there is a Markov decision process (MDP) equivalent to the LMP with respect to the formulas. On the MDP, we can apply known algorithms and tools for probabilistic systems with finite or countable state space. The MDP is constructed in such a way that we can infer that the LMP satisfies the reachability property if and only if the MDP also satisfies it. Theoretically, our approach gives a precise final result. The initial goal of the present work was to improve CISMO [17], a model checker on LMPs, and there is indeed an implementation of the proposed technique in this tool. However the idea is theoretical and could be used in other tools. We show how the result can be extended to other distributions than uniform under some conditions.

At implementation, since the numerical computation for the MDP is subjected to numerical errors, the result can be inexact, as expected. In the implementation, we use a numerical integration method to determine the transition probabilities in the MDP. Despite the errors that can affect the outcome of a verification, we show that our approach makes sense at implementation by quantifying the errors. We show on the one hand that numerical errors are always bounded from above in the MDP and we establish, on the other hand, approximate bisimilarity between the MDP constructed theoretically and the MDP generated algorithmically with errors.

Other approximations have been defined on LMPs but they are general and hence cannot guarantee, even theoretically, equivalent satisfaction of formulas between the LMP and its approximant [11,13]. Those that quotient with respect to logical formulas do so for a simple logic without reachability properties [9]. There is much work on the verification of Stochastic Hybrid Systems, but as far as we know, the exact discretization that we obtain in this paper has not been considered before.

The plan of the paper is as follows. In the next section, we define the basic model and logic that we will work with; in Sect. 3, we show how the discretization is defined; in Sect. 4, we analyze the properties of the construction, including its limitation. We conclude in Sect. 5.

2   Preliminaries

The model of continuous-state systems we consider is a restriction of LMPs. Throughout the years, LMPs have been studied a lot, mostly on their theoretical aspect, and in full generality. When comes the need to actually check properties of an LMP, some restrictions must be chosen. In a general LMP [12], the set of states S is arbitrary and equipped with a sigma-algebra Σ, forming an analytic space. To model the probabilities associated with the change of states, there are transition functions indexed with actions: $\mu_a : S \times \Sigma \to [0,1]$ such that for all $s \in S$, the function $\mu_a(s,.)$ is a probability measure and for all $E \in \Sigma$, the function $\mu_a(.,E)$ is measurable. LMPs are deterministic: there is only one function μ_a for any action a; consequently, given a state and an action, only one distribution with action a exists. Even if LMPs can be as general as possible, when one wants to actually describe a system, one is forced to use a finite list of states and transitions, with the help of parametrized functions, most likely. In particular, the set of states will most of the time be the reals, or $\mathbb{R}^n$. Most of the time, parametrization of transitions will force to split the state space into intervals or their equivalent [3]. This explains why our restriction specifically exhibits a set of intervals. Now because of the construction we propose further on, we also assume that the transitions follow the uniform law (this can be relaxed, as we will argue later on). In the following definition we use the notation $U(I)$ for the uniform law on the interval I.

Definition 1. *An LMP_U is a tuple $\mathcal{S} = (S, I, i, AP, Act, Label, \{\mu_a\}_{a \in Act})$ where*

- *$S \subseteq \mathbb{R}$ is a set of states; $i \in S$ is the initial state;*
- *I is a finite set of pairwise disjoint intervals, a partition of $S = \cup I$;*
- *AP is a countable set of atomic propositions (on states);*
- *Act is a countable set of actions;*
- *$Label : AP \to \mathbb{B}(S)$ is a function that returns the (Borel) set of states that satisfy a label;*
- *A transition with action a has the form $\mu_a : S \times \mathbb{B}(S) \to [0,1]$, with the following restrictions. For each action a and pair $\iota_1, \iota_2 \in I$, a measurable function $f : \iota_1 \to [0,1]$ is associated to parametrize the (uniform) distribution on ι_2 and this is emphasised by the notation $\overline{\mu}_a(s \in \iota_1, \iota_2) \sim f(s)U(\iota_2)$. Because the total probability out of a state must be at most one, we require $\sum_{\iota \in I} \mu_a(s, \iota) \leq 1$ for all $s \in S$. Finally, we require that only one action is enabled in any state, that is, if $\mu_a(s, S) > 0$ and $\mu_b(s, S) > 0$, then $a = b$.*

It is straightforward to see that defining μ by pieces on the intervals of I is sufficient for its extension to $\mathbb{B}(S)$, and hence that an LMP_U is an LMP. In general LMPs, the labelling function of states is usually omitted. To lighten up the writing, instead of LMP_U we often write LMP and we assume an LMP $\mathcal{S}$ has set of states S.

In this definition, we write $\overline{\mu}_a(s \in \iota_1, \iota_2)$ as a notation for the behaviour of $\mu_a(s, -)$ inside the interval ι_2 (because $\mu_a(s, \iota_2)$ is actually a value). The interpretation of $\overline{\mu}_a(s \in \iota_1, \iota_2) \sim f(s)U(\iota_2)$ is that $f(s)$ multiplies the value of the uniform distribution on ι_2. So, if $\overline{\mu}_a(s \in [1,2], [3,5)) \sim (s-1)^2 U([3,5])$, then we get the following computations:

$$\mu_a(s, [3.5, 4]) = (s-1)^2 \frac{|[3.5, 4]|}{|[3, 5]|} = (s-1)^2 \frac{0.5}{2} \quad \text{for } s \in [1, 2]$$

$$\mu_a(1.1, [3.5, 4]) = (1.1 - 1)^2 (0.25) = 0.01 \cdot 0.25$$

$$\mu_a(2, [3.5, 4]) = (2 - 1)^2 (0.25) = 0.25.$$

The probability $\frac{0.5}{2}$ to a subset of $[3, 5]$ is relative to its length, because the distribution is uniform. Finally, as expected, for a set $S_0 \in \mathbb{B}(S)$ overlapping more than one element of I, the probability $\mu_a(s, S_0)$ is computed by partitioning S_0 with respect to the sets in I and then summing up the probabilities:

$$\mu_a(s, S_0) = \sum_{\iota \in I} \mu_a(s, \iota \cap S_0).$$

Example 1. *An LMP_U is illustrated in Fig. 1; $S = [0, 7] \cup \{10\}$, $Act = \{a, b, e\}$; $i = 0$. The arrows must be interpreted as follows, the variable involved in the label of an arrow belonging to the set in the source of the arrow.*

- *The arrow from $\{0\}$ to $]0, 1]$ is for $\overline{\mu}_a(0,]0, 1]) \sim \frac{1}{4}U(0, 1)$.*
- *The arrow from $]1, 2]$ to $]5, 7]$ is for $\overline{\mu}_b(z \in]1, 2],]5, 7]) \sim \frac{z^2}{4}U(5, 7)$*
- *The arrow from $]5, 7]$ to $\{10\}$ means that $\mu_e(k \in]5, 7], \{10\}) = \frac{k^2}{100}$*
- *For all other values of x, $\mu_a(x, \cdot)$ and $\mu_b(x, \cdot)$ have value 0.*

We see that the total probability out of 0 is $\frac{1}{2}$, the total probability out of any $z \in$ $]1, 2]$ is $\frac{z^2}{4} + \frac{4-z^2}{4} = 1$. Note that $\{0\}$ could be merged with the interval state $]0, 1]$ and follow the parametrized transitions out of $]0, 1]$, as $\overline{\mu}_a(0,]0, 1]) \sim \frac{0+1}{4}U(0, 1)$ and $\overline{\mu}_a(0,]2, 3]) \sim \frac{0}{4}U(2, 3)$.

This model is neither a restriction of continuous-time Markov chains nor of hybrid systems[1]. However, it is a discrete time stochastic hybrid system.

The discretization that we will define from LMP_U are finite state LMPs, a model very similar to MDPs [6,19].

Definition 2. *A tuple $(S, i, AP, Act, Label, P)$ is a Markov decision process (MDP) if S is a countable set of states, $i \in S$, AP, Act and $Label$ are as in Definition 1, $P : S \times Act \times S \to [0, 1]$ is the transition probability function; for all states $s \in S$ and actions $a \in Act : \sum_{s' \in S} P(s, a, s') \leq 1$.*

Finally, we define the logic, denoted $\mathbf{L_1}$, a branch of PCTL [2,14].

Definition 3. *The logic $\mathbf{L_1}$ is a subset of the logic PCTL, restricting to state formulas ϕ without nested Until operators. PCTL follows the following syntax:*

$$\phi := T \mid p \mid \neg\phi \mid \phi \vee \phi \mid \mathbb{P}_q(\Psi) \qquad \Psi := \langle a \rangle \phi \mid \phi U \phi$$

where $p \in AP$, $q \in [0, 1]$, $a \in Act$.

Examples of formulas of $\mathbf{L_1}$ are $\mathbb{P}_q(\langle a \rangle \mathbb{P}_r(\langle b \rangle T))$ and $\mathbb{P}_q(a U b)$; the latter is the typical one for which we propose a finite equivalent MDP. The logic $\mathbf{L_1}$ does not contain formula $\mathbb{P}_r(\langle a \rangle \mathbb{P}_q(a U b))$, but it accepts formulas $\mathbb{P}_q(a U \mathbb{P}_r(\langle a \rangle \langle b \rangle)T)$ and $\mathbb{P}_q(a U b) \wedge \mathbb{P}_r(T U c)$.

The semantics is as usual (see [2,14]), we summarize it here. A path $\pi = s_0 a_0 s_1...$ (an alternation of states and actions) satisfies $\langle a \rangle \phi$ if and only if $a_0 = a \wedge s_1 \models \phi$. A path satisfies $\phi U \varphi$ as for the well-known logic LTL. A state

---

[1] See [4] for a comparison between hybrid systems and LMPs on $\mathbb{R}$ with a finite list of transitions (similar but a little more general than LMP_U's).

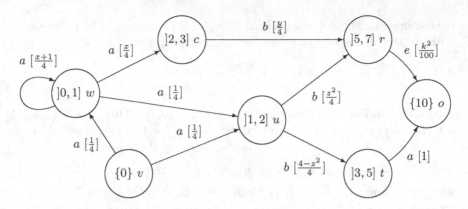

Fig. 1. An LMP$_U$ where $AP = \{w, v, c, u, r, t, o\}$, $x \in [0, 1]$, $y \in]2, 3]$, $k \in]5, 7]$, $z \in]1, 2]$.

satisfies formula $\mathbb{P}_q(\Psi)$ if the probability of the set of paths starting in this state that satisfy Ψ is greater than or equal to q:

$$s \models \mathbb{P}_q(\Psi) \text{ if } Prob_S(\{\pi \in \text{Path}(s) \mid \pi \models \Psi\}) \geq q. \qquad (1)$$

Finally, a process satisfies a formula if its initial state does. Note that Eq. (1) is well defined, because there is only one action enabled in any state.

3   The Mean MDP

Our goal is to verify, for an LMP, attainability properties of the form $\mathbb{P}_q(\phi U \psi)$ where ϕ and ψ are properties without Until operators. We prove that our result is theoretically exact; of course, during implementation, digital inaccuracies may distort the result, but this is part of any system analysis with numeric values. We limit ourselves to the LMPs having distributions of uniform probabilities on their states. The choice of the uniform law simplifies the use of the mean-value theorem. This provides us with an exact result, which ensures that the constructed MDP satisfies the formula if and only if the initial LMP also does (from its initial state, or states).

Here is our strategy. In the description of an LMP, since the set of intervals is finite, there is a finite number of transitions, which calls for a finite structure of transition system, as we can see in Fig. 1. If we can simply replace these transition functions by well-chosen probability values, we will get an MDP on which we can possibly check our property with a finite-state probabilistic model checker, like PRISM [15] (more precisely, on the initial state of the MDP). We will do just that, but we will first split the states along the formula $\phi U \psi$. We proceed in the following way:

1. We first determine the state intervals of the LMP that satisfy ϕ or ψ, using known strategy for the sublogic of PCTL without U-operator. If the initial state does not belong to any such interval, the LMP does not satisfy the property;

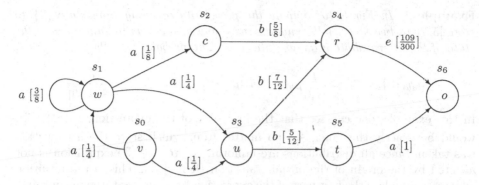

Fig. 2. MDP corresponding to the LMP of Fig. 1, for formula $T U o$.

2. we then construct an MDP from these states, the transition probabilities values being defined by the mean-value theorem;
3. the resulting MDP is fed to a model-checker like PRISM which will determine whether or not $\mathbb{P}_q(\phi U \psi)$ is satisfied.

Figure 2 illustrates the MDP resulting from the first two steps of our approach, for verifying formula $T U o$ on the LMP illustrated in Fig. 1 (there is no need to split the states here). In this example, s_1 represents the LMP's state interval $]0, 1]$ and $\frac{3}{8}$ is the mean of the transition function $\frac{x+1}{4}$ over $]0, 1]$. There are infinitely many paths satisfying $T U o$; for example $s_0 a s_1 a s_2 b s_4 e s_6$ and $s_0 a s_1 a s_1 a s_1 a s_2 b s_4 e s_6$.

3.1 Mean of a Probability Transition

The *mean-value theorem for definite integrals* is a known result in calculus saying that the mean-value of a function on a given interval is the image of some point.

Theorem 1 ([8]). *For any real-valued function f, defined and continuous on a segment $[a, b] \subset \mathbb{R}$, there exists $c \in]a, b[$, satisfying $f(c) = \frac{1}{b-a} \int_a^b f(x) dx$. We denote by $M_f^{[a,b]}$ the mean of f on $[a, b]$. In general, if $f : [a, b] \to R$ is continuous and g is an integrable function that does not change sign on $[a, b]$, then there exists $c \in (a, b)$ such that $\int_a^b f(x) g(x) dx = f(c) \int_a^b g(x) dx$.*

For example, consider $f : [0, 2] \to \mathbb{R}$, $f(x) = 2x$. Since f is integrable on $[0, 2]$ and $\int_0^2 f(x) = 4$ then the mean-value of f is $M_{2x}^{[0.2]} = \frac{1}{2} \int_0^2 f(x) = 2$ and $c = 1$.

Remark 1. At this point, the reader may wonder why we would work through the computation of an integral if we can avoid it. The point is that we cannot. When evaluating a reachability property, one must sum up through all possible paths. Even just to compute a path of length two, one has to evaluate an integral, as illustrated in the following example.

Example 2. *In Fig. 1, to compute the probability of going from state $\{0\}$ to state $]5, 7]$ in two steps, following actions a and b, one has to sum up over the state of the intermediate interval state $]1, 2]$. The probability is then*

$$\mu_{ab}(\{0\},]5, 7]) := \int_1^2 \mu_b(z,]5, 7])\mu_a(\{0\}, dz) = \int_1^2 \frac{z^2}{4} \cdot \frac{dz}{4|]1, 2]|}$$

In this example, one can see that the first part of the expression, $\mu_b(z,]5, 7])$, would be exactly the same, $\frac{z^2}{4}$, no matter from which state the a-transition was taken. Since all distributions are uniform, the rest of the expression is not affected by the origin of the initial transition (here $\{0\}$). This is the intuition why we can replace the first part of the expression by a constant, the mean value.

If we want to ultimately check the property on the LMP in practice, we will need to actually compute the mean value, which forces us to evaluate an integral. Usually, an integral is evaluated using classical mathematical formulas for determining primitives. For example, the primitive of $f(x) = \frac{1}{x}$, $x \neq 0$, is $\ln|x|$. Generally it is difficult to construct algorithmically a program that derives or calculates the primitive of mathematical functions. Due to this difficulty, we chose the numerical evaluation of an integral. The numerical integration techniques are numerous and diverse [18]. Two fundamental aspects are observed in their use: the desired precision and the calculation time. In terms of numerical integral evaluation, there are two main categories of methods: deterministic methods, like the rectangle method, the trapezoidal rule, Romberg's method, Simpson's rule, Gauss', etc; there are also probabilistic methods, like the Monte Carlo techniques. In our implementation, we have chosen the trapezoidal rule.

3.2   From LMP to MDP

We construct an MDP, from an LMP $\mathcal{S}$, that will satisfy the reachability property $\beta = \mathbb{P}_q(\phi U \psi)$ if and only if $\mathcal{S}$ also does. As is usually done when verifying a property of the form $\phi U \psi$, the first step is to keep in $\mathcal{S}$ only the states that satisfy ϕ or ψ. These formulas come from a syntax without reachability properties, and they already taken care of with direct techniques, as is done in CISMO, for example, so we do not give anymore detail on the matter. W.l.o.g., we can thus assume that we want to check a property $\beta = \mathbb{P}_q(\phi U \psi)$ in an LMP $\mathcal{S}$, where the states satisfying ϕ and ψ have already been computed. In the following, the sets S_ϕ and S_ψ have this role (and hence, they could be any sets of states).

To define the states of the MDP, we refine the partition I of the state space S with respect to the sets S_ϕ and S_ψ. The following definition is similar to the function $\lceil . \rceil$ presented in [16].

Definition 4. *Let $\mathcal{S} = (S, I, i, Act, Label, \{\mu_a\}_{a \in Act})$, $S_\phi, S_\psi \subseteq S$. The set of intervals states(I, S_ϕ, S_ψ) is a partition of $S_\phi \cup S_\psi$ that satisfies*
1. *each interval of states(I, S_ϕ, S_ψ) is included in (exactly) one interval of I*
2. *states(I, S_ϕ, S_ψ) contains a partition of $S_\phi \setminus S_\psi$ and a partition of S_ψ*
3. *it is minimal: replacing any two intervals $I_1, I_2 \in states(I, S_\phi, S_\psi)$ with their union $I_1 \cup I_2$ violates item 1 or 2.*

Note that Condition 3, on minimality, is not necessary, but it gives a unique construction, and will yield an MDP with the least number of state. No proof relies on this condition.

Example 3. *Consider the LMP of Fig. 1. We have* $S = [0,7] \cup [10,10]$, $I = \{[0,0], [0,1],]1,2],]2,3],]3,5],]5,7], [10,10]\}$ *with initial state 0. Suppose* $S_\phi = [0,0] \cup]1,2] \cup [3,4]$ *and* $S_\psi =]2.5,3.5]$, *then* $S_\phi \setminus S_\psi = [0,0] \cup]1,2] \cup]3.5,4]$ *and the minimal refinement is given by*

$$states(I, S_\phi, S_\psi) = \{[0,0],]1,2],]2.5,3],]3,3.5],]3.5,4]\}.$$

We are now ready to define the MDP from the LMP.

Definition 5. *Let* $S = (S, I, i, AP, Act, Label, \{\mu_a\}_{a \in Act})$ *be an LMP, and let* $S_\phi, S_\psi \subseteq S$, *with* $i \in S_\phi \cup S_\psi$. *The mean MDP associated to* S, S_ϕ, S_ψ *is defined as:*

$$S_{\phi U \psi} = (states(I, S_\phi, S_\psi), I_{init}, AP, Act, Label, \mu')$$

where I_{init} *is the unique interval of* $states(I, S_\phi, S_\psi)$ *containing* i, *and* μ' *is defined as follows: for* $I_1, I_2 \in states(I, S_\phi, S_\psi)$, *let* $\iota_1, \iota_2 \in I$ *be the unique intervals of* I *such that* $I_k \subseteq \iota_k$, $k = 1, 2$. *Then* S *contains some* $\overline{\mu}_a(x \in \iota_1, \iota_2) \sim f(x) U(\iota_2)$ *and we set*

$$\mu'(I_1, a, I_2) = \frac{|I_2|}{|\iota_2|} \frac{1}{|I_1|} \int_{I_1} f(x) dx.$$

$$= \frac{|I_2|}{|\iota_2|} M_f^{I_1},$$

where $M_f^{I_1}$ *is the mean value of* f *in interval* I_1 *(see Theorem 1).*

Note that the states of the MDP $S_{\phi U \psi}$ are intervals given by the minimal refinement, each interpreted as a single state, or label, or equivalence class. The value $\frac{|I_2|}{|\iota_2|}$ normalizes the uniform distribution value: the distribution μ_a, defined on ι_2 in S, is applied on the interval $I_2 \subseteq \iota_2$. Note that as for LMPs, the probability of transiting from one state of the mean MDP to the other states may not sum up to 1. However, they sum up to at most one.

Theorem 2. *For any LMP, the mean MDP* $S_{\phi U \psi}$ *associated to* S, S_ϕ, S_ψ *is indeed an MDP.*

Proof. First observe that since ϕ and ψ do not contain an Until operator, the determination of S_ϕ and S_ψ can be done for a general LMP_U. Now, the only thing that needs to be checked is that the total probability out of a state I_1 is less than or equal to 1 for every action.

$$\sum_{I_2 \in states(I,S_\phi,S_\psi)} \mu'(I_1,a,I_2) = \sum_{I_2 \in states(I,S_\phi,S_\psi)} \frac{|I_2|}{|\iota_2|} M_f^{I_1}$$

$$= \sum_{I_2 \in states(I,S_\phi,S_\psi)} \frac{|I_2|}{|\iota_2|} \frac{1}{|I_1|} \int_{I_1} f(x) dx.$$

$$= \frac{1}{|I_1|} \int_{I_1} \sum_{I_2 \in states(I,S_\phi,S_\psi)} f(x) \frac{|I_2|}{|\iota_2|} dx.$$

$$= \frac{1}{|I_1|} \int_{I_1} \sum_{I_2 \in states(I,S_\phi,S_\psi)} \mu_a(x,I_2) dx.$$

$$= \frac{1}{|I_1|} \int_{I_1} \mu_a(x, S_\phi \cup S_\psi) dx.$$

$$\leq \frac{1}{|I_1|} \int_{I_1} (1) dx.$$

$$= 1.$$

□

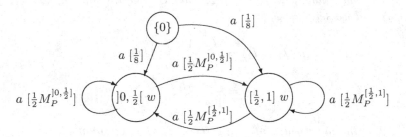

Fig. 3. Mean MDP of Fig. 1 with $S_\phi = [0, \frac{3}{4}]$ and $S_\psi = [\frac{1}{2}, 1]$.

Example 4. *Consider the LMP S of Fig. 1, with initial state 0. Suppose that $S_\phi = [0, \frac{3}{4}]$ and $S_\psi = [\frac{1}{2}, 1]$. Then the states of $S_{\phi U \psi}$ are illustrated in Fig. 3. Where P is the function defining the transitions of S of Fig. 1, so that M_P^I is the mean value of P in I. Let us make an observation about Rule 2 of the construction of $states(I, S_\phi, S_\psi)$. If this rule was omitted, there would be no partitioning of states in this example, because $(S_\phi \cup S_\psi) \setminus \{0\} =]0, 1]$ is entirely included in an interval of I. It would result in a less precise MDP with only one state apart from the initial state. Since states satisfying ϕ and ψ would not even be split, then we would not preserve the satisfiability of formula $\beta = \phi U \psi$. We illustrate in Fig. 4 a finer partition where S_ϕ is split (thus dropping the third condition of Definition 4). This discretization will have the same properties we are seeking in our approach. The downside of it is that there are more mean values to compute. The upside is that it may be useful for other reachability properties, for example for $\mathbb{P}_q(\psi U \phi)$.*

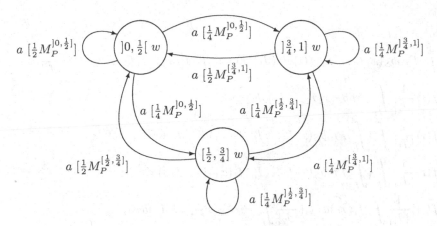

Fig. 4. A finer partition than in Fig. 3. (Initial state $\{0\}$ is omitted).

3.3 The Reachability Property in $\mathcal{S}$ and $\mathcal{S}_{\phi U \psi}$

We are at the end of the second stage of the approach. The aim is to verify $\beta = \phi U \psi$ in $\mathcal{S}$ through the mean MDP, so we must calculate the probability of the set of paths in $\mathcal{S}_{\phi U \psi}$ that satisfy $\phi U \psi$ and compare it to q. In the following theorem, the sets of paths $\text{Path}_{\mathcal{S}}$ and $\text{Path}_{\mathcal{S}_{\phi U \psi}}$ are initial paths, that is, starting at initial states.

Theorem 3. *Let $\mathcal{S}$ be an LMP and $\mathcal{S}_{\phi U \psi}$ a corresponding mean MDP. Then*

$$Prob_{\mathcal{S}}(\{\pi \in \text{Path}_{\mathcal{S}} \mid \pi \models \phi U \psi\}) = Prob_{\mathcal{S}_{\phi U \psi}}(\{\pi \in \text{Path}_{\mathcal{S}_{\phi U \psi}} \mid \pi \models \phi U \psi\}).$$

Proof. Recall that we assume that $\mathcal{S}$ is deterministic in the sense that only one action is possible in any state. We will discuss a generalisation later on. Let $I_1, I_2, \ldots, I_n$ be interval states of $\mathcal{S}_{\phi U \psi}$ and $a_1, a_2, \ldots, a_{n-1}$ be actions. The transition probabilities between these intervals are obtained from $\mathcal{S}$ as follows. Let $J_1, J_2, \ldots, J_n \in I$ be such that $I_k \subseteq J_k$, and $f, f_1, f_2, f_3 \ldots, f_{n-1}$ be the functions that satisfy $\overline{\mu}_{a_k}(x, J_{k+1}) \sim f_k(x) U(J_{k+1})$, $x \in J_k$ representing the transition function from J_k to J_{k+1} in $\mathcal{S}$, for $k = 1, \ldots, n$ and let $fU(J_1)$ be the transition function from i to I_1.

We define the sets of paths that pass through a sequence of intervals of I in $\mathcal{S}$ (where a path goes through individual states $i_k \in S$):

$$i \, a_1 I_1 a_2 I_2 a_2 I_3 \ldots a_{n-1} I_n := \{i \, a_1 \, i_1 \, a_2 \, i_2 \ldots a_n \, i_n \in \text{Path}_{\mathcal{S}}(s) \mid i_j \in J_j\}.$$

We say that a set of paths satisfy a property ϕ, and we write $X \models \phi$ if all the paths of the set satisfy it. To simplify the notation, we omit the actions below; they determine the functions f_k from I_k. We have:

$$Prob_S(iI_1I_2\ldots I_n)$$

$$= \int_{I_1} Prob_S(x_1I_2\ldots I_n)f(i)dU(J_1)(x_1) = \int_{I_1} Prob_S(x_1I_2\ldots I_n)f(i)\frac{1}{|J_1|}dx_1$$

$$= f(i)\frac{1}{|J_1|}\int_{I_1} Prob_S(x_1I_2\ldots I_n)\,dx_1$$

$$= f(i)\frac{1}{|J_1|}\int_{I_1}\left(\int_{I_2} Prob_S(x_2I_3\ldots I_n)f_1(x_1)dU(J_2)(x_2)\right)dx_1$$

$$= f(i)\frac{1}{|J_1|}\int_{I_1}\left(\int_{I_2} Prob_S(x_2I_3\ldots I_n)f_1(x_1)\frac{1}{|J_2|}dx_2\right)dx_1$$

$$= f(i)\left(\frac{1}{|J_1|}\int_{I_1} f_1(x_1)dx_1\right)\left(\frac{1}{|J_2|}\int_{I_2} Prob_S(x_2I_3\ldots I_n)dx_2\right)$$

$$\vdots$$

$$= f(i)\left(\frac{1}{|J_1|}|I_1|M_{f_1}^{I_1}\right)\left(\frac{1}{|J_2|}\int_{I_2} f_2(x_2)dx_2\right)\ldots\left(\frac{1}{|J_{n-2}|}\int_{I_{n-2}} f_{n-2}(x_{n-2})dx_{n-2}\right)$$

$$\cdot\left(\int_{I_{n-1}} Prob_S(x_{n-1}I_n)dU(J_{n-1})(x_{n-1})\right)$$

$$= f(i)\frac{|I_1|}{|J_1|}M_{f_1}^{I_1}\frac{|I_2|}{|J_2|}M_{f_2}^{I_2}\ldots\frac{|I_{n-2}|}{|J_{n-2}|}M_{f_{n-2}}^{I_{n-2}}\cdot\left(\frac{1}{|J_{n-1}|}\int_{I_{n-1}} f_{n-1}(x_{n-1})\frac{|I_n|}{|J_n|}dx_{n-1}\right)$$

$$= f(i)\frac{|I_1|}{|J_1|}M_{f_1}^{I_1}\frac{|I_2|}{|J_2|}M_{f_2}^{I_2}\ldots\frac{|I_{n-2}|}{|J_{n-2}|}M_{f_{n-2}}^{I_{n-2}}\left(\frac{|I_{n-1}|}{|J_{n-1}|}M_{f_{n-1}}^{I_{n-1}}\frac{|I_n|}{|J_n|}\right)$$

$$= \mu'(i,I_1)\,\mu'(I_1,I_2)\,\cdots\,\mu'(I_{n-1},I_n)$$

$$= Prob_{S_{\phi U\psi}}(iI_1I_2\ldots I_n).$$

The first equality above is based on the fact that the probability of the paths starting in i is equal to the probability of the paths starting on x_1 in I_1 weighted by the probability from i to these x_1 (i.e., $f(i)dU(J_1)(x_1)$), those of which there are uncountably many: we thus have to integrate. The uniform distribution $dU(J_1)(x_1)$ must take into account the size of J_1, resulting in the factor $\frac{1}{|J_1|}$ at the second equality, followed by the usual Lebesgue measure dx_1.

Note that $Prob_S(x_{n-1}I_n) = \mu_{a_{n-1}}(x_{n-1},I_n) = f_{n-1}(x_{n-1})\frac{|I_n|}{|J_n|}$ because the function from I_{n-1} to I_n is $f_{n-1}U(J_n)$.

We can now conclude the demonstration. The first equality below comes from the fact that the paths can be grouped according to the intervals they traverse in S. These sets are all disjoint, hence the second equality.

$$Prob_S(\{\pi \in Paths(L) \mid \pi \models \phi U\psi\}$$

$$= Prob_S\left(\bigcup_{iI_1I_2\ldots I_n \models (\phi\wedge\neg\psi)U\psi} iI_1I_2\ldots I_n\right)$$

$$= \sum_{i I_1 I_2 \ldots I_n \models (\phi \wedge \neg \psi) \, U \psi} Prob_S(i I_1 I_2 \ldots I_n)$$

$$= \sum_{i I_1 I_2 \ldots I_n \models (\phi \wedge \neg \psi) \, U \psi} Prob_{S_{\phi U \psi}}(i I_1 I_2 \ldots I_n)$$

$$= Prob_{S_{\phi U \psi}}(\{\pi \in \mathrm{Path}(S_{\phi U \psi}) \mid \pi \models \phi U \psi\}.$$

□

This theorem shows that as far as properties ϕ and ψ are computed exactly, then the computed MDP contains the same information as the initial LMP, with regards to reachability formulas involving ϕ and ψ. Hence we can use a model-checker for MDP to check if the property is satisfied.

Remark 2. Even if we do not enter into the details, let us comment on the restriction we impose, for simplicity, on LMPs. In general, starting for any LMP S, the MDP $S_{\phi U \psi}$ could have some states with multiple outgoing transitions (with different actions), each summing up to at most one. Thus, as is usually done, reachability properties must be evaluated with respect to some determination of the MDP, usually through schedulers, for example. The statement in the theorem, as well as the argument in most of the proof, do not detail on this matter. However, any fixed scheduler determines a *sub*-LMP that respects the condition of the theorem. Accordingly, the schedulers of S and $S_{\phi U \psi}$ are in one-to-one correspondence, the pair satisfying the theorem.

4 Analysis of the Mean MDP

4.1 Possible Extension, Simplification and Reusability

The technique is suited for a very specific type of formula and we have limited the analysis to probabilistic transition functions defined by the uniform law. In this subsection, we discuss how to extend and take advantage of the construction. First, the mean-value theorem also applies to real-valued functions of multiple variables, provided the domain is convex. This is why for our setting, we have split the state space in intervals; for higher dimensions, the state-space will have to be split with analogue structures, called hyper-rectangles. Second, it is interesting to observe an exact discretization for reachability properties has more impact than for finite depth formulas. Formulas are usually short, so the number of computations when evaluating a formula without reachability operator is relatively small, and hence so is the error generated. The errors do not accumulate that much, as the cumulation of errors comes from the depth of the formula – the number of $\langle a \rangle$ nested operators, more precisely. However, when a reachability property is evaluated, then we are forced to compute the probability of sets of *paths*, along which the errors accumulate. Hence, having on hand a discretization of the continuous system that is exact for some reachability properties can very much improve the quality of the verification.

Reusability of MDP We have assumed a property $\phi U \psi$. The same discretization would be suitable for another reachability property $\phi' U \psi'$ as long as $states(I, S_\phi, S_\psi)$ is a more refined partition than $states(I, S_{\phi'}, S_{\psi'})$. Figure 5 shows an example. Let $\beta = (u \wedge v)Uw$ and $\beta' = vUw$. Here, $S_{(u \wedge v)Uw}$ cannot be used to evaluate β'. However, if $\beta'' = uUw$ we have $S_{(u \wedge v)Uw}$ equal to S_{uUw}. In general, one could perform the first step of the approach by considering a set of reachability properties. The refinement of I would then be done in such a way that it also refines all the state subformulas of this set of properties, and not only S_ϕ and S_ψ. The discretization could then be used to check all properties. Alternatively, after a discretization with respect to some properties, one could adapt the result locally to take another reachability property into account, instead of computing all of it from scratch. For example, in Fig. 5 one could just add a state $]4, 8]$ to $S_{(u \wedge v)Uw}$ and obtain an MDP that would have the same reachability properties as S_{vUw} has. All previously computed transitions would still be correct. Indeed, the computation of the MDP is a local computation, from an interval to another interval. If an interval is split, one has to compute the probabilities of its ingoing and outgoing edges, but nothing more.

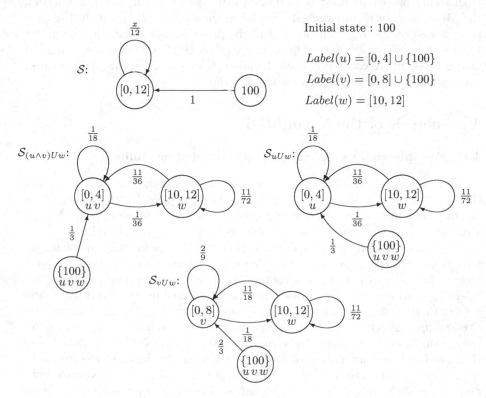

Fig. 5. LMP and corresponding MDPs for formulas $(u \wedge v)Uw$, uUw and vUw .

Initial State(s). The theorem is stated for sets of paths that start at the initial state. The requirement is necessary because, in the MDP, the probability from a (regrouped) state I to state I' is an aggregate from all states in the original interval I. If the latter is a singleton, the value is exact, else it may not be for a particular $s \in I$. The probability is exact only when used for path computations. More generally, states of the MDP whose original outgoing transition functions f are constant in $\mathcal{S}$ (restricted to S_ϕ and S_ψ) are also suitable as initial states in the theorem; this would be the case if this state interval contained all bisimilar states (w.r.t. $\mathcal{S}$ restricted to S_ϕ and S_ψ). Finally, all states that must be considered as initial could be extracted from the intervals they belong to in the partitioning of the states when constructing the MDP.

Uniform Distributions. We discuss the role of the assumption of uniform distributions. It is used in the proof of Theorem 3 when the integral $\int_{I_k} f_k(x_k) dU(J_k)(x_k)$ is factored out of the integral and computed independently, and then replaced by its value in the MDP. This is not so much because the distribution is uniform, but because the distribution into the interval I_k is *always the same*, no matter from where this jump is made! We observed this partially in Example 2. The partial integral above would become $\int_{I_k} f_k(x_k) d\mathcal{D}_k(x_k)$ for a general distribution $\mathcal{D}_k$ on I_k. As long as this distribution can be expressed with a density function g such as stated in the mean theorem (1), then the proof of Theorem 3 stays correct. So the result should apply to a generalisation of LMP_U where the transitions are restricted as follows

A transition with action a has the form $\mu_a : S \times \mathbb{B}(S) \to [0,1]$, with the following restrictions. For each pair $\iota_1, \iota_2 \in I$, we have $\overline{\mu}_a(s \in \iota_1, \iota_2) \sim f(s)\mathcal{D}_{\iota_2}$; where the distribution $\mathcal{D}_{\iota_2}$ on the interval ι_2 is fixed.

It is still restrictive, but note that if one wants to model two different distributions to interval ι_2, it can be done with a bisimilar LMP, where the latter interval is recopied twice, one copy for each of the two different distributions.

4.2   Error Analysis at Implementation – Comparison of $\mathcal{S}_\beta$ and $\widehat{\mathcal{S}}_\beta$

We now evaluate the numerical errors that can arise when *computing* in practice the mean MDP. In this section, the notation $\mathcal{S}_\beta$ refers to the theoretical MDP with exact computations; we write $\widehat{\mathcal{S}}_\beta$ for its actual computed version. The errors come from the computations of the mean values M_f^J, using integrals. We focus on the errors generated by our construction for $\beta = \phi U \psi$, and ignore the numerical errors that are due to the computation of the subformulas ϕ and ψ, and those from arithmetic calculations. These errors can be considered negligible, as they are mainly due to the floating point non recursive computations.

Consider a transition from $[a, b]$ to $[c, d]$ with distribution $fU[c, d]$ in $\mathcal{S}$. To compute the mean value, the errors depend on the method for numerical integration. In our implementation, we have chosen the trapezoidal rule. In $\mathcal{S}_\beta$, both intervals $[a, b]$ and $[c, d]$ may have been split. The errors only depend on the first

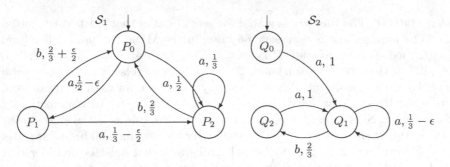

Fig. 6. $R = \{(P_0, Q_0), (P_0, Q_2), (P_1, Q_1), (P_2, Q_1)\}$ is an ϵ-bisimilation

interval, and it is in $\Theta(\frac{(b-a)^3}{12N^2} \sup_{[a,b]} f'')$ where N is the number of subdivisions of $[a, b]$. Hence, when the functions f are of degree less than or equal to one, the computations are exact. If set $[a, b]$ has been split, the error is smaller. If set $[c, d]$ has been split, there is no cumulation of errors: the mean is computed once on the interval at the starting point of the transition, it is then weighted by the size of the sub-interval of its arriving point. For example, consider $S_{(a \wedge bUc)}$ in Fig. 5; the means from state $[0, 4]$ to itself and to state $[10, 12]$ are obtained using the integral $\int_0^4 x/3 dx = 1/6$, a value that must then be multiplied by the ratio of the starting state $[0, 4]$ over the arriving states, which are, respectively, $1/3$ and $1/6$, yielding $1/18$ and $1/36$.

We now compare $\widehat{S_\beta}$, S_β and S through a notion of approximate bisimulation from [10]. Of course, none of these processes are equivalent (bisimilar), but we want to establish how close they are. Informally, two probabilistic systems are ϵ-bisimilar if their transition probabilities are within ϵ.

Definition 6. ([10]) *Let* $S = (S, I, i, AP, Act, Label, \{\mu_a\}_{a \in Act})$ *and* S' *be two LMPs. Let* $\epsilon \in [0, 1]$. *A symmetric relation* R *on* $S \cup S'$ *is an* ϵ-*bisimulation if for all* $s \in S$, $s' \in S'$, *such that* $s \, R \, s'$, *we have that* s *and* s' *are labelled with the same atomic proposition, and*

$$\forall a \in Act, \forall C \in \mathbb{B}(R), |\mu_a(s, C) - \mu'_a(s', C)| \leq \epsilon.$$

where we write $\mathbb{B}(R)$ *for the Borel sets* X *that are* R-*closed, that is, if* $x \in X$ *and* xRx', *then* $x' \in X$. *If there is an* ϵ-*bisimulation between* S *and* S', *we write* $S \sim_\epsilon S'$.

In Fig. 6, we give a simple example of ϵ-bisimilar finite LMPs, to help intuition. The notion captures, in particular, the similarities between a system and an approximation of its probabilities (which may have been obtained by observation or standard assumptions). Note that, as expected, 0-bisimulation gives the usual notion of bisimulation for LMPs.

Theorem 4. *Let* S *be an LMP, let* β *be the reachability property under study. Let* $\varepsilon_{s,a}$ *be the (absolute value of the) error at state* s *of* $\widehat{S_\beta}$ *for action* a, *that is,* *the sum of the errors of transitions from s with action a. Let $\varepsilon = \max_{s,a} \varepsilon_{s,a}$. Then $\mathcal{S}_\beta \sim_\varepsilon \widehat{\mathcal{S}_\beta}$.*

Moreover, if K is the maximal number of branching in LMP $\mathcal{S}$, if l is the maximal length of interval states in $\mathcal{S}_\beta$, if γ is $\sup f''$ (see note[2]) then ε is bounded above by $\frac{Kl^3\gamma}{12N^2}$, where N is the number of subdivisions in the trapezoidal rule.

Proof. The first part of the theorem is straightforward from the definition of ϵ-bisimulation. The relation R is given by the identity relation, considering that $\mathcal{S}_\beta$ and $\widehat{\mathcal{S}_\beta}$ have the same state space. Let us write μ^β and $\widehat{\mu}_a^\beta$ for the probabilistic transition functions of $\mathcal{S}_\beta$ and $\widehat{\mathcal{S}_\beta}$. Then, for any state s, action a and R-closed set X (of interval states), we have

$$|\mu_a^\beta(s, X) - \widehat{\mu}_a^\beta(s, X)| = |\sum_{I \in X} \mu_a^\beta(s, I) - \sum_{I \in X} \widehat{\mu}_a^\beta(s, I)|$$

$$\leq \sum_{I \in X} |\mu_a^\beta(s, I) - \widehat{\mu}_a^\beta(s, I)|$$

$$\leq \varepsilon_{s,a}.$$

For the second part of the theorem, it is well known that the trapezoidal rule applied to approximate $\int_b^c f(x)\,dx$ has error $\frac{|c-b|^3 f''(\xi)}{12N^2}$, where ξ is some value in the interval $[b, c]$ and N is the number of subdivisions used in the method. Thus from state s with action a, each outgoing transition generates this error, so $\varepsilon_{s,a}$ is bounded above by $\frac{\sum_i l_i^3 f''(\xi_i)}{12N^2}$, where l_i is the length of the target interval of the ith transitions, and ξ_i is the value where the mean is obtained. Since the number of outgoing a-transitions from s is bounded by K and $f''(\xi_i) \leq \gamma$, the result follows.   □

The relation between $\mathcal{S}$ and $\mathcal{S}_\beta$ is not as tight, because here $\varepsilon_{s,a}$ will have to compare the mean value with the exact value, and because $\mathcal{S}_\beta$ is constructed specifically for formula β; of course information about $\mathcal{S}$ is lost. So $\mathcal{S}_\beta$ is not a very "good" approximation of $\mathcal{S}$ in general and we do not explore the subject further.

When computing a reachability property on an MDP, there are always numerical errors on the way. The errors are at many stages, among which: abstraction of an actual system; floating-point rounding errors; fixpoint computation for $\phi U \psi$, involving a stopping condition on iteration. We expect that with a discretization that is exact for a reachability property, our precision is better.

5   Conclusion

We have developed a technique to construct a finite system from an LMP$_U$ in such a way that a family of reachability properties can be checked exactly. The transition functions of these systems are functions of uniform distribution.

---

[2] The supremum is taken for every f over the interval of $\mathcal{S}$ on which it is defined.

This particular restriction allows us to use the *mean-value theorem* to construct a discretization of LMP_U in such a way that some reachability properties are preserved exactly. On the discretized system, an MDP, we can use known tools for probabilistic systems, like PRISM, to determine the probability of the property. Even if the systems considered are limited, we believe that an exact result is interesting, especially for reachability properties.

We have implemented the technique. From an LMP_U, we use CISMO [17] to determine states that satisfy formulas ϕ and ψ (formulas without reachability operators), we then compute the MDP using the technique of the mean-value and we pass the result to PRISM.

At implementation, since the numerical computation for the MDP is subjected to numerical errors, the result can be inexact, as expected. Despite the errors that can affect the outcome of a verification, we have shown that the theoretical and the computed discretization are within a small error, using the notion of ϵ-bisimulation.

Acknowledgements. The authors thank the reviewers for their helpful comments. This research has been supported by NSERC grants RGPIN-239294 and RGPIN-262067.

References

1. Abate, A., Kwiatkowska, M., Norman, G., Parker, D.: Probabilistic model checking of labelled Markov processes via finite approximate bisimulations. In: van Breugel, F., Kashefi, E., Palamidessi, C., Rutten, J. (eds.) Horizons of the Mind. A Tribute to Prakash Panangaden. LNCS, vol. 8464, pp. 40–58. Springer, Cham (2014). https://doi.org/10.1007/978-3-319-06880-0_2
2. de Alfaro, L.: Formal verification of probabilistic systems. Ph.D. thesis, Stanford University. Technical Report STAN-CS-TR-98-1601 (1997)
3. Amin, S., Abate, A., Prandini, M., Lygeros, J., Sastry, S.: Reachability analysis for controlled discrete time stochastic hybrid systems. In: Hespanha, J.P., Tiwari, A. (eds.) HSCC 2006. LNCS, vol. 3927, pp. 49–63. Springer, Heidelberg (2006). https://doi.org/10.1007/11730637_7
4. Assouramou, J., Desharnais, J.: Continuous time and/or continuous distributions. In: Aldini, A., Bernardo, M., Bononi, L., Cortellessa, V. (eds.) EPEW 2010. LNCS, vol. 6342, pp. 99–114. Springer, Heidelberg (2010). https://doi.org/10.1007/978-3-642-15784-4_7
5. Baier, C., Haverkort, B.R., Hermanns, H., Katoen, J.-P.: Model-checking algorithms for continuous-time Markov chains. IEEE Trans. Software Eng. (2003)
6. Baier, C., Katoen, J.-P.: Principles of Model Checking (Representation and Mind Series). The MIT Press, Cambridge (2008)
7. Blute, R., Desharnais, J., Edalat, A., Panangaden, P.: Bisimulation for labelled Markov processes. In: Proceedings of the Twelfth IEEE Symposium on Logic in Computer Science (LICS), Warsaw, Poland, Test-of-time award in 2017 (1997)
8. Comenetz, M.: Calculus: The Elements, 1st edn. World Scientific, Singapore (2002)
9. Danos, V., Desharnais, J., Panangaden, P.: Labelled Markov processes: stronger and faster approximations. Electron. Notes Theor. Comput. Sci. **87**, 157–203 (2004)

10. Desharnais, J., Laviolette, F., Tracol, M.: Approximate analysis of probabilistic processes: logic, simulation and games. In: Fifth International Conference on the Quantitative Evaluaiton of Systems, QEST: 14–17 September 2008, Saint-Malo, France (2008)
11. Desharnais, J., Panangaden, P., Jagadeesan, R., Gupta, V.: Approximating labeled Markov processes. In: Proceedings of the 15th Annual IEEE Symposium on Logic in Computer Science, LICS 2000, p. 95 (2000)
12. Desharnais, J., Abbas, E., Panangaden, P.: Bisimulation for labelled Markov processes. Inf. Comput. **179**(2), 163–193 (2002)
13. Desharnais, J., Gupta, V., Jagadeesan, R., Panangaden, P.: Approximating labeled Markov processes. Inf. Comput. **184**(1), 160–200 (2003)
14. Hansson, H., Jonsson, B.: A logic for reasoning about time and reliability. Formal Aspects Comput. **6**(5), 512–535 (1994)
15. Kwiatkowska, M., Norman, G., Parker, D.: PRISM 4.0: verification of probabilistic real-time systems. In: Gopalakrishnan, G., Qadeer, S. (eds.) CAV 2011. LNCS, vol. 6806, pp. 585–591. Springer, Heidelberg (2011). https://doi.org/10.1007/978-3-642-22110-1_47
16. Kwiatkowska, M., Norman, G., Sproston, J.: Symbolic computation of maximal probabilistic reachability. In: Larsen, K.G., Nielsen, M. (eds.) CONCUR 2001. LNCS, vol. 2154, pp. 169–183. Springer, Heidelberg (2001). https://doi.org/10.1007/3-540-44685-0_12
17. LSFM: CISMO. http://www.ift.ulaval.ca/~jodesharnais/cismo/. Accessed 14 Aug 2018
18. Press, W.H., Teukolsky, S.A., Vetterling, W.T., Flannery, B.P.: Numerical Recipes 3rd Edition: The Art of Scientific Computing, 3rd edn. Cambridge University Press, New York (2007)
19. Vardi, M.Y.: Automatic verification of probabilistic concurrent finite state programs. In: Proceedings of the 26th Annual Symposium on Foundations of Computer Science, SFCS 1985, pp. 327–338. IEEE Computer Society, Washington, DC (1985). https://doi.org/10.1109/SFCS.1985.12

Learning and Verification

Bayes-Adaptive Planning
for Data-Efficient Verification
of Uncertain Markov Decision Processes

Viraj Brian Wijesuriya[✉] and Alessandro Abate

Department of Computer Science, University of Oxford, Oxford, UK
viraj.wijesuriya@cs.ox.ac.uk

Abstract. This work concerns discrete-time parametric Markov decision processes. These models encompass the uncertainty in the transitions of partially unknown probabilistic systems with input actions, by parameterising some of the entries in the stochastic matrix. Given a property expressed as a PCTL formula, we pursue a data-based verification approach that capitalises on the partial knowledge of the model and on experimental data obtained from the underlying system: after finding the set of parameters corresponding to model instances that satisfy the property, we quantify from data a measure (a confidence) on whether the system satisfies the property. The contribution of this work is a novel Bayes-Adaptive planning algorithm, which synthesises finite-memory strategies from the model allowing Bayes-Optimal selection of actions. Actions are selected for collecting data, with the goal of increasing its information content that is pertinent to the property of interest: this active learning goal aims at increasing the confidence on whether or not the system satisfies the given property.

1 Introduction

Markov Decision Processes (MDPs) [23] have been successfully employed to solve many demanding decision making problems in complex environments. A fully-specified MDP can be leveraged to provide quantitative guarantees for correct behaviour of intricate engineering systems. Formal methods provide mathematically rigorous machinery to obtain such guarantees [3], but their applicability might fall short in the case of incomplete knowledge of the underlying system. Available knowledge of a partially unknown system can be encompassed by a parametric MDP (pMDP) [13], where a set of parameters is used to account for imperfect knowledge.

We are interested in performing data-efficient verification of partially unknown systems with input actions, which can be modelled using pMDPs. Input actions represent nondeterministic choices available for planning. We reason about system properties expressed in probabilistic computational tree logic (PCTL [15]). In this paper, we assume that full data can be gathered from the system to reason about these properties.

© Springer Nature Switzerland AG 2019
D. Parker and V. Wolf (Eds.): QEST 2019, LNCS 11785, pp. 91–108, 2019.
https://doi.org/10.1007/978-3-030-30281-8_6

Our verification approach is both model-based and data-driven [20,21]: on the one hand, we classify the pMDP model into those MDPs satisfying the given property and those that do not; on the other, we augment the knowledge about the pMDP with the information content derived from limited amount of data actively gathered from the system; we finally quantify a confidence on whether the system satisfies the property.

In this work, we perform active learning [1] by seeking optimal strategies to gather data from the system, with the objective of performing verification with the greatest degree of accuracy. The novelty of our contribution is to extend [21], where memoryless strategies were synthesised towards this task, which are highly sub-optimal for maximising the confidence estimate [23]. Here we tackle the requirement of memory dependency by formalising the verification problem as a model-based reinforcement learning task, which is cast into a framework that augments the pMDP with the information acquired through histories of interactions with the system: this results in a model formulation known as Bayes-Adaptive MDP (BAMDP) [8]. We also introduce a new algorithm called *Bayes-Adaptive Temporal Difference Search* for planning with BAMDPs. A reward function, related to the confidence estimate, is introduced over the BAMDP to set up an optimisation task. Optimal strategies help to steer the interaction with the underlying system to ultimately attain the most accurate confidence estimate.

1.1 Related Work

The parameter synthesis problem [13] aims at formulating a range of possible valuations for a set of parameters corresponding to the satisfaction of a property of interest. Recent works [6,13,24] perform synthesis utilising increasingly efficient techniques that scale well on larger state and parameter spaces. Parameter synthesis alone does not answer the question whether the underlying, partly known system satisfies the property. Instead, some information about the parameters also needs to be inferred from the system. We not only perform parameter synthesis but also *parameter inference*, which draws valuations for model parameters from measurement data from the system. When measurement data is readily available (i.e. need not gather from the underlying system as part of the planning process), approaches such as [2] proceed to find *parameter-independent* strategies where the expected reachability probability is optimised.

Depending entirely on data, [5,18] attempt to learn a completely deterministic representation of an MDP model from the system and to subsequently verify properties over the learnt model. Unlike our approach, they do not take into account prior knowledge available to the learner through the incomplete model at hand and the property given, leading to a single model fitting the underlying system. Characterising the transition model of an MDP, [1,25] aim at learning representations from sequences of data using Bayesian learning. The lack of information from a partial model and without the ability to exploit known relationships between parameters themselves, renders these approaches data-inefficient.

In this work we incorporate Bayesian inference [25] when *planning* over pMDPs. Simplicity and analytical behaviour of Bayesian inference has motivated its use herein. Statistical Model Checking (SMC) techniques [26] perform verification over fully-specified models by generating simulation data, or by gathering data from the underlying system if no model is specified. While we do not solve the same problem as SMC techniques do, our work strikes a balance between the two mentioned alternatives, allowing for a substantial reduction in data gathered owing to the knowledge encompassed in the partially known model (i.e., the pMDP). Notice that SMC techniques for MDPs solve nondeterminism via memoryless strategies [16], or employ history-dependent strategies from only a subset of possible strategies [17]. On the other hand, in this work, we have the ability to construct memory-efficient strategies that are focused on the specific objective of asserting whether or not the system satisfies the property.

Computing confidence estimates in formal verification is seen as a meaningful approach in the presence of uncertainty in models. Cognate to this work, [21] utilises ideas from *experiment design* to calculate memoryless schedulers for pMDPs to ultimately compute a confidence estimate for the satisfaction of properties. Similarly, [20] computes confidence estimates for parametric Markov chains (PMCs) and [12] performs data-driven verification over *linear, time-invariant dynamical systems* encompassing measurement uncertainty.

2 Rudiments

2.1 Markov Decision Processes

Let $\mathbb{P}(H)$ denote a probability measure over an *event* H while $\mathbb{E}[X]$ denote the expectation for any given random variable X. We use $P(\cdot)$ to denote a discrete probability distribution over a finite set S where $P\colon S \to [0,1]$ and $\sum_{s \in S} P(s) = 1$.

We consider a discrete-time Markov decision process (MDP), represented as a tuple $\mathcal{M} = (S, A, \mathcal{T}, \iota)$, where S is the finite set of states, A is the finite set of actions. $\mathcal{T}\colon S \times A \times S \to [0,1]$ is a transition probability function such that $\forall s \in S$ and $\forall a \in A\colon \sum_{s' \in S} \mathcal{T}(s, a, s') \in \{0,1\}$. $\iota \subseteq S$ denotes the set of initial states. Any action that belongs to the set $A(s) = \{a \in A \mid \sum_{s' \in S} \mathcal{T}(s, a, s') = 1\}$ is said to be *enabled* at state s.

Consider an underlying data-generating system that allows to observe and collect finite traces of data in the form of sequences of visited states and chosen actions. We take into account the case where an MDP model that exactly represents the system is unknown but is assumed belonging to a class of uncertain models comprised in a parametric Markov decision process (pMDP).

A pMDP is a tuple $\mathcal{M}_p = (S, A, \mathcal{T}_p, \iota, \Theta)$, where the previous definition of an MDP is lifted, ceteris paribus, to include a transition function $\mathcal{T}_p$ with a target set specified using parameters found in an n-dimensional vector θ. All possible *valuations* of θ are held in $\Theta \subseteq [0,1]^n$, with $n \in \mathbb{N}_{>0}$. For any $\theta \in \Theta$, we enforce that $\forall s \in S$ and $\forall a \in A(s)\colon \sum_{s' \in S} \mathcal{T}_p(s, a, s') = 1$. Hence, each valuation of θ induces a single MDP $\mathcal{M}(\theta)$ with a transition function that can be represented using a *stochastic matrix*. Whilst the probabilities of all non-parameterised transitions of $\mathcal{M}_p$ are assumed to be known, we allow (unknown) probabilities of parameterised transitions to be linearly related, as in [21].

2.2 Strategies

A *strategy* (a.k.a. *policy* or *scheduler*) designates an agent's behaviour within its environment. For an MDP, a strategy is a distribution over actions at a particular state. A *deterministic* strategy selects a single action at a particular state and a *deterministic memoryless* (a.k.a. *stationary*) strategy $\pi\colon S \to A$ where $\forall s \in S$: $\pi(s) \in A(s)$, always selects the same action per state, regardless of any available *memory* of previously selected actions and/or visited states, hence allowing for a time-independent choice.

In this work, we are compelled to introduce the notion of memory (a.k.a. *history*) with respect to strategies, since memoryless strategies fail to be adequate with optimality in the choice of actions [2,23]. We call a memory $\mathbf{m}$, a sequence of states and actions, namely $\mathbf{m} = s_0 a_0 s_1 a_1 s_2 a_2 \ldots$, where $s_i \in S$, $a_i \in A(s_i)$ and $\mathcal{T}(s_i, a_i, s_{i+1}) > 0$. A memory $\mathbf{m}_t = s_0 a_0 s_1 a_1 s_2 a_2 \ldots a_{t-1} s_t$ is finite if it covers a finite-time horizon $t \in \mathbb{N}_{>0}$. Let $\mathcal{M}$ represent the set of possible finite memories. A *deterministic finite-memory* strategy $\hat{\pi}\colon S \times \mathbf{M} \to A$ has finitely many modes $\mathbf{M} \subseteq \mathcal{M}$, such that a single action a_t is chosen at time t, namely $a_t = \hat{\pi}(s_t, \mathbf{m}_t)$, $\forall t > 0$ where $\mathbf{m}_t \in \mathbf{M}$ and $a_t \in A(s_t)$. Obviously, s_t is the last state in memory $\mathbf{m}_t$: the redundant emphasis on pairs $(s_t, \mathbf{m}_t)$ is a notational convenience inherited from literature that will be further justified in Sect. 4.1.

2.3 Bayesian Reinforcement Learning (Bayesian RL)

Model-based Bayesian RL for pMDPs relies on an explicit model, which is learnt assuming priors over model parameters and by updating a posterior distribution using Bayesian inference [25] as more data is gathered from the underlying system. Subsequently (and possibly iteratively), the MDP with parameters sampled from the current posterior is employed to find an optimal policy that maximises the long-term expected reward.

We consider a Bayes-Adaptive RL formulation [8], with a model that allows to encode memory as part of the state space. A Bayes-Adaptive MDP is a tuple $\mathcal{M}_{ba} = (\hat{S}, A, \hat{\mathcal{T}}, \hat{\iota}, \mathcal{R})$, where $\hat{S} = S \times \mathbf{M}$ is the state space encompassing memory, A is as defined in Sect. 2.1 for an MDP, the transition function $\hat{\mathcal{T}}(s, \mathbf{m}, a, s', \mathbf{m}')$ designates transitions between *belief states* $(s, \mathbf{m}) \in \hat{S}$ and $(s', \mathbf{m}') \in \hat{S}$ after choosing an action a. Further, $\hat{\iota} \subseteq \hat{S}$ where $(s, \mathbf{m}) \in \hat{\iota}$ if $s \in \iota$. $\mathcal{R}\colon \hat{S} \times A \times \hat{S} \to \mathbb{R}$ is the newly introduced *transition reward*. The transition (s, a, s') plus the *information state* $\mathbf{m}$ affect the next information state $\mathbf{m}'$, thereby preserving the *Markov property* among transitions between belief states.

When an action a is selected in a belief state $(\mathbf{s}, \mathbf{m})$, a transition occurs to the successive belief state $(s', \mathbf{m}') \sim \hat{\mathcal{T}}(s, \mathbf{m}, a, \cdot, \cdot)$ and a transition reward r is received. With a slight abuse of notation, the function $\mathcal{R}(s, \mathbf{m}, a) = \mathbb{E}[r \mid s_t =$

$s, \mathbf{m}_t = \mathbf{m}, a_t = a]$ shall denote the expected reward for the pair: belief state $(s, \mathbf{m})$ and action a.

Given a finite-time horizon $T \in \mathbb{N}_{>0}$ and strategy $\hat{\pi}$, the *action-value* function $Q^{\hat{\pi}}(s, \mathbf{m}, a) = \mathbb{E}_{\hat{\pi}}[\Sigma_{j=t}^T r_j \mid s_t = s, \mathbf{m}_t = \mathbf{m}, a_t = a]$ is the expected cumulative transition reward up to the horizon T after action a is chosen over belief state $(s, \mathbf{m})$ and thereafter following strategy $\hat{\pi}$. Solving a BAMDP in our context boils down to finding a finite-memory, deterministic strategy $\hat{\pi}^*$ that maps belief states from the augmented state space $\hat{S}$ to actions in A and which maximises an expected cumulative transition reward over a given finite horizon T.

2.4 PCTL Properties

We consider *non-nested* properties expressed in a fragment of probabilistic computational tree logic (PCTL) [15]. For a PCTL formula ϕ interpreted over states of a given MDP $\mathcal{M}$, a formula φ interpreted over the paths [3], $\bowtie \in \{<, \leq, \geq, >\}$ and $b \in [0, 1]$, the probabilistic operator $\phi = \mathbf{P}_{\bowtie b}(\varphi)$ in a state $s \in S$ expresses the probability for paths starting at s that satisfy φ meet the bounds given by $\bowtie b$. We consider path formulae both bounded: $\varphi = \phi_i \, \mathcal{U}^{\leq k} \, \phi_j$ (with a finite-time bound $k \in \mathbb{N}_{>0}$) and unbounded time: $\varphi = \phi_i \, \mathcal{U} \, \phi_j$. Denote by $\mathbb{P}^{\pi}(\varphi \mid s)$ the probability of satisfying φ along the paths of MDP $\mathcal{M}$ that start from $s \in S$ and follow a given strategy π. The satisfaction of formula $\mathbf{P}_{\bowtie b}(\phi_i \, \mathcal{U}^{\leq k} \, \phi_j)$ over $\mathcal{M}$ is thus given by:

$$\mathcal{M} \models \mathbf{P}_{\bowtie b}(\phi_i \, \mathcal{U}^{\leq k} \, \phi_j) \iff \forall s \in \iota: \underset{\pi \in Str_{\mathcal{M}}}{\mathfrak{A}} \, \mathbb{P}^{\pi}(\phi_i \, \mathcal{U}^{\leq k} \, \phi_j \mid s) \bowtie b,$$

where $\mathbb{P}^{\pi}$ is the measure over the events corresponding to the formula ϕ for the MDP under strategy π and $Str_{\mathcal{M}}$ is the set of all strategies of $\mathcal{M}$ and $\mathfrak{A} \in \{\inf, \sup\}$ with following choices for $\bowtie$: inf if $\geq$ or $>$ and sup if $<$ or $\leq$. The satisfaction of $\mathbf{P}_{\bowtie b}(\phi_i \, \mathcal{U} \, \phi_j)$ over $\mathcal{M}$ can be derived similarly. Considering a pMDP $\mathcal{M}_p$, we let Θ_ϕ denote the set of valuations of θ for which the formula ϕ is satisfied: $\theta \in \Theta_\phi \Leftrightarrow \mathcal{M}(\theta) \models \phi$ and we call Θ_ϕ the *feasible set*.

3 Verification and Learning

We present here our integrated verification and learning approach [20].

3.1 Parameter Synthesis

The aim of parameter synthesis is to classify induced MDP models of the corresponding pMDP, between those that satisfy a given property ϕ and those that do not. This is achieved by producing an output that maps regions corresponding to parameter valuations to truth values [13]. Regions that map to "true" are considered belonging to the feasible set Θ_ϕ, while those that maps to "false" are guaranteed to not contain valuations that satisfy the property ϕ. This step handles actions pessimistically, namely models are considered to verify property ϕ

regardless of the action selection. This is because we plan to fully utilise actions for learning at a later step. Evidently, a different trade-off on actions could be struck, which we delegate to future work. We employ the probabilistic model checker Storm [7] to perform synthesis because it supports the PCTL properties of interest.

3.2   Bayesian Inference

Bayesian inference allows to determine a posterior probability distribution over the likely values of the parameters $\theta \in \Theta$, based on data gathered from the underlying system and to update this probability distribution as more data is collected [25]. It is also possible to incorporate any subjective beliefs about the parameters using a prior distribution $P(\theta)$.

Denote by $\mathscr{D}$ the set of *finite traces* gathered from the system so far, comprising a count $C_{s,s'}^a$ of how many times a particular transition (s, a, s') has been observed. We limit parameterisation of pMDP transitions to two cases of interest: (a) each transition of the pMDP is parameterised either with a single parameter (e.g. θ_i or $1 - \theta_i$) or with a constant $k \in (0, 1]$; (b) the pMDP includes transitions whose transition probabilities are expressed as affine functions of the form $g_{s,a}^{s'}(\theta) = k_0 + k_1\theta_1 + \ldots + k_n\theta_n$. For any instance of (b), [21] suggests two transformations that produce a pMDP that contains only transition probabilities of the form given in (a). Therefore, without loss of generality and for the purpose of succinctness, we assume pMDPs with parameterisation corresponding to the form in (a) herein. For transitions having identical parameterisations, their transition counts can be grouped together using *parameter tying* [22].

Denote by $C(\theta_i)$, the number of times transitions parameterised by θ_i have been observed in $\mathscr{D}$: $C(\theta_i) = \sum C_{s,s'}^a$ for $\mathcal{T}_p(s, a, s') = \theta_i$. Similarly, we define $C'(\theta_i)$ to count transitions parameterised by $1 - \theta_i$, i.e., transitions not parameterised by θ_i given that there exists a transition parameterised by θ_i under the same action $a \in A(s)$. We collect both $C(\theta_i)$ and $C'(\theta_i)$ in $\bar{C}(\theta_i)$ for brevity.

Assuming a prior distribution $P(\theta_i)$ over each component parameter $\theta_i \in \theta$, the posterior distribution over θ_i can be expressed using Bayes' rule as:

$$P(\theta_i \mid \mathscr{D}) \propto P(\theta_i)\theta_i^{C(\theta_i)}(1 - \theta_i)^{C'(\theta_i)}.$$

The counts $C_{s,s'}^a$ follow a multinomial distribution [25]. Selecting a Beta distribution: $\mathrm{Beta}(\theta_i; (\alpha_{\theta_i}, \beta_{\theta_i}))$ as a conjugate prior, the posterior distribution has a closed-form expression, allowing it to be updated by adding respective parameter counts to the hyper-parameter pair $(\alpha_{\theta_i}, \beta_{\theta_i})$ [25]:

$$P(\theta_i \mid \mathscr{D}) = \mathrm{Beta}(\theta_i; \bar{C}(\theta_i) + (\alpha_{\theta_i}, \beta_{\theta_i})).$$

Note that hyper-parameters α_{θ_i} and β_{θ_i} are the parameters of the (Beta) prior distribution over θ_i. We denote by $U_i(\mathbf{m})$ the update on hyper-parameter counts $\bar{C}(\theta_i) + (\alpha_{\theta_i}, \beta_{\theta_i})$, corresponding to an information state $\mathbf{m}$. Marginals $P(\theta_j \mid \mathscr{D})$ can be combined to form the posterior for the entire vector $\theta \in \Theta$ under the assumption that each component parameter θ_j is independent over those independent state-action pairs in the pMDP. Thus, the posterior $P(\theta \mid \mathscr{D})$ is given by: $P(\theta \mid \mathscr{D}) = \prod_{\theta_j \in \theta} P(\theta_j \mid \mathscr{D})$. Whenever an analytical update is impossible, we can resort to *sampling* techniques to obtain posterior realisations [21].

3.3   Confidence Computation

Given a specification ϕ and the posterior distribution $P(\theta \mid \mathscr{D})$ for $\theta \in \Theta$ obtained from a system of interest S, the confidence on whether $S \models \phi$, can be quantified according to [20]:

$$\mathscr{C} = \mathbb{P}(S \models \phi \mid \mathscr{D}) = \int_{\Theta_\phi} P(\theta \mid \mathscr{D})d\theta = \int_{\Theta_\phi} \prod_{\theta_j \in \theta} P(\theta_j \mid \mathscr{D})d\theta. \quad (1)$$

This quantity in general can be computed using Monte Carlo integration [20].

3.4   Overview of the Approach

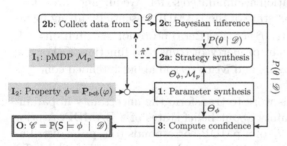

Fig. 1. Verification and Learning. $\mathbf{I}_1$, $\mathbf{I}_2$ are inputs and output **O** is the confidence estimate $\mathscr{C}$.

The different phases of our approach are shown in Fig. 1. We assume that the available parametric model $\mathcal{M}_p$ best represents the underlying system together with its uncertain dynamics. We first perform parameter synthesis over the given parametric model to find the feasible set of parameter values Θ_ϕ that satisfy the specification at hand. We then collect data from the system and employ Bayesian inference to update the posterior distribution over the likely values of the parameters with respect to the gathered data. Finally, we output a quantification of the confidence that the underlying system satisfies the specification over the data gathered so far.

In this work, the feasible set and the current distribution over the parameter space are propagated through a BAMDP model and used to plan (viz. synthesise a strategy) for gathering valuable data from the underlying system. We synthesise strategies to sequentially collect data to optimise the measure in Eq. (1) which will be further elaborated in Sect. 4.2.

4   Active Learning

We introduce a new technique for model-based Bayesian RL to synthesise a finite-memory strategy to further explore the system from data.

4.1 Bayes-Adaptive Model

In order to collect maximally useful data from the underlying system, we take into account the importance of (both past and expected) information to decrease the uncertainty associated with model parameters with respect to property satisfaction. Our confidence quantification (cf. Eq. (1)) is a proxy for this uncertainty: if the property is satisfied over the underlying system, one would expect the confidence to be as high as possible and, conversely, as low possibly if the property is not satisfied (essentially, in either case, one ideally wants to be away from the value 0.5).

We lift the BAMDP model described in Sect. 2 to support uncertainty described by parameterised transitions assuming a Beta-Binomial representation for the posterior distribution (as in Sect. 3.2). Using this uncertainty representation, we encompass the information state $\mathbf{m}$ by a joint probability distribution $i_{\mathbf{m}}$ over the hyper-parameters, which are collectively denoted by the pair $(\boldsymbol{\alpha}, \boldsymbol{\beta}) = \{U_j(\mathbf{m}) \mid \forall \theta_j \in \theta\}$, namely, $\boldsymbol{\alpha} = \langle C(\theta_1) + \alpha_{\theta_1}, C(\theta_2) + \alpha_{\theta_2}, \ldots, C(\theta_n) + \alpha_{\theta_n} \rangle$ and $\boldsymbol{\beta} = \langle C'(\theta_1) + \beta_{\theta_1}, C'(\theta_2) + \beta_{\theta_2}, \ldots, C'(\theta_n) + \beta_{\theta_n} \rangle$ where $n = |\theta|$. The hyper-parameters for θ_j are thus denoted by $i_{\mathbf{m},j}$. The distribution $i_{\mathbf{m}}$ acts as a statistic for $\mathbf{m}$ that summarises all information accumulated so far. We furthermore adapt the pair $(s, \mathbf{m})$ to $(s, i_{\mathbf{m}})$ as a belief state of the BAMDP, essentially lifting memories $\mathbf{M}$ to the hyper-parameters $(\boldsymbol{\alpha}, \boldsymbol{\beta})$ of the (Beta) posterior distributions, $\widehat{\mathbf{M}}$. This lifting preserves the Markovian property of transition function $\hat{\mathcal{T}}$. For the remainder of the paper, we employ $i_{\mathbf{m}}$ and $\widehat{\mathbf{M}}$ to reason about intended concepts over $\mathbf{m}$ and $\mathbf{M}$, respectively.

We formulate BAMDPs in this work to transform the uncertainty in parameters $\theta \in \Theta$ of a pMDP into certainty about belief states of a BAMDP. A prior $P(\mathcal{T}_p)$ on the transition function $\mathcal{T}_p$ of a pMDP corresponds to a prior distribution $P(\theta)$ over parameters θ. Accordingly, we can define a posterior belief $b(\mathcal{T}_p)$ over the transition function given data $\mathcal{D}$, so that $b(\mathcal{T}_p) = P(\theta \mid \mathcal{D})$. For a memory $\mathbf{m}$, this belief can be quantified as $b(\mathcal{T}_p) = P(\mathcal{T}_p \mid i_{\mathbf{m}}) \propto P(i_{\mathbf{m}} \mid \mathcal{T}_p) P(i_{\mathbf{m}})$. The transition dynamics for the BAMDP can be formulated as:

$$\hat{\mathcal{T}}(s, i_{\mathbf{m}}, a, s', i_{\mathbf{m}'}) = \int_{\mathcal{T}_p} \mathcal{T}_p(s, a, s') P(\mathcal{T}_p \mid i_{\mathbf{m}}) \mathrm{d}\mathcal{T}_p, \tag{2}$$

where $\mathbf{m}'$ is the updated memory after the transition (s, a, s') is witnessed. The RHS of Eq. 2 is the expectation of $\mathcal{T}_p(s, a, s')$, which corresponds to the expectation of the posterior $P(\theta \mid \mathcal{D})$, hence $\hat{\mathcal{T}}$ can be expressed as:

$$\hat{\mathcal{T}}(s, i_{\mathbf{m}}, a, s', i_{\mathbf{m}'}) = \begin{cases} \mathbb{1}_{i_{\mathbf{m}a s'}}(i_{\mathbf{m}'}) \frac{\alpha_{\theta_j}}{\alpha_{\theta_j} + \beta_{\theta_j}} & \text{if } \mathcal{T}_p(s, a, s') = \theta_j, \\ \mathbb{1}_{i_{\mathbf{m}a s'}}(i_{\mathbf{m}'}) \frac{\beta_{\theta_j}}{\alpha_{\theta_j} + \beta_{\theta_j}} & \begin{array}{l} \text{if } \mathcal{T}_p(s, a, s') \neq \theta_j \text{ and} \\ \exists s_k \in S \colon \mathcal{T}_p(s, a, s_k) = \theta_j, \end{array} \\ \mathcal{T}_p(s, a, s') & \text{otherwise,} \end{cases} \tag{3}$$

where $\mathbb{1}_b(a)$ is 1 if a equals b, else 0.

4.2   Synthesis of Bayes-Adaptive Strategies

The reward function $\mathcal{R}$ is used in the BAMDP to designate expected confidence updates resulting from transitions and thereby to guide the learning process. We work with finite-horizon problems. Positive rewards defined at the horizon (and zero elsewhere) require the learner to consider complete trajectories spanning the horizon in order to accumulate non-zero rewards. This can be computationally expensive should the horizon be large. Therefore, we focus on obtaining immediate rewards at each step of the learning process.

We define belief-dependent rewards [19] based on the difference between the confidence estimate at the given time step and that at the successive time step. As a learning process is designed to maximise rewards [28], here it is focussed on maximising the information content from the system's data to compute the most accurate confidence estimate possible. In order to achieve this, we need to synthesise strategies that maximise the deviation between *future confidence* and the base case $\mathcal{K} = 0.5$.

Denote by $\mathscr{C}_t$, the confidence estimate at current time step t, where we have set $\mathscr{C}_0 = \mathcal{K}$. After selecting action a_t over the belief state $(s_t, \mathrm{i}_{\mathrm{m}_t})$, the next-step confidence $\mathscr{C}_{t+1}$ can be used to define an immediate confidence gain r_{t+1} as:

$$r_{t+1} = |\mathcal{K} - \mathscr{C}_{t+1}| - |\mathcal{K} - \mathscr{C}_t|.$$

The reward function $\mathcal{R}(s, \mathrm{i}_{\mathrm{m}}, a)$ is thus defined as $\mathcal{R}(s, \mathrm{i}_{\mathrm{m}}, a) = \mathbb{E}[r_{t+1} \mid s_t = s, \mathrm{i}_{\mathrm{m}_t} = \mathrm{i}_{\mathrm{m}}, a_t = a]$, where clearly $\mathcal{R}(s, \mathrm{i}_{\mathrm{m}}, a) = 0$ if there is no associated parameterised transition. An interesting observation about $\mathcal{R}$ is that corresponding rewards might converge to zero in the limit, i.e., go to zero as the agent is left with nothing more to learn.

With respect to a prior distribution $P(\theta)$, the *Bayes-Optimal* policy $\hat{\pi}^*$ that maximises the expected cumulative reward over a finite horizon T is given by

$$\hat{\pi}^*(s, \mathrm{i}_{\mathrm{m}}) = \arg \max_{a \in A(s)} Q^*(s, \mathrm{i}_{\mathrm{m}}, a),$$

where the corresponding Bayes-Optimal *action-value* function is given by

$$Q^*(s, \mathrm{i}_{\mathrm{m}}, a) = \sup_{\hat{\pi}} \mathbb{E}_{\hat{\pi}}[\Sigma_{j=t}^T r_j \mid s_t = s, \mathrm{i}_{\mathrm{m}_t} = \mathrm{i}_{\mathrm{m}}, a_t = a, \theta \sim P(\theta)].$$

Note that the Bayes-Optimal policy $\hat{\pi}^*(s, \mathrm{i}_{\mathrm{m}})$ depends on prior beliefs and is consistent with the way in which $\mathscr{C}_{t+1}$ is calculated. According to Eq. (1), this value corresponds to the *expected values* of the parameter counts after transitioning from state s_t by selection of action a_t. Starting from the expected values of transition counts, as described in Sect. 3.2, we can collect the expected parameter counts in $\bar{C}(\theta)$. The expected transition counts correspond to the Binomial distribution over the transitions under a chosen action. The expected values of the transition probabilities can be calculated via the expected values of the parameters. For instance, the expected value of a given parameter θ_j is $\mathbb{E}[\theta_j] = \dfrac{\alpha_{\theta_j}}{\alpha_{\theta_j} + \beta_{\theta_j}}$. The expected transition probability for the transition (s, a, s') is hence $g_{s,a}^{s'}(\mathbb{E}[\theta_j])$.

4.3   Bayes-Adaptive Temporal Difference Search

The learning algorithm we introduce is based on learning from simulated episodes of experiences gathered from the BAMDP. However, a BAMDP can be sizeable, even for a corresponding simple concrete MDP [8]. The information space grows exponentially with the number of state-action pairs in the concrete MDP and the horizon T of exploration. In our setting, T directly relates to the length of the traces drawn from the system and T can be chosen arbitrarily but needs to be large enough to witness several state transitions. However, when the PCTL property imposes a finite-time bound k on satisfaction, T should not exceed k. Even though there exist exact solutions to BAMDPs, for instance, via dynamic programming using *Gittins indices* [11], in most practical cases they are intractable. Let us recall that we denote by $Q(s, i_m, a)$ the expected cumulative transition reward, when action a is selected at belief state (s, m). It is in practice not possible to store all values of $Q(s, i_m, a)$ in memory and learning exact values might be too slow. One way of reducing these computational burdens is to observe that distinct memories may yield the same (or a similar) belief [8], hence generalisation of Q values among related paths can be helpful. [10] proposes a Monte Carlo simulation algorithm to estimate Q values with a function approximator, which allows generalisation between states, actions and beliefs. However, such methods require to evaluate the final step of the simulation to update all corresponding Q values.

In this work, we follow an approach based on temporal difference (TD) learning [28], which can update the estimate of the Q value after every step of a simulation. This is helpful when the time horizon is very long (or non-terminating). Furthermore, the ability to learn step by step helps in estimating Q values with low variance and to plan via subsequent decisions that can be correlated in time. Temporal Difference Search (TD Search) is a simulation-based algorithm that employs value function approximation. Initially used for planning in Computer Go [27], we extend TD search to the context of Bayes-Adaptive models.

A new Bayes-Adaptive temporal difference search algorithm is outlined in Algorithm 1. It gathers episodes of simulated trajectories starting from the current belief state (s_t, i_{m_t}) according to an ϵ–*greedy* strategy. An ϵ–*greedy* strategy selects an action that maximises the local Q value with a probability equal to $1 - \epsilon$ or outputs a random action with probability ϵ. Rather than exploring the whole BAMDP, our algorithm commits to solving a sub-BAMDP that starts at the current belief state and spans a given time horizon T. Once Algorithm 1 synthesises a strategy (based on the current posterior $P(\theta \mid \mathscr{D})$), we roll it out up to the designated time horizon T and collect data (in the form of traces of length T) from the underlying system. We then update the BAMDP model via Bayesian inference using the collected data: the current posterior distributions of each parameter $\theta_i \in \theta$ is updated using the new data. Next, we synthesise a new strategy to further gather data. We continue in this fashion until we have gathered an arbitrary allowed number of traces from the system. We then output the eventual confidence estimate that asserts whether the system satisfies the property (cf. Fig. 1).

Value Function Approximation. In order to approximate the value function, we lift the action-value function with a weight matrix β of learnable parameters: $Q(s, i_m, a; \beta)$. The goal of the learning process is to then find β that minimises the mean-squared error between approximate and true Q functions: $\mathbb{E}[(Q(s, i_m, a; \beta) - Q(s, i_m, a))^2]$.

Algorithm 1. Bayes-Adaptive Temporal Difference Search

```
 1: Inputs:
 2:     s_t, i_m_t
 3: Initialize:
 4:     β ← 0, ρ ← 0
 5: procedure SEARCH(s_t, i_m_t)
 6:     while time remaining do                    ▷ Start episode
 7:         s ← s_t, i_m ← i_m_t, ť ← t
 8:         a ← π̂_ε-greedy(s, i_m; Q)
 9:         ξ ← 0, ξ' ← 0
10:         while ť < T do
11:             s' ~ T̂(s, i_m, a, ·, i_ma·)
12:             R ← R(s, i_m, a)
13:             a' ← π̂_ε-greedy(s', i_mas'; Q)
14:             δ ← R + Q(s', i_mas', a'; β) - Q(s, i_m, a; β)
15:             ξ = λξ + y(i_m) ⊗ x(s, a)
16:             ξ' = λξ + y(i_mas') ⊗ x(s', a')
17:             β ← β + αδξ - αξ'(ξ^T ρ)
18:             ρ ← ρ + ω(δJ - ξ^T ρ)ξ
19:             s ← s', a ← a', ť ← ť + 1
20:         end while
21:     end while
22:     return arg max_{a_t} Q(s_t, i_m_t, a_t; β)
23: end procedure
```

We use the backward view of SARSA(λ) [28], with ϵ-greedy strategy improvement, to help learn parameters β at each step of simulation sequences. The parameter λ designates up to how far in time should *bootstrapping* occur (bootstrapping refers to updating an estimated value with one or more estimated values of the same kind). To implement the backward view, it is required to maintain an eligibility trace ξ^a_{s, i_m} [1] for each tuple (s, i_m, a). An *eligibility trace* [28] temporarily remembers the occurrence of an activity, for instance, visiting a state and choosing an action. The trace signals the learning process that the credit associated to the activity is eligible for change. The trace assigns credit to eligible activities based on combing assignments from two common heuristics: frequency heuristic (where credit is assigned to most frequent activities) and recency heuristic (where credits is assigned to most recent activities).

A BAMDP entails a convex action-value function [8] as a function of the information state: this function becomes piecewise linear if the horizon is finite and if the state-action space is discrete. Therefore, linear value function approximation is appropriate to represent the true convex action-value function for a particular state-action pair. Since the transition rewards we receive correspond to a gain in confidence, the values are bounded within $[0, 1]$. As such, the introduced function approximation is truncated using the sigmoid function $\sigma(x) = \frac{1}{1+e^{-x}}$ [27].

Feature Representation. The quality of the value function approximation $Q(s, i_m, a; \beta)$ greatly depends on employed features: we use the feature triple (s, i_m, a). A good approximation procedure should generalise well for those memories that lead to similar information states (or beliefs). The feature representation should facilitate likewise representations for such memories. We propose the following representation for $Q(s, a, i_m; \beta)$ [10]:

$$Q(s, i_m, a; \beta) = \mathbf{y}(i_m)^\top \beta \mathbf{x}(s, a).$$

---

[1] Note that in Algorithm 1, we actually maintain an eligibility trace matrix ξ.

This form encodes the feature triple into the Q approximation, with vector $\mathbf{x}(s, a)$ concerning state-action pairs and vector $\mathbf{y}(i_m)$ representing information states.

$\mathbf{x}(s, a)$ indicates which state-action pair is currently involved. Therefore, for a particular pair, this representation of Q is linear as a function of i_m which approximates the true convex action-value function. State-action pairs with similar features will be considered to be similar. We associate $\mathbf{x}(s, a)$ with binary features by assigning it a column vector of size $\mathcal{Z}$, with value *one* assigned to the location of the element corresponding to (s, a) and to any entry corresponding to parameter similar state-action pairs of (s, a), while other entries are assigned the value zero[2]. By parameter similar state-action pairs, we mean those pairs with outgoing transitions having identical parameterisation.

The construction of the vector $\mathbf{y}(i_m)$ requires representing beliefs in a coordinate vector form. However, as beliefs are not finite-dimensional objects, a finite-dimensional approximation is therefore required. [10] proposes a sampling mechanism based on a *sequential importance sampling particle filter*. We construct $\mathbf{y} \colon \widehat{\mathbf{M}} \to \mathbb{R}^{\mathcal{Z}}$ as follows, assuming that $\mathcal{Z}$ is the degree of the finite-dimensional approximation. We initialise $\mathbf{y}(i_m)$, a column vector with $\mathcal{Z}$ elements, to $\frac{1}{\mathcal{Z}}$ at the beginning of each episode of Algorithm 1. We then modify $\mathbf{y}(i_m)$ by updating each entry j at the current step, using the probability given by $\hat{T}(s, i_m, a, s', i_{mas'})$, as:

$$\mathbf{y}_j(i_{mas'}) = \mathbf{y}_j(i_m)\hat{T}(s, i_m, a, s', i_{mas'}).$$

Notice how this scheme allows different memories leading to same belief to be mapped as identical representations, i.e., $\mathbf{y}(i_{m'}) = \mathbf{y}(i_m)$ if $b(\mathbf{m'}) = b(\mathbf{m})$. With this construction, it is not required to explicitly update the information of belief states that are not directly traversed in the simulation, since these updates are implicitly reflected in the finite-dimensional representation. The two updates (cf. Algorithm 1, lines: 15 and 16) on eligibility traces capture the joint effect of the introduced feature vectors $\mathbf{x}$ and $\mathbf{y}$ (in the algorithm, $\otimes$ denotes the standard outer product).

Feature vectors $\mathbf{x}(s, a)$ and $\mathbf{y}(i_m)$ effectively generalise from states, actions, and memories already seen to those unseen. As such, the rolled-out simulations will achieve the generalisation without the need to traverse all possible states of the BAMDP, making the algorithm much more efficient.

We run Algorithm 1 episodically (note that an episode starts from line 6 and ends in line 21) to learn β using simulated traces from the BAMDP model. We roll-out these simulations (cf. from line 10 to 20) up to the horizon T. The propagation of knowledge from one step of the algorithm to the other is fundamental to learning good representations from past experiences. When rolling-out simulations, one can use $\mathbf{y}(i_m)$ from previous time step $\tilde{t} - 1$ for learning in the current step $\tilde{t}$ but $\mathbf{y}(i_m)$ may degenerate, e.g. leaving one of its entries $\mathbf{y}_j(i_m) = 1$ and rest being zero. Given a threshold Y, we simply re-initialise all entries if $\frac{1}{\sum_j \mathbf{y}_j(i_{m_{\tilde{t}}})^2} < Y$ or else, we reuse $\mathbf{y}$ from the previous β update.

---

[2] This scheme is sometimes called *one-hot encoding*.

Convergence. In the context of our action-value function approximation, convergence means that entries of β reach a fixed point. Linear SARSA (λ) (which is the underlying learning algorithm used in TD search) is sensitive to initial values of β and does not always converge in view of chattering [9]. Convergence guarantees for standard SARSA(λ), requiring a *greedy in the limit with infinite exploration* (GLIE) sequence of strategies and *Robbins-Monro conditions* for the learning rate α [28], are not in general enough to guarantee convergence for linear SARSA (λ), since it is not a true gradient-descent method [29]. Policy gradient methods [28] on the other hand can be used in place if one desires guaranteed convergence. This motivates the use of stochastic gradient descent algorithms. Since we roll-out a complete strategy before the next stage of data gathering, we are essentially performing open loop control together with the current beliefs we possess. Therefore, our planning stage is based on *off-policy learning*: training on outcomes from an ϵ-*greedy* strategy in order to learn the value of another. *Linear TD with gradient correction* (TDC) [29] can be used to force SARSA to follow the true gradient. We adopt TDC to support the form of Q and the presence of matrix β. Based on the *mean-square projected Bellman error (MSPBE)* objective function [29], TDC employs two additional parameters: matrix ρ and scalar ω and updates β and ρ accordingly on each state transition (cf. lines 17–18 of Algorithm 1, where J is a properly-sized *all-ones* matrix).

5 Experiments

5.1 Setup

We evaluate our approach over three case studies. We consider a range of simulated underlying systems (corresponding to different instantiations of the parameters) and compare obtained confidence results against the corresponding ground truths, via mean-squared error (MSE) metric. We compare strategies generated by our synthesis algorithm, Algorithm 1 (denoted *BA strategy*) against other strategies: a strategy synthesised in [21] (denoted *Synth strategy*), a given probabilistic memoryless strategy (denoted *RS strategy*), and a strategy that randomly select actions at each state (denoted *No strategy*).

First case study involves the pMDP given in [21], endowed with 6 states, 12 transitions, and 2 parameters and the PCTL property $\mathbf{P}_{\geq 0.5}(\text{true } \mathcal{U} \text{ complete})$ (complete is the label associated to one of the 6 states).

For the second case study (cf. Fig. 2), we extend an MDP model for a smart buildings application [4] to a pMDP with action space $A = \{f_{\text{off}}, f_{\text{on}}\}$ and parameter vector $\theta = \{\theta_1, \theta_2\}$. Actions correspond to the on/off state of a fan inside a room. We verify the satisfaction of the PCTL property $\mathbf{P}_{\geq 0.35}\neg(\text{true } \mathcal{U}^{\leq 20} (\text{E}, \text{O}))$. Beyond the comparison between strategies described above, we use traces generated by our algorithm and by other strategies with a Bayesian Statistical Model Checking (Bayesian SMC) [30] implementation. SMC collects trajectories from the system, checks whether trajectories satisfy a given property and subsequently uses statistical methods such as *hypothesis testing* to determine whether the system satisfies the property.

Thirdly, we carry out experiments with a well-known pMDP benchmark from [14], the *Randomised Consensus Protocol*. This case study allows to show-case the efficiency and effectiveness of our approach over large MDPs. We consider an instance of the problem with 4112 states and 7692 transitions, where we fix the number of processes $N = 2$ (i.e. two parameters) and the protocol constant $K = 32$ to check the PCTL property $\mathbf{P}_{\geq 0.25}$(true $\mathcal{U}$ (finished & allCoinsEqualToOne)).

Like [21], for convenience of presentation, we have selected all simulated underlying systems such that a single parameter θ_* is responsible for the satisfaction/falsification of the property: $\theta_* = (\theta_1 = \theta_2)$. Intervals over θ_* (namely, feasible sets) corresponding to the system verifying the corresponding property are: $I_1 = [0.369, 0.75]$ (for case 1), $I_2 = [0.0, 0.16]$ (for case 2) and $I_3 = [0.2, 0.5]$ (for case 3), respectively.

Our approach has been implemented in C++. We consider non-informative priors for all parameters involved i.e., uniform ones ($\alpha = \beta = 1$). Over different values of θ_*, data from the simulated underlying system is collected as traces containing state-action pairs visited over time. We gather confidence estimates through n number of runs. If $\mathscr{C}_j$ is the confidence estimate at the j-th run of case study i, then MSE is computed as $= \frac{1}{n}\Sigma_{j=1}^{n}(\mathbb{1}_{I_i}(\theta_*) - \mathscr{C}_j)^2$. For the experiments, we set algorithm parameters (cf. Algorithm 1) as follows. λ is set to 0.8, ω to 0.9, α to $\frac{1}{(\bar{t}+1)^{0.65}}$ and ϵ to $\frac{1}{\bar{t}+1}$. We train β on 1000 episodes for all case studies and set $\mathcal{Z} = 50$ for the large model in case study 3.

5.2   Results

The MSE outcomes for each case study are summarised in Figs. 3a, b and c. The horizontal axis (system parameter) represents values of θ_* for the simulated underlying systems. The intervals I_i above allow to separate systems that satisfy the property from those that do not, by a clear edge/boundary (e.g., for Fig. 3a this edge is rooted at 0.369). The MSE results have been drawn by experiments carried out under limited amount of data (e.g. for the third case study, we have used 10 traces, each of length 10, namely $(t10, l10)$). These results show that our approach clearly outperforms other strategies.

It is important to note that we incur a comparably higher error at system parameters that are very close to the mentioned edge. This is due to the nature of the confidence computation that we perform. For a point closer to the edge, this could yield a posterior distribution that will have its peak centred at the point with probability mass falling almost equally in both the feasible set and outside it. As more data is gathered, the posterior distribution may grow taller and thinner, but with the slightest shift in its peak, a large proportion of the mass may fall on either side, resulting in an increase of the MSE. This increased sensitivity near the edge soon subsides as we move away from the edge, where the mass can now fall in either part of the interval.

Note that we achieve a significant performance for the large pMDP model in case study 3 (cf. Fig. 3c). For a model of this magnitude, the corresponding

Bayes-Adaptive model is enormous, making it impossible to search/traverse it entirely. The proposed generalisation approach embedded in our search algorithm allows to tackle this otherwise intractable problem. Unfortunately, [21] times out (in 1 h after going out of memory) when attempting synthesis, i.e., explicitly evaluating memoryless strategies does not scale well for large models.

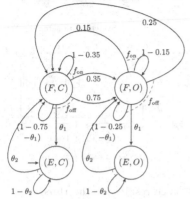

Fig. 2. Extended pMDP model for case study in [4].

Figures 3d and e present results on case study 2. These show that our method is able to gather more useful data than *Synth* and to rapidly converge to the ground truth. Maintaining good performance even at very low amounts of data (e.g. 10 traces or traces of length 5) shows that our approach is robust to the nature of the gathered data. The major reason behind this is that the current strategy constantly looks for parameterised transitions as much as possible: it is over these transitions that confidence gain may happen. This is in stark contrast to techniques like SMC, where the length of horizon of the trace needs to be long enough to either reach a designated state or find counterexamples for the given property.

Figure 3f provides results for an experiment conducted over case study 2, and shows two significant aspects of our approach: first, the information content of the traces that we have generated from our approach, by comparing them against those generated from other strategies; and secondly, the demonstration that SMC can be problematic in situations where one has access to only a limited amount of data. Running traces generated by different strategies (*BA* vs *Synth* vs *No*) through a Bayesian SMC algorithm, demonstrates that our (*BA*) approach converges rapidly to the ground truth, faster than other methods (*Synth* and *No*). This shows that our traces encompass much richer information content to compute better confidence estimates to decide the satisfiability of the property.

SMC provides outcomes that are usually much faster than canonical model checking tools. However, for case study 2, the property we have selected is a negative *bounded-time* property that requires falsification by reaching a specific state (E, O). This is a tricky property to ascertain via SMC, due to the lack of counterexamples with trace lengths much shorter than the formula horizon. On the other hand, such traces processed with our approach (i.e. confidence computation using the posterior distribution) yield much better results than Bayesian SMC. This shows that we can work with much shorter horizons than the formula horizon and are still able to accurately verify properties. Performance at shorter trace lengths is an important performance criterion for large models, like the one in case study 3, where you would need fairly longer trace lengths (e.g. 1000 or more) for SMC to work, whereas our approach is able to verify the property with a couple of orders of magnitudes lower trace lengths (e.g. 10).

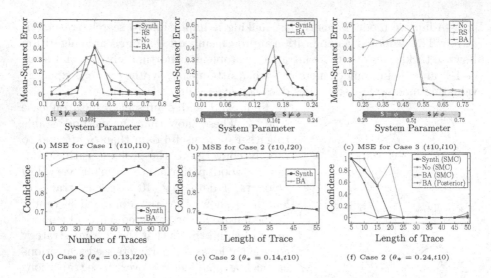

(a) MSE for Case 1 ($t10, l10$)   (b) MSE for Case 2 ($t10, l20$)   (c) MSE for Case 3 ($t10, l10$)

(d) Case 2 ($\theta_* = 0.13, l20$)   (e) Case 2 ($\theta_* = 0.14, t10$)   (f) Case 2 ($\theta_* = 0.24, t10$)

Fig. 3. Mean-squared error (MSE) and average confidence results for the three case studies. The number of generated traces or the length of generated traces is indicated by values prefixed with t or l, respectively. Translucent bars along the x-axes of a, b and c designate the simulated underlying systems that satisfy the property ϕ and those which do not. Respective edges/boundaries are followed by a $\ddagger$ symbol.

6   Conclusions and Future Work

We have a data-based efficient verification approach to assert whether a partially unknown probabilistic system satisfies a given property expressed as a logical specification. Our approach takes into account memory in calculating optimal strategies to gather data from the underlying system so as to derive the most accurate confidence estimates possible.

As future work, based on the updated confidence value, one could *tune/repair* the parametric model until a decisive confidence is achieved. For instance, if the output confidence value is 0.5, then there exists an equal chance that the property is either satisfied or not over the system. If this value has been obtained after gathering a substantial amount of data, this may mean that the employed parametric model was not supportive enough to gauge the satisfaction of the property, hence it could be adjusted until a substantial judgement about the satisfiability can be made.

Furthermore, in this work, actions are exclusively selected for learning tasks. Instead, one might choose them in the context of model classification (i.e., parameters selection), in order to steer the system towards property satisfaction.

References

1. Araya-López, M., Buffet, O., Thomas, V., Charpillet, F.: Active learning of MDP models. In: Sanner, S., Hutter, M. (eds.) EWRL 2011. LNCS (LNAI), vol. 7188, pp. 42–53. Springer, Heidelberg (2012). https://doi.org/10.1007/978-3-642-29946-9_8
2. Arming, S., Bartocci, E., Chatterjee, K., Katoen, J.-P., Sokolova, A.: Parameter-independent strategies for pMDPs via POMDPs. In: McIver, A., Horvath, A. (eds.) QEST 2018. LNCS, vol. 11024, pp. 53–70. Springer, Cham (2018). https://doi.org/10.1007/978-3-319-99154-2_4
3. Baier, C., Katoen, J.P.: Principles of Model Checking. MIT Press, Cambridge (2008)
4. Cauchi, N., Abate, A.: StocHy: automated verification and synthesis of stochastic processes. In: 25th International Conference on Tools and Algorithms for the Construction and Analysis of Systems (TACAS) (2019)
5. Chen, Y., Nielsen, T.: Active learning of Markov decision processes for system verification. In: 2012 11th International Conference on Machine Learning and Applications (ICMLA), vol. 2, pp. 289–294, December 2012
6. Cubuktepe, M., et al.: Sequential convex programming for the efficient verification of parametric MDPs. In: Legay, A., Margaria, T. (eds.) TACAS 2017. LNCS, vol. 10206, pp. 133–150. Springer, Heidelberg (2017). https://doi.org/10.1007/978-3-662-54580-5_8
7. Dehnert, C., Junges, S., Katoen, J.-P., Volk, M.: A STORM is Coming: a modern probabilistic model checker. In: Majumdar, R., Kunčak, V. (eds.) CAV 2017. LNCS, vol. 10427, pp. 592–600. Springer, Cham (2017). https://doi.org/10.1007/978-3-319-63390-9_31
8. Duff, M.O.: Optimal learning: computational procedures for Bayes-Adaptive Markov decision processes. Ph.D. thesis (2002)
9. Gordon, G.J.: Chattering in SARSA(λ) - a CMU learning lab internal report. Technical report (1996)
10. Guez, A., Heess, N., Silver, D., Dayan, P.: Bayes-adaptive simulation-based search with value function approximation. In: Ghahramani, Z., Welling, M., Cortes, C., Lawrence, N.D., Weinberger, K.Q. (eds.) Advances in Neural Information Processing Systems, Curran Associates, Inc., vol. 27, pp. 451–459 (2014)
11. Guez, A., Silver, D., Dayan, P.: Efficient Bayes-adaptive reinforcement learning using sample-based search. In: Advances in Neural Information Processing Systems 25: 26th Annual Conference on Neural Information Processing Systems 2012, Proceedings of a Meeting Held 3–6 December 2012, Lake Tahoe, Nevada, United States, pp. 1034–1042 (2012)
12. Haesaert, S., Van den Hof, P.M., Abate, A.: Data-driven property verification of grey-box systems by Bayesian experiment design. In: American Control Conference (ACC), 2015, IEEE, pp. 1800–1805 (2015)
13. Hahn, E.M., Han, T., Zhang, L.: Synthesis for PCTL in parametric markov decision processes. In: Bobaru, M., Havelund, K., Holzmann, G.J., Joshi, R. (eds.) NFM 2011. LNCS, vol. 6617, pp. 146–161. Springer, Heidelberg (2011). https://doi.org/10.1007/978-3-642-20398-5_12
14. Hahn, E.M., Hermanns, H., Wachter, B., Zhang, L.: PARAM: a model checker for parametric Markov models. In: Touili, T., Cook, B., Jackson, P. (eds.) CAV 2010. LNCS, vol. 6174, pp. 660–664. Springer, Heidelberg (2010). https://doi.org/10.1007/978-3-642-14295-6_56

15. Hansson, H., Jonsson, B.: A logic for reasoning about time and reliability. Formal Aspects Comput. **6**(5), 512–535 (1994)
16. Henriques, D., Martins, J.G., Zuliani, P., Platzer, A., Clarke, E.M.: Statistical model checking for Markov decision processes. In: Proceedings of the 2012 Ninth International Conference on Quantitative Evaluation of Systems, QEST 2012, IEEE Computer Society, Washington, DC, USA. pp. 84–93 (2012)
17. Legay, A., Sedwards, S., Traonouez, L.-M.: Scalable verification of Markov decision processes. In: Canal, C., Idani, A. (eds.) SEFM 2014. LNCS, vol. 8938, pp. 350–362. Springer, Cham (2015). https://doi.org/10.1007/978-3-319-15201-1_23
18. Mao, H., Chen, Y., Jaeger, M., Nielsen, T.D., Larsen, K.G., Nielsen, B.: Learning Markov decision processes for model checking. In: Proceedings Quantities in Formal Methods, QFM 2012, Paris, France, 28 August 2012, pp. 49–63 (2012)
19. Marom, O., Rosman, B.: Belief reward shaping in reinforcement learning. In: Thirty-Second AAAI Conference on Artificial Intelligence (2018)
20. Polgreen, E., Wijesuriya, V.B., Haesaert, S., Abate, A.: Data-efficient Bayesian verification of parametric Markov chains. In: Agha, G., Van Houdt, B. (eds.) QEST 2016. LNCS, vol. 9826, pp. 35–51. Springer, Cham (2016). https://doi.org/10.1007/978-3-319-43425-4_3
21. Polgreen, E., Wijesuriya, V.B., Haesaert, S., Abate, A.: Automated experiment design for data-efficient verification of parametric Markov decision processes. In: Bertrand, N., Bortolussi, L. (eds.) QEST 2017. LNCS, vol. 10503, pp. 259–274. Springer, Cham (2017). https://doi.org/10.1007/978-3-319-66335-7_16
22. Poupart, P., Vlassis, N., Hoey, J., Regan, K.: An analytic solution to discrete Bayesian reinforcement learning. In: Proceedings of the 23rd International Conference on Machine Learning, ICML 2006, ACM, New York, NY, USA, pp. 697–704 (2006)
23. Puterman, M.L.: Markov Decision Processes: Discrete Stochastic Dynamic Programming, 1st edn. Wiley, New York (1994)
24. Quatmann, T., Dehnert, C., Jansen, N., Junges, S., Katoen, J.-P.: Parameter synthesis for Markov models: faster than ever. In: Artho, C., Legay, A., Peled, D. (eds.) ATVA 2016. LNCS, vol. 9938, pp. 50–67. Springer, Cham (2016). https://doi.org/10.1007/978-3-319-46520-3_4
25. Ross, S., Pineau, J., Chaib-draa, B., Kreitmann, P.: A Bayesian approach for learning and planning in partially observable Markov decision processes. J. Mach. Learn. Res. **12**(May), 1729–1770 (2011)
26. Sen, K., Viswanathan, M., Agha, G.: Statistical model checking of black-box probabilistic systems. In: Alur, R., Peled, D.A. (eds.) CAV 2004. LNCS, vol. 3114, pp. 202–215. Springer, Heidelberg (2004). https://doi.org/10.1007/978-3-540-27813-9_16
27. Silver, D., Sutton, R.S., Müller, M.: Temporal-difference search in Computer Go. Mach. Learn. **87**(2), 183–219 (2012)
28. Sutton, R.S., Barto, A.G.: Introduction to Reinforcement Learning, 1st edn. MIT Press, Cambridge (1998)
29. Sutton, R.S., Maei, H.R., Precup, D., Bhatnagar, S., Silver, D., Szepesvári, C., Wiewiora, E.: Fast gradient-descent methods for temporal-difference learning with linear function approximation. In: Proceedings of the 26th Annual International Conference on Machine Learning, ICML 2009, ACM, New York, NY, USA, pp. 993–1000 (2009)
30. Zuliani, P., Platzer, A., Clarke, E.M.: Bayesian statistical model checking with application to Stateflow/Simulink verification. Form. Methods Syst. Des. **43**(2), 338–367 (2013)

Strategy Representation by Decision Trees with Linear Classifiers

Pranav Ashok[1], Tomáš Brázdil[2], Krishnendu Chatterjee[3], Jan Křetínský[1], Christoph H. Lampert[3], and Viktor Toman[3(✉)]

[1] Technical University of Munich, Munich, Germany
[2] Masaryk University, Brno, Czech Republic
[3] IST Austria, Klosterneuburg, Austria
viktor.toman@ist.ac.at

Abstract. Graph games and Markov decision processes (MDPs) are standard models in reactive synthesis and verification of probabilistic systems with nondeterminism. The class of ω-regular winning conditions; e.g., safety, reachability, liveness, parity conditions; provides a robust and expressive specification formalism for properties that arise in analysis of reactive systems. The resolutions of nondeterminism in games and MDPs are represented as strategies, and we consider succinct representation of such strategies. The decision-tree data structure from machine learning retains the flavor of decisions of strategies and allows entropy-based minimization to obtain succinct trees. However, in contrast to traditional machine-learning problems where small errors are allowed, for winning strategies in graph games and MDPs no error is allowed, and the decision tree must represent the entire strategy. In this work we propose decision trees with linear classifiers for representation of strategies in graph games and MDPs. We have implemented strategy representation using this data structure and we present experimental results for problems on graph games and MDPs, which show that this new data structure presents a much more efficient strategy representation as compared to standard decision trees.

1 Introduction

Graph Games and MDPs. Graph games and Markov decision processes (MDPs) are classical models in reactive synthesis. In graph games, there is a finite-state graph, where the vertices are partitioned into states controlled by the two players, namely, player 1 and player 2, respectively. In each round the state changes according to a transition chosen by the player controlling the current state. Thus, the outcome of the game being played for an infinite number of rounds, is an infinite path through the graph, which is called a play. In MDPs, instead of an adversarial player 2, there are probabilistic choices. An objective specifies a subset of plays that are satisfactory. A strategy for a player is a recipe to specify the choice of the transitions for states controlled by the player. In games, given an objective, a winning strategy for a player from a state ensures

© Springer Nature Switzerland AG 2019
D. Parker and V. Wolf (Eds.): QEST 2019, LNCS 11785, pp. 109–128, 2019.
https://doi.org/10.1007/978-3-030-30281-8_7

the objective irrespective of the strategy of the opponent. In MDPs, given an objective, an *almost-sure winning strategy* from a state ensures the objective with probability 1.

Reactive Synthesis and Verification. The above models play a crucial role in various areas of computer science, in particular analysis of reactive systems. In reactive-system analysis, the vertices and edges of a graph represent the states and transitions of a reactive system, and the two players represent controllable versus uncontrollable decisions during the execution of the system. The reactive synthesis problem asks for construction of winning strategies in adversarial environment, and almost-sure winning strategies in probabilistic environment. The reactive synthesis for games has a long history, starting from the work of Church [15,18] and has been extensively studied [16,27,38,47], with many applications in synthesis of discrete-event and reactive systems [44,48], modeling [1,22], refinement [28], verification [4,20], testing [7], compatibility checking [19], etc. Similarly, MDPs have been extensively used in verification of probabilistic systems [6,21,33]. In all the above applications, the objectives are ω-regular, and the ω-regular sets of infinite paths provide an important and robust paradigm for reactive-system specifications [37,49].

Strategy Representation. The strategies are the most important objects as they represent the witness to winning/almost-sure winning. The strategies can represent desired controllers in reactive synthesis and protocols, and formally they can be interpreted as a lookup table that specifies for every controlled state of the player the transition to choose. As a data structure to represent strategies, there are some desirable properties, which are as follows: (a) *succinctness*, i.e., small strategies are desirable, since smaller strategies represent efficient controllers; (b) *explanatory*, i.e., the representation explains the decisions of the strategies. While one standard data structure representation for strategies is binary decision diagrams (BDDs) [2,14], recent works have shown that decision trees [39,45] from machine learning provide an attractive alternative data structure for strategy representation [10,12]. The two key advantages of decision trees are: (a) Decision trees utilize various predicates to make decisions and thus retain the inherent flavor of the decisions of the strategies; and (b) there are entropy-based algorithmic approaches for decision tree minimization [39,45]. However, one of the key challenges in using decision trees for strategy representation is that while in traditional machine-learning applications errors are allowed, for winning and almost-sure winning strategies errors are not permitted.

Our Contributions. While decision trees are a basic data structure in machine learning, their various extensions have been considered. In particular, they have been extended with linear classifiers [13,25,35,46]. Informally, a linear classifier is a predicate that checks inequality of a linear combination of variables against a constant. In this work, we consider decision trees with linear classifiers for strategy representation in graph games and MDPs, which has not been considered before. First, for representing strategies where no errors are permitted, we present a method to avoid errors both in decision trees as well as in linear classification. Second, we present a new method (that is not entropy-based) for choosing predicates in the decision trees, which further improves the succinctness of decisions trees with linear classifiers. We have implemented our approach, and applied it to examples of reactive synthesis from SYNTCOMP benchmarks [30], model-checking examples from PRISM benchmarks [34], and synthesis of randomly generated LTL formulae [43]. Our experimental results show significant improvement in succinctness of strategy representation with the new data structure as compared to standard decision trees.

2   Stochastic Graph Games and Strategies

Stochastic Graph Games. We denote the set of probability distributions over a finite set X as $\mathcal{D}(X)$. A *stochastic graph game* is a tuple $G = \langle S_1, S_2, A_1, A_2, \delta \rangle$, where:

- S_1 and S_2 is a finite set of states for player 1 and player 2, respectively, and $S = S_1 \cup S_2$ denotes the set of all states;
- A_1 and A_2 is a finite set of actions for player 1 and player 2, respectively, and $A = A_1 \cup A_2$ denotes the set of all actions; and
- $\delta \colon (S_1 \times A_1) \cup (S_2 \times A_2) \to \mathcal{D}(S)$ is a transition function that given a player 1 state and a player 1 action, or a player 2 state and a player 2 action, gives the probability distribution over the successor states.

We consider two special cases of stochastic graph games, namely:

- *graph games*, where for each (s,a) in the domain of δ, $\delta(s,a)(s') = 1$ for some $s' \in S$.
- *Markov decision processes (MDPs)*, where $S_2 = \emptyset$ and $A_2 = \emptyset$.

We consider stochastic graph games with several classical objectives, namely, *safety* (resp. its dual *reachability*), *Büchi* (resp. its dual *co-Büchi*), and *parity* objectives.

Stochastic Graph Games with Variables. Consider a finite subset of natural numbers $X \subseteq \mathbb{N}$, and a finite set *Var* of variables over X, partitioned into state-variables and action-variables $Var = Var_S \uplus Var_A$ ($\uplus$ denotes a disjoint union). A *valuation* is a function that assigns values from X to the variables. Let X^{Var_S} (resp., X^{Var_A}) denote the set of all valuations to the state-variables (resp., the action-variables). We associate a stochastic graph game $G = \langle S_1, S_2, A_1, A_2, \delta \rangle$ with a set of variables *Var*, such that (i) each state $s \in S$ is associated with a unique valuation $val_s \in X^{Var_S}$, and (ii) each action $a \in A$ is associated with a unique valuation $val_a \in X^{Var_A}$.

Example 1. Consider a simple system that receives requests for two different channels A and B. The requests become pending and at a later point a response handles a request for the respective channel. A controller must ensure that (i) the request-pending queues do not overflow (their sizes are 2 and 3 for channels A and

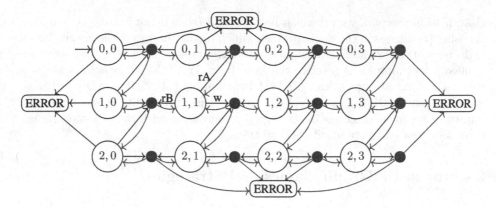

Fig. 1. A reactive system with two request channels.

B, respectively), and that (ii) no response is issued for a channel without a pending request. The system can be modeled by the graph game depicted in Fig. 1. The states of player 1 (controller issuing responses) are labeled with valuations of state-variables capturing the number of pending requests for channel A and B, respectively. For brevity of presentation, the action labels (corresponding to valuations of a single action-variable) are shown only outgoing from one state, with a straightforward generalization for all other states of player 1. Further, for clarity of presentation, the labels of states and actions for player 2 (environment issuing requests, with filled blue-colored states and actions) are omitted. The controller must ensure the safety objective of avoiding the four error states.

Strategy Representation. The algorithmic problem treated in this work considers representation of memoryless almost-sure winning strategies for stochastic graph games with variables. Given a stochastic graph game and an objective, a *memoryless* strategy for player $i \in \{1, 2\}$ is a function $\pi \colon S_i \to A_i$ that resolves the nondeterminism for player i by choosing the next action based on the currently visited state. Further, a strategy is *almost-sure winning* if it ensures the given objective irrespective of the strategy of the other player. In synthesis and verification of reactive systems, the problems often reduce to computation of memoryless almost-sure winning strategies for stochastic graph games, where the state space and action space is represented by a set of variables. In practice, such problems arise from various sources, e.g., AIGER specifications [29], LTL synthesis [43], PRISM model checking [33].

We refer to our technical report [5] for detailed description of the technical concepts regarding games and strategies.

3   Decision Trees and Decision Tree Learning

Here we recall decision trees (DT), representing strategies by DT, and learning DT.

Decision tree (DT) over $\mathbb{N}^d$ is a tuple $\mathcal{T} = (T, \rho, \theta)$ where T is a finite rooted binary (ordered) tree, ρ assigns to every inner node an (in)equality predicate comparing arithmetical expressions over variables $\{x_1, \ldots, x_d\}$, and θ assigns to every leaf a value *YES* or *NO*. The language $\mathcal{L}(\mathcal{T}) \subseteq \mathbb{N}^d$ of the tree is defined as follows. For a vector $\boldsymbol{x} = (x_1, \ldots, x_d) \in \mathbb{N}^d$, we find a path p from the root to a leaf such that for each inner node n on the path, $\rho(n)(\boldsymbol{x}) = \mathbf{true}$ (i.e., the predicate $\rho(n)$ is satisfied with valuation $\boldsymbol{x}$) iff the first child of

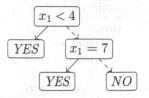

Fig. 2. A decision tree for $\{0, 1, 2, 3, 7\} \subseteq \mathbb{N}^1$.

n is on p. Denote the leaf on this particular path by ℓ. Then $\boldsymbol{x}$ is in the language $\mathcal{L}(\mathcal{T})$ of $\mathcal{T}$ iff $\theta(\ell) = YES$. Intuitively, $\mathcal{L}(\mathcal{T})$ captures the set of vectors *accepted* by the tree $\mathcal{T}$, i.e., vectors with accepting path in the tree (ending with *YES*). An example is illustrated in Fig. 2 with the first children connected with unbroken arrows and the second children with dashed ones.

The (usually finite) set of predicates in the co-domain of ρ is denoted by *Pred*. In the example above *Pred* are comparisons of variables to constants.

Representing strategies by DT has been introduced in [10]. The dimension of data points here is $d = |Var|$. The data points are natural tuples representing state-action pairs, thus we also write them as (s, a). The strategy induced by a decision tree $\mathcal{T}$ allows to play a in s iff $(s, a) \in \mathcal{L}(\mathcal{T})$.

A given input strategy $\pi \colon S_i \to A_i$ for player $i \in \{1, 2\}$ defines the sets (i) $Good = \{\langle s, \pi(s)\rangle \in S_i \times A_i\}$, (ii) $Bad = \{\langle s, a\rangle \in S_i \times A_i \mid a \neq \pi(s)\}$, and (iii) $Train = Good \uplus Bad$ ($\uplus$ denotes a disjoint union). Further, given a subset $data \subseteq Train$, we define $maxclass(data)$ as (i) *YES* if $|data \cap Good| \geq |data \cap Bad|$, and (ii) *NO* otherwise. When strategies need to be represented exactly, as in the case of games, the trees have to classify all decisions correctly [12]. This in turn causes difficulties not faced in standard DT learning [39], as described below.

Example 2. Consider the reactive system and the corresponding game described in Example 1. Consider a strategy π for the controller (player 1) in this system that (i) waits in state $(0, 0)$, (ii) issues a response for channel B when there are more pending requests for channel B than pending requests for channel A, and (iii) issues a response for channel A in all other cases. Then, the strategy π induces: $Good = \{(0, 0, w), (0, 1, rB), (0, 2, rB), (0, 3, rB), (1, 0, rA), (1, 1, rA), (1, 2, rB), (1, 3, rB), (2, 0, rA), (2, 1, rA), (2, 2, rA), (2, 3, rB)\}$, and $Bad = \{(p_A, p_B, act) \in \{0, 1, 2\} \times \{0, 1, 2, 3\} \times \{w, rA, rB\} \mid (p_A, p_B, act) \notin Good\}$. The task is to represent π exactly, i.e., to accept all *Good* examples and reject all *Bad* examples.

Learning DT from the set *Good* of positive examples and the set *Bad* of negative examples is described in Algorithm 1. A node with all the data points is gradually split into offsprings until the point where each leaf contains only elements of *Good* or only *Bad*. Note that in the classical DT learning algorithms such as ID3 [45], one can also stop this process earlier to prevent overfitting, which induces smaller trees with a classification error, unacceptable in the strategy representation.

Algorithm 1. Basic decision-tree learning algorithm

Input: $Train \subseteq \mathbb{N}^{|Var|}$ partitioned into subsets $Good$ and Bad.
Output: A decision tree $\mathcal{T}$ such that $\mathcal{L}(\mathcal{T}) \cap Train = Good$.
/* train $\mathcal{T}$ on positive set $Good$ and negative set Bad */
1: $\mathcal{T} \leftarrow (T = \{root\}, \rho = \emptyset, \theta = \emptyset)$
2: $\mathsf{q} \leftarrow \{(root, Train)\}$
3: **while** q nonempty **do**
4:    $(\ell, data_\ell) \leftarrow pop_\mathsf{q}$
5:    **if** $data_\ell \subseteq Good$ or $data_\ell \subseteq Bad$ **then**
6:        $\theta(\ell) \leftarrow maxclass(data_\ell)$
7:    **else**
8:        $\rho(\ell) \leftarrow$ predicate selected by a split procedure $Split(data_\ell)$
9:        create children ℓ_{sat} and ℓ_{unsat} of ℓ
10:       $push_\mathsf{q}((\ell_{sat}, data_\ell[\rho(\ell)])), push_\mathsf{q}((\ell_{unsat}, data_\ell[\neg\rho(\ell)]))$
11: **return** $\mathcal{T}$

Algorithm 2. Split procedure – information gain

Input: $data \subseteq \mathbb{N}^{|Var|}$ partitioned into subsets $data_G$ and $data_B$.
Output: A predicate pr maximizing information gain on $data$.
1: $ig \leftarrow \emptyset$
2: **for** $pr \in Pred$ **do**
3:    $ig(pr) \leftarrow$ information gain$(data, pr)$
4: **if** $\max_{pr}\{ig(pr)\} = 0$ **then**   ▷ condition checks if information gain failed
5:    **for** $pr \in Pred$ **do**
6:        $ig(pr) \leftarrow \max \left\{ \frac{|data_B[\neg pr]|}{|data[\neg pr]|} + \frac{|data_G[pr]|}{|data[pr]|}, \frac{|data_G[\neg pr]|}{|data[\neg pr]|} + \frac{|data_B[pr]|}{|data[pr]|} \right\}$
7: **return** $\arg\max_{pr}\{ig(pr)\}$

The choice of the predicate to split a node with is described in Algorithm 2. From the finite set $Pred$[1] we pick the one which maximizes *information gain* (i.e., decrease of entropy [39]). Again, due to the need of fully expanded trees with no error, we need to guarantee that we can split all nodes with mixed data even if none of the predicates provides any information gain in one step. This issue is addressed in [12] as follows. Whenever no positive information gain can be achieved by any predicate, a predicate is chosen according to a very simple different formula using a heuristic that always returns a positive number. One possible option suggested in [12] is captured on Line 6.

4   Decision Trees with Linear Classifiers

In this section, we develop an algorithm for constructing decision trees with linear classifiers in the leaf nodes. As we are interested in representation of

---

[1] The set of considered predicates $Pred$ is typically domain-specific, and finitely restricted in a natural way. In this work, we consider (in)equality predicates that compare values of variables to constants. A natural finite restriction is to consider only constants that appear in the dataset.

winning and almost-sure winning strategies, we have to address the challenge of allowing no error in the strategy representation. Thus we consider an algorithm that provably represents a given strategy in its entirety. Furthermore, we present a split procedure for decision-tree algorithms, which aims to propose predicates leading into small trees with linear classifiers.

4.1 Linear Classifiers in the Leaf Nodes

During the construction of a decision tree for a given dataset, each node corresponds to a certain subset of the dataset. This subset exactly captures the data points from the dataset that would reach the node starting from the root and progressing based on the predicates visited along the travelled path (as explained in Sect. 3). Notably, there might be other data points also reaching this node from the root, however, they are not part of the dataset, and thus their outcome on the tree is irrelevant for the correct dataset representation. This insight allows us to propose a decision-tree algorithm with more expressive terminal (i.e., leaf) nodes, and in this work we consider linear classifiers as the leaf nodes.

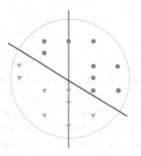

Fig. 3. *Good* (triangles) and *Bad* (circles). No horizontal or vertical classifier can separate *Train*, but *Train* is linearly separable (by a slanted classifier).

Given two vectors $\boldsymbol{a}, \boldsymbol{b} \in \mathbb{R}^d$, their dot product (or scalar product) is defined as $\boldsymbol{a} \cdot \boldsymbol{b} = \sum_{i=1}^{d} a_i b_i$. Given a weight vector $\boldsymbol{w} \in \mathbb{R}^d$ and a bias term $b \in \mathbb{R}$, a *linear classifier* $c_{\boldsymbol{w},b} \colon \mathbb{R}^d \to YES, NO$ is defined as

$$c_{\boldsymbol{w},b}(\boldsymbol{x}) = \begin{cases} YES & \boldsymbol{w} \cdot \boldsymbol{x} \geq b \\ NO & \text{otherwise.} \end{cases}$$

Informally, a linear classifier checks whether a linear combination of vector values is greater than or equal to a constant. Intuitively, we consider strategies as good and bad vectors of natural numbers, and we use linear classifiers to decide for a given vector whether it is good or bad. On a more general level, a linear classifier partitions the space $\mathbb{R}^d$ into two half-spaces, and a given vector gets classified based on the half-space it belongs to.

Consider a finite dataset $Train \subseteq \mathbb{N}^d$ partitioned into subsets $Good$ and Bad. A linear classifier $c_{\boldsymbol{w},b}$ *separates* $Train$, if for every $\boldsymbol{x} \in Train$ we have that $c_{\boldsymbol{w},b}(\boldsymbol{x}) = YES$ iff $\boldsymbol{x} \in Good$. The corresponding decision problem asks, given a dataset $Train \subseteq \mathbb{N}^d$, for existence of a weight vector $\boldsymbol{w} \in \mathbb{R}^d$ and bias $b \in \mathbb{R}$ such that the linear classifier $c_{\boldsymbol{w},b}$ separates $Train$. In such a case we say that $Train$ is linearly separable. Figure 3 provides an illustration. There are efficient oracles for the decision problem of linear separability, e.g., linear-programming solvers.

Example 3. We illustrate the idea of representing strategies by decision trees with linear classifiers. Consider the game described in Example 1 and the controller strategy π for this game described in Example 2. An example of a decision

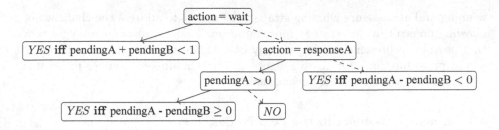

Fig. 4. A decision tree for the system's controller.

tree that represents the strategy π is displayed in Fig. 4. The input samples with action w (*wait*) end in and get classified by the leftmost linear classifier, and the samples with action rB (*responseB*) get classified by the rightmost linear classifier. Finally, the samples with action rA (*responseA*) are rejected if there are no pending requests to channel A, and otherwise they get classified by the bottommost linear classifier. Note that the decision tree accepts each sample from *Good* and rejects each sample from *Bad*, and thus indeed represents the strategy π.

Algorithm 3. Learning algorithm for decision trees with linear classifiers

---

Input: *Train* $\subseteq \mathbb{N}^{|Var|}$ partitioned into subsets *Good* and *Bad*.
Output: A decision tree $\mathcal{T}$ such that $\mathcal{L}(\mathcal{T}) \cap \textit{Train} = \textit{Good}$.
/* train $\mathcal{T}$ on positive set *Good* and negative set *Bad* */
1: $\mathcal{T} \leftarrow (T = \{root\}, \rho = \emptyset, \theta = \emptyset)$
2: $\mathsf{q} \leftarrow \{(root, \textit{Train})\}$
3: **while** q nonempty **do**
4:     $(\ell, \textit{data}_\ell) \leftarrow pop_{\mathsf{q}}$
5:     **if** $\textit{data}_\ell \subseteq \textit{Good}$ or $\textit{data}_\ell \subseteq \textit{Bad}$ **then**
6:         $\theta(\ell) \leftarrow maxclass(\textit{data}_\ell)$
7:     **else if** $\textit{data}_\ell$ is linearly separable by a classifier $c_{w,b}$ **then**
8:         $\theta(\ell) \leftarrow c_{w,b}$
9:     **else**
10:        $\rho(\ell) \leftarrow$ predicate selected by a split procedure $Split(\textit{data}_\ell)$
11:        create children ℓ_{sat} and ℓ_{unsat} of ℓ
12:        $push_{\mathsf{q}}((\ell_{sat}, \textit{data}_\ell[\rho(\ell)])), \ push_{\mathsf{q}}((\ell_{unsat}, \textit{data}_\ell[\neg\rho(\ell)]))$
13: **return** $\mathcal{T}$

We are now ready to describe our algorithm for representing strategies as decision trees with linear classifiers. Algorithm 3 presents the pseudocode. At the beginning, in Line 2 the queue is initiated with the root node and the whole training set *Train*. Intuitively, the queue maintains the tree nodes that are to be processed, and in every iteration of the loop (Line 3) one node ℓ gets processed. First, in Line 4 the node ℓ gets popped together with $\textit{data}_\ell$, which is the subset of *Train* that would reach ℓ from the root node. If $\textit{data}_\ell$ contains only samples from *Good* (resp., only samples from *Bad*), then ℓ becomes a leaf node with *YES* (resp., *NO*) as the answer (Line 6). If $\textit{data}_\ell$ contains samples from both, but is linearly separable by some classifier, then ℓ becomes a leaf node with this classifier (Line 8). Otherwise, ℓ becomes an inner node. In Line 10 it gets assigned a predicate by an external split procedure and in Line 11 two children of ℓ are created. Finally, in Line 12, $data_\ell$ is partitioned into the subset that satisfies the chosen predicate of ℓ and the subset that does not, and the two children of ℓ are pushed into the queue with the two subsets, to be processed in later iterations. Once there are no more nodes to be processed, the final decision tree is returned. For further intuition on Algorithm 3, we refer to our technical report [5] that contains an example displaying a step-by-step construction of a decision tree from a given dataset.

Correctness. We now prove the correctness of Algorithm 3. In other words, we show that given a strategy in the form of a training set, Algorithm 3 can be used to provably represent the training set (i.e., the strategy) without errors.

Theorem 1. *Let G be a stochastic graph game, and let $\pi\colon S_i \to A_i$ be a memoryless strategy for player $i \in \{1,2\}$ that defines a training set Train partitioned into Good and Bad. Consider an arbitrary split procedure that considers only predicates from Pred which produce nonempty sat- and unsat-partitions. Given Train as input, Algorithm 3 using the split procedure outputs a decision tree $T = (T,\rho,\theta)$ such that $\mathcal{L}(T) \cap Train = Good$, which means that for all $s \in S_i$ we have that $\langle s,a\rangle \in \mathcal{L}(T)$ iff $\pi(s) = a$. Thus T represents the strategy π.*

Proof. We consider stochastic graph games with variables *Var* over a finite domain $X \subseteq \mathbb{N}$, thus $Train \subseteq X^{|Var|}$. Recall that given a decision tree $T = (T,\rho,\theta)$ constructed by Algorithm 3, ρ assigns to every inner node a predicate from *Pred*, and θ assigns to every leaf either *YES*, or *NO*, or a linear classifier $c_{w,b}$ that classifies elements from $\mathbb{R}^{|Var|}$ into *YES* resp. *NO*.

Partial Correctness. Consider Algorithm 3 with input *Train*, and let $T = (T,\rho,\theta)$ be the output decision tree. Consider an arbitrary $\langle s,a\rangle \in S_i \times A_i$, note that it belongs to *Train*. Consider the leaf ℓ corresponding to $\langle s,a\rangle$ in T. There is a unique path for $\langle s,a\rangle$ down the tree T from its root, induced by the predicates in the inner nodes given by ρ. Thus ℓ is well-defined. At some point during the algorithm, ℓ was popped from the queue q in Line 4, together with a dataset $data_\ell$, and note that $\langle s,a\rangle \in data_\ell$. Since ℓ is a leaf, there are three cases to consider:

1. $\theta(\ell) = YES$. Then $data_\ell \subseteq \mathcal{L}(T)$, which implies $\langle s,a\rangle \in \mathcal{L}(T)$. The assignment happened in Line 6, so (i) the condition in Line 5 was satisfied, and (ii) $maxclass(data_\ell) = YES$. Thus $data_\ell \subseteq Good$, which implies $\langle s,a\rangle \in Good$. By the definition of *Good*, we have $\pi(s) = a$.
2. $\theta(\ell) = NO$. Then $data_\ell \cap \mathcal{L}(T) = \emptyset$, which implies $\langle s,a\rangle \notin \mathcal{L}(T)$. The assignment happened in Line 6, so (i) the condition in Line 5 was satisfied, and (ii) $maxclass(data_\ell) = NO$. Thus $data_\ell \subseteq Bad$, which implies $\langle s,a\rangle \in Bad$. By the definition of *Bad*, we have $\pi(s) \neq a$.
3. $\theta(\ell) = c_{w,b}$. This assignment happened in Line 8. Thus the condition in Line 7 was satisfied, and hence $c_{w,b}$ linearly separates $data_\ell$. As $\langle s,a\rangle \in data_\ell$, we have that $c_{w,b}(\langle s,a\rangle) = YES$ iff $\langle s,a\rangle \in Good$. This gives that $\langle s,a\rangle \in \mathcal{L}(T)$ iff $\pi(s) = a$.

The desired result follows.

Total Correctness. Algorithm 3 uses a split procedure that considers only predicates from *Pred* which produce nonempty sat- and unsat-partitions. Thus the algorithm maintains the following invariant for every path $\bar{p}$ in $\mathcal{T}$ starting from the root: For each predicate $pr \in Pred$, there is at most one inner node $\bar{n}$ in the path $\bar{p}$ such that $\rho(\bar{n}) = pr$. This invariant is indeed maintained, since any predicate considered the second time in a path inadvertently produces an empty data partition, and such predicates are not considered by the split procedure that selects predicates for ρ (in Line 10 of Algorithm 3).

From the above we have that the length of any path in $\mathcal{T}$ starting from the root is at most $|Pred| \leq 2 \cdot |Var| \cdot |X|$, i.e., twice the number of variables times the size of the variable domain. We prove that the number of iterations of the loop in Line 3 is finite. The branch from Line 9 happens finitely many times, since it adds two vertices (in Line 11) to the decision tree $\mathcal{T}$ and we have the bound on the path lengths in $\mathcal{T}$. Since only the branch from Line 9 pushes elements into the queue q, and each iteration of the loop pops an element from q in Line 4, the number of loop iterations (Line 3) is indeed finite. This proves termination, which together with partial correctness proves total correctness.   □

4.2   Splitting Criterion for Small Decision Trees with Classifiers

During construction of decision trees, the predicates for the inner nodes are chosen based on a supplied metric, which heuristicly attempts to select predicates leading into small trees. The entropy-based *information gain* is the most prevalent metric to construct decision trees, in machine learning [39,45] as well as in formal methods [3,10,26,41]. Algorithm 2 presents a split procedure utilizing information gain, supplemented with a stand-in metric proposed in [12].

In this section, we propose a new metric and we develop a split procedure around it. When selecting predicates for the inner nodes, we exploit the knowledge that in the descendants the data will be tested for linear separability. Thus for a given predicate, the metric tries to estimate, roughly speaking, how well-separable the corresponding data partitions are. While the metric is well-studied in machine learning, to the best of our knowledge, the corresponding decision-tree-split procedure is novel, both in machine learning and in formal methods.

True/False Positive/Negative. Consider a fixed linear classifier c, and a sample $x \in Train$ such that $c(x) = YES$. If $x \in Good$, then x is a *true positive* (*TP*) w.r.t. the classifier c, otherwise $x \in Bad$ and thus x is a *false positive* (*FP*). Consider a different sample $\bar{x} \in Train$ such that $c(\bar{x}) = NO$. If $\bar{x} \in Bad$, then $\bar{x}$ is a *true negative* (*TN*), otherwise $\bar{x} \in Good$ and $\bar{x}$ is a *false negative* (*FN*). Figure 5 summarizes the terminology.

x ╲ c	YES	NO
Good	TP	FN
Bad	FP	TN

Fig. 5. True/False Positive/Negative.

True/False Positive Rate. Consider a fixed linear classifier c and a fixed dataset $Train = Good \uplus Bad$. We denote by $|TP|$ the number of true positives within $Train$ w.r.t. the classifier c. Similarly we denote $|FP|$ for false positives. Then, the *true positive rate* (*TPR*) is defined as $|TP|/|Good|$, and the *false positive rate* (*FPR*) is $|FP|/|Bad|$. Intuitively, *TPR* describes the fraction of good samples that are correctly classified, whereas *FPR* describes the fraction of bad samples that are misclassified as good.

Area Under the Curve. Consider a fixed dataset $Train = Good \uplus Bad$ and a fixed weight vector $w \in \mathbb{R}^d$. In what follows we describe a metric that evaluates w w.r.t. $Train$. First, consider a set of *boundaries*, which are the dot products of w with samples from $Train$. Formally, $bnd = \{w \cdot x \mid x \in Train\}$. Further, consider $b_{none} = \max bnd + \varepsilon$ for some $\varepsilon > 0$. Then, consider the set of linear classifiers that "hit" the boundaries, plus a classifier that rejects all samples. Formally, $cl = \{c_{w,b} \mid b \in bnd \cup \{b_{none}\}\}$. Now, the *receiver operating characteristic* (ROC) is a curve that plots *TPR* against *FPR* for the classifiers in cl. Intuitively, the ROC curve captures, for a fixed set of weights, how changing the bias term affects *TPR* and *FPR* of the resulting classifier. Ideally, we want the *TPR* to increase rapidly when bias is weakened, while the *FPR* increases as little as possible. We consider the area under the ROC curve (denoted $auc \in [0,1]$) as the metric to evaluate the weight vector w w.r.t. the dataset $Train$. Intuitively, the faster the *TPR* increases, and the slower the *FPR* increases, the bigger the area under the ROC curve (auc) will be.

Figure 6 provides an intuitive illustration of the concept, where the weight vector is fixed as $w = (1,0)$. The classifiers cl are then shown on the left subfigure, and the corresponding ROC curve (with the shaded area under the curve – auc) is shown on the right subfigure. Note that the points in the ROC curve correspond to the classifiers from cl, and they capture their (*FPR*, *TPR*). The extra point $(0/2, 0/3)$ corresponds to the classifier that rejects all samples.

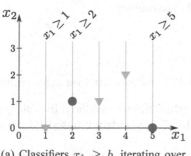

(a) Classifiers $x_1 \geq b$, iterating over the bias b from 5 down to 1.

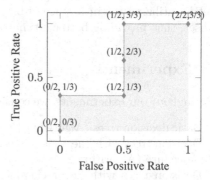

(b) ROC curve and the shaded auc.

Fig. 6. Area under the curve for $w = (1,0)$ w.r.t. *Good* (triangles) and *Bad* (circles).

Algorithm 4. Split procedure – area under the curve (auc)

Input: $data \subseteq \mathbb{N}^{|Var|}$ partitioned into subsets $data_G$ and $data_B$.

Output: A predicate pr maximizing area under the sat and $unsat$ ROC curves.

1: areas $\leftarrow \emptyset$
2: **for** $pr \in Pred$ **do**
3:      $w_{sat} \leftarrow LinearLeastSquares(data[pr])$
4:      $w_{unsat} \leftarrow LinearLeastSquares(data[\neg pr])$
5:      areas$(pr) \leftarrow$ auc$(w_{sat}, data[pr]) +$ auc$(w_{unsat}, data[\neg pr])$
6: **return** $\arg\max_{pr}\{$areas$(pr)\}$

Algorithm 4 presents a split procedure that uses auc as the metric to select predicates. Each considered predicate partitions input $data$ into the subset that satisfies the predicate and the subset that does not. Then, in Lines 3 and 4, two weight vectors are obtained by solving the linear least squares problem on the data partitions. This is a classical problem in statistics with a known closed-form solution, and [5] provides detailed description of the problem. Finally, the score for the predicate equals the sum of auc for the two weight vectors with respect to their corresponding data partitions (Line 5). At the end, in Line 6 the predicate with maximum score is selected.

The choice of auc as the split metric is motivated by heuristicly estimating well-separability of data in the setting of strategy representation. A simpler metric of *accuracy* (i.e., the fraction of correctly classified samples) may seem as a natural choice for the estimate of well-separability. However, in strategy representation, the data is typically very inbalanced, i.e., the sizes of *Good* are typically much smaller than the sizes of *Bad*. As a result, for all considered predicates the corresponding proposed classifiers focus heavily on the *Bad* samples and neglect the few *Good* samples. Thus all classifiers achieve remarkable accuracy, which gives us little information on the choice of a predicate. This is a well-known insight, as in machine learning, the accuracy metric is notoriously problematic in the case of disproportionate classes. On the other hand, the auc metric, utilizing the invariance of bias, is able to focus also on the sparse *Good* subset, thus providing better estimates on well-separability.

5   Experiments

Throughout our experiments, we consider the following construction algorithms:

- Basic decision trees (Algorithm 1 with Algorithm 2), as considered in [12]. ($\star$)
- Decision trees with linear classifiers (Algorithm 3) and entropy-based splitting procedure (Algorithm 2). ($\dagger$)
- Decision trees with linear classifiers (Algorithm 3) and auc-based splitting procedure (Algorithm 4). ($\ddagger$)

For the experimental evaluation of the construction algorithms, we consider multiple sources of problems that arise naturally in reactive synthesis, and reduce to stochastic graph games with Integer variables. These variables provide semantical information about the states (resp., actions) they identify, so a strategy-representation method utilizing predicates over the variables produces naturally interpretable output. Moreover, there is an inherent internal structure in the states and their valuations, which machine-learning algorithms can exploit to produce more succinct representation of strategies.

Given a game and an objective, we use an explicit solver to obtain an almost-sure winning strategy. Then we consider the strategy as a list of played (*Good*) and non-played (*Bad*) actions for each state, which can be used directly as an input training set (*Train*). We evaluate the construction algorithms based on succinctness of representation, which we express as the number of non-pure nodes (i.e., nodes with either a predicate or a linear classifier). Further experimental details are presented in [5].

5.1   Graph Games and Winning Strategies

We consider two sources of problems reducible to strategy representation in graph games, namely, AIGER safety synthesis [29] and LTL synthesis [43].

AIGER – Scheduling of Washing Cycles. The goal of this problem is to design a centralized controller for a system of washing tanks running in parallel. The system is parametrized by the number of tanks, the time limit to fill a tank with water after a request, the delay after which the tank has to be emptied again, and a number of tanks per one shared water pipe. The controller has to ensure that all requests are satisfied within the specified time limit.

The problem has been introduced in the second year of SYNTCOMP [30], the most important and well-known synthesis competition. The problem is implicitly described in the form of AIGER safety specification [29], which uses circuits with input, output, and latch Boolean variables. This reduces directly to graph games with $\{0, 1\}$-valued Integer variables and safety objectives. The state-variables represent for each tank whether it is currently filled, and the current deadline for filling (resp., emptying). The action-variables capture environment requests to fill water tanks, and the controller commands to fill (resp., empty) water tanks. We consider 364 datasets, where the sizes of *Train* range from 640 to 1024000, and the sizes of *Var* range from 16 to 62.

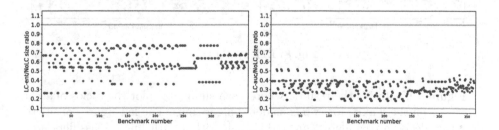

Fig. 7. Scheduling of washing cycles.

We illustrate the results in Fig. 7. Both subfigures plot the ratios of sizes for two considered algorithms. Each dot represents a dataset, the y-axis captures the ratios, and the two red lines represent equality and order-of-magnitude improvement, respectively. The left figure considers the size ratios of the basic decision-tree algorithm and the algorithm with linear classifiers and entropy-based splits ($\star/\dagger$). The arithmetic, geometric, and harmonic means of the ratios are 59%, 57%, and 55%, respectively. The right figure considers the basic algorithm and the algorithm with linear classifiers and auc-based splits ($\star/\ddagger$). The arithmetic, geometric, and harmonic means of the ratios are 33%, 31%, and 30%, respectively.

LTL Synthesis. In reactive synthesis, most properties considered in practice are ω-regular objectives, which can be specified as linear-time temporal logic (LTL) formulae over input/output signals [43]. Given an LTL formula and input/output signal partitioning, the controller synthesis for this specification is reducible to solving a graph game with parity objective.

In our experiments, we consider LTL formulae randomly generated using the tool SPOT [24]. Then, we use the tool Rabinizer [31] to translate the formulae into deterministic parity automata. Crucially, the states of these automata contain semantic information retained by Rabinizer during the translation. We consider an encoding of the semantic information (given as sets of LTL formulae and permutations) into binary vectors. The encoding aims to capture the inherent structure within automaton states, which can later be exploited during strategy representation. Finally, for each parity automaton we consider various input/output partitionings of signals, and thus we obtain parity graph games with $\{0, 1\}$-valued Integer variables. The whole pipeline is described in detail in [12].

We consider graph games with liveness (parity-2) and strong fairness (parity-3) objectives. In total we consider 917 datasets, with sizes of *Train* ranging from 48 to 8608, and sizes of *Var* ranging from 38 to 128.

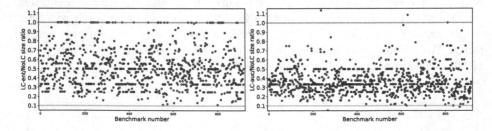

Fig. 8. LTL synthesis.

Figure. 8 illustrates the results, where both subfigures plot the ratios of sizes (captured on the y-axis) for two considered algorithms. The left figure considers the basic decision-tree algorithm and the algorithm with linear classifiers and entropy-based splits (⋆/†). The arithmetic, geometric, and harmonic means of the ratios are 51%, 47%, and 43%, respectively. The right figure considers the basic decision-tree algorithm and the algorithm with linear classifiers and auc-based splits (⋆/‡). The arithmetic, geometric, and harmonic means of the ratios are 36%, 34%, and 31%, respectively.

5.2 MDPs and Almost-Sure Winning Strategies

LTL Synthesis with Randomized Environment. In LTL synthesis, given a formula and an input/otput signal partitioning, there may be no controller that satisfies the LTL specification. In such a case, it is natural to consider a different setting where the environment is not antagonistic, but behaves randomly instead. There are LTL specifications that are unsatisfiable, but become satisfiable when randomized environment is considered. Such special case of LTL synthesis reduces to solving MDPs with almost-sure parity objectives [17]. Note that in this setting, the precise probabilities of environment actions are immaterial, as they have no effect on the existence of a controller ensuring an objective almost-surely (i.e., with probability 1).

We consider 414 instances of LTL synthesis reducible to graph games with *co-Büchi* (i.e., parity-2) objective, where the LTL specification is unsatisfiable, but becomes satisfiable with randomized environment (which reduces to MDPs with almost-sure co-Büchi objective). The examples have been obtained by the same pipeline as the one described in the previous subsection. In the examples, the sizes of *Train* range from 80 to 26592, and the sizes of *Var* range from 38 to 74.

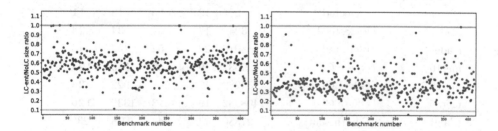

Fig. 9. LTL synthesis with randomized environment.

The experimental results are summarized in Fig. 9. The two subfigures plot the ratios of sizes (captured on the y-axis) for two considered algorithms. The left figure considers the basic decision-tree algorithm and the algorithm with linear classifiers and entropy-based splits (⋆/†). The arithmetic, geometric, and harmonic means of the ratios are 58%, 56%, and 54%, respectively. The right figure considers the basic decision-tree algorithm and the algorithm with linear classifiers and auc-based splits (⋆/‡). The arithmetic, geometric, and harmonic means of the ratios are 38%, 36%, and 34%, respectively.

Table 1. PRISM model checking.

Model	Specification	|Train|	|Var|	NoLC	LC-ent	LC-auc
coin2_K1	F[finished&agree]	1820	7	142	135	**45**
coin2_K2	F[finished&agree]	3484	7	270	261	**55**
coin2_K3	F[finished&agree]	5148	7	386	373	**60**
coin2_K4	F[finished&agree]	6812	7	536	520	**55**
coin2_K9	F[finished&agree]	15132	7	1137	1123	**68**
coin3_K1	F[finished&agree]	27854	9	772	713	**298**
coin3_K2	F[finished&agree]	51566	9	1142	1074	**316**
coin3_K3	F[finished&agree]	75278	9	1580	1500	**378**
coin3_K4	F[finished&agree]	98990	9	2047	1967	**388**
coin4_K0	F[finished&agree]	52458	11	742	632	**221**
coin5_K0	F[finished&agree]	451204	13	2572	1626	**566**
csma2_2	F[succ_min_bo≤2]	8590	13	70	52	**32**
csma2_2	F[max_col≥3]	10380	13	65	**54**	54
csma2_3	F[succ_min_bo≤3]	25320	13	66	48	**35**
csma2_3	F[max_col≥4]	28730	13	63	**48**	59
csma2_4	F[succ_min_bo≤4]	73110	13	60	42	**40**
csma2_4	F[max_col≥5]	79580	13	54	**41**	59
firewire_abst	F[exists_leader]	2535	4	12	10	**8**
firewire_impl_01	F[exists_leader]	22633	12	99	86	**71**
firewire_impl_02	F[exists_leader]	37180	12	101	85	**81**
firewire_impl_05	F[exists_leader]	90389	12	102	85	**72**
leader2	F[elected]	204	12	25	18	**11**
leader3	F[elected]	3249	17	61	34	**23**
leader4	F[elected]	38016	22	152	92	**45**
mer10	G[!err_G]	499632	19	552	510	**124**
mer20	G[!err_G]	954282	19	963	922	**124**
mer30	G[!err_G]	1408932	19	1373	1332	**126**
wlan0	F[both_sent]	27380	14	244	**198**	232
wlan1	F[both_sent]	81940	14	272	**200**	286
wlan2	F[both_sent]	275140	14	288	**206**	353
zeroconf	F[configured]	268326	24	413	**330**	376

PRISM Model Checking. We consider model checking of probabilistic systems in the model checker PRISM [33]. Given an implicit description of a probabilistic system in PRISM, and a reachability/safety LTL formula as a specification, the model checking problem of the model and the specification reduces to construction of an almost-sure winning strategy in an MDP with nonnegative Integer variables. The state-variables correspond to the variables in the implicit

PRISM model description, i.e., local states of the moduli, counter values, etc. The action-variables capture the id of the module performing an action, and the id of the action performed by the module.

Table 1 presents the PRISM experimental results, where we consider various case studies available from the PRISM benchmark suite [34] (e.g., communication protocols). The columns of the table represent the considered model and specification, the sizes of *Train* and *Var*, and the decision-tree sizes for the three considered construction algorithms ($\star, \dagger, \ddagger$).

In this set of experiments, we have noticed several cases where the split heuristic based on auc achieves significantly worse results. Namely, in csma, wlan, and zeroconf, it is mostly outperformed by the information-gain split procedure, and sometimes it is outperformed even by standard decision trees without linear classifiers. This was caused by certain variables repeatedly having high auc scores (for different thresholds) when constructing some branches of the tree, even though subsequent choices of the predicates did little progress to linearly separate the data. We were able to mitigate the cases of bad predicate suggestions, e.g., by penalizing the predicates on the variables that already appear in the path to the current node (that is about to be split), however, the inferior overall performance in these benchmarks persists. This discovery motivates to consider various combinations of auc and information-gain methods, e.g., using information gain as a stand-in metric, in cases where auc yields poor scores for all considered predicates.

6 Related Work

Strategy Representation. Previous non-explicit representation of strategies for verification or synthesis purposes typically used BDDs [50] or automata [40, 42] and do not explain the decisions by the current valuation of variables. Classical *decision trees* have been used a lot in the area of machine learning as a classifier that naturally explains a decision [39]. They have also been considered for representation of values and thus implicitly strategies for MDP in [8,9]. In the context of verification, this approach has been modified to capture strategies guaranteed to be ε-optimal, for MDPs [10], partially observable MDPs [11], and (non-stochastic) games [12]. Learning a compact decision tree representation of an MDP strategy was also investigated in [36] for the case of body sensor networks.

Linear extensions of decision trees have been considered already in [23] for combinatoric optimization problems. In the field of machine learning, combinations of decision trees and linear models have been proposed as interpretable models for classification and regression [13,25,35,46]. A common feature of these works is that they do not aim at classifying the training set without any errors, as in classification tasks this would bear the risk of overfitting. In contrast, our usage requires to learn the trees so that they fully fit the data.

The closest to our approach is the work of Neider et al. [41], which learns decision trees with linear classifiers in the leaves in order to capture functions with generally non-Boolean co-domains. Since the aim is not to classify, but represent fully a function, our approach is better tailored to representing strategies. Indeed, since the trees and the lines in the leaves of [41] are generated from counterexamples in the learning process, the following issues arise. Firstly, each counterexample has to be captured exactly using a generated line. With the geometric intuition, each point has to lie on a line, while in our approach we only need to separate positive and negative points by lines, clearly requiring less lines. Secondly, the generation of lines is done online and based on the single discussed point (counterexample). As a result, lines that would work for more points are not preferred, while our approach maximizes the utility of a generated line with respect to the complete data set and thus generally prefers smaller solutions. Unfortunately, even after discussing with the authors of [41] there is no compilable version of their implementation at the time of writing and no experimental confirmation of the above observations could be obtained.

7   Conclusion and Future Work

In this work, we consider strategy representation by an extension of decision trees. Namely, we consider linear classifiers as the leaf nodes of decision trees. We note that the decision-tree framework proposed in this work is more general. Consider an arbitrary data structure $\mathscr{D}$, with an efficient decision oracle for existence of an instance of $\mathscr{D}$ representing a given dataset without error. Then, our scheme provides a straightforward way of constructing decision trees with instances of $\mathscr{D}$ as the leaf nodes.

Besides representation algorithms that provably represent entire input strategy, one can consider models where an error may occur and the data structure is refined into a more precise one only when the represented strategy is not winning. Here we can consider more expressive models in the leaves, too. This could capture representation of controllers exhibiting more complicated functions, e.g. quadratic polynomial capturing that a robot navigates closely (in Euclidean distance) to a given point, or deep neural networks capturing more complicated structure difficult to access directly [32].

Acknowledgments. This work has been partially supported by DFG Grant No KR 4890/2-1 (SUV: Statistical Unbounded Verification), TUM IGSSE Grant 10.06 (PARSEC), Czech Science Foundation grant No. 18-11193S, Vienna Science and Technology Fund (WWTF) Project ICT15-003, the Austrian Science Fund (FWF) NFN Grants S11407-N23 (RiSE/SHiNE) and S11402-N23 (RiSE/SHiNE).

References

1. Abadi, M., Lamport, L., Wolper, P.: Realizable and unrealizable specifications of reactive systems. In: ICALP, pp. 1–17 (1989)
2. Akers, S.B.: Binary decision diagrams. IEEE Trans. Comput. **C–27**(6), 509–516 (1978)

3. Alur, R., et al.: Syntax-guided synthesis. In: Dependable Software Systems Engineering, pp. 1–25 (2015)
4. Alur, R., Henzinger, T., Kupferman, O.: Alternating-time temporal logic. J. ACM **49**, 672–713 (2002)
5. Ashok, P., Brázdil, T., Chatterjee, K., Křetínský, J., Lampert, C.H., Toman, V.: Strategy representation by decision trees with linear classifiers. arXiv.org.1906.08178 (2019)
6. Baier, C., Katoen, J.: Principles of Model Checking. MIT Press, New York (2008)
7. Blass, A., Gurevich, Y., Nachmanson, L., Veanes, M.: Play to test. In: FATES, pp. 32–46 (2005)
8. Boutilier, C., Dearden, R.: Approximate value trees in structured dynamic programming. In: ICML, pp. 54–62 (1996)
9. Boutilier, C., Dearden, R., Goldszmidt, M.: Exploiting structure in policy construction. In: IJCAI, pp. 1104–1113 (1995)
10. Brázdil, T., Chatterjee, K., Chmelík, M., Fellner, A., Křetínský, J.: Counterexample explanation by learning small strategies in Markov decision processes. In: CAV, pp. 158–177 (2015)
11. Brázdil, T., Chatterjee, K., Chmelík, M., Gupta, A., Novotný, P.: Stochastic shortest path with energy constraints in POMDPs: (extended abstract). In: AAMAS, pp. 1465–1466 (2016)
12. Brázdil, T., Chatterjee, K., Křetínský, J., Toman, V.: Strategy representation by decision trees in reactive synthesis. In: TACAS, pp. 385–407 (2018)
13. Breiman, L., Friedman, J.H., Olshen, R.A., Stone, C.J.: Classification and Regression Trees. Chapman and Hall/CRC, Boca Raton (1984)
14. Bryant, R.: Graph-based algorithms for Boolean function manipulation. IEEE Trans. Comput. **C−35**(8), 677–691 (1986)
15. Büchi, J.: On a decision method in restricted second-order arithmetic. In: International Congress on Logic, Methodology, and Philosophy of Science, pp. 1–11 (1962)
16. Büchi, J., Landweber, L.: Solving sequential conditions by finite-state strategies. Trans. AMS **138**, 295–311 (1969)
17. Chatterjee, K., Henzinger, T.A., Jobstmann, B., Singh, R.: Measuring and synthesizing systems in probabilistic environments. J. ACM **62**(1), 9:1–9:34 (2015)
18. Church, A.: Logic, arithmetic, and automata. In: International Congress of Mathematicians, pp. 23–35 (1962)
19. de Alfaro, L., Henzinger, T.: Interface automata. In: FSE, pp. 109–120 (2001)
20. de Alfaro, L., Henzinger, T., Mang, F.: Detecting errors before reaching them. In: CAV, pp. 186–201 (2000)
21. Dehnert, C., Junges, S., Katoen, J., Volk, M.: A storm is coming: a modern probabilistic model checker. In: CAV, pp. 592–600 (2017)
22. Dill, D.: Trace Theory for Automatic Hierarchical Verification of Speed-independent Circuits. MIT Press, Cambridge (1989)
23. Dobkin, D.P.: A nonlinear lower bound on linear search tree programs for solving knapsack problems. J. Comput. Syst. Sci. **13**(1), 69–73 (1976)
24. Duret-Lutz, A., Lewkowicz, A., Fauchille, A., Michaud, T., Renault, E., Xu, L.: Spot 2.0 - a framework for LTL and ω-automata manipulation. In: ATVA, pp. 122–129 (2016)
25. Frank, E., Wang, Y., Inglis, S., Holmes, G., Witten, I.H.: Using model trees for classification. Mach. Learn. **32**(1), 63–76 (1998)
26. Garg, P., Neider, D., Madhusudan, P., Roth, D.: Learning invariants using decision trees and implication counterexamples. In: POPL, pp. 499–512 (2016)

27. Gurevich, Y., Harrington, L.: Trees, automata, and games. In: STOC, pp. 60–65 (1982)
28. Henzinger, T., Kupferman, O., Rajamani, S.: Fair simulation. Inf. Comput. **173**, 64–81 (2002)
29. Jacobs, S.: Extended AIGER format for synthesis. CoRR, abs/1405.5793 (2014)
30. S. Jacobs, et al.: The second reactive synthesis competition (SYNTCOMP 2015). In: SYNT, pp. 27–57 (2015)
31. Komárková, Z., Křetínský, J.: Rabinizer 3: safraless translation of LTL to small deterministic automata. In: ATVA, pp. 235–241 (2014)
32. Kontschieder, P., Fiterau, M., Criminisi, A., Bulò, S.R.: Deep neural decision forests. In: IJCAI, pp. 4190–4194 (2016)
33. Kwiatkowska, M.Z., Norman, G., Parker, D.: PRISM: probabilistic symbolic model checker. In: TOOLS, pp. 200–204 (2002)
34. Kwiatkowska, M.Z., Norman, G., Parker, D.: The PRISM benchmark suite. In: QEST, pp. 203–204 (2012)
35. Landwehr, N., Hall, M., Frank, E.: Logistic model trees. In: ECML, pp. 241–252 (2003)
36. Liu, S., Panangadan, A., Raghavendra, C.S., Talukder, A.: Compact representation of coordinated sampling policies for body sensor networks. In: Advances in Communication and Networks, pp. 6–10 (2010)
37. Manna, Z., Pnueli, A.: The Temporal Logic of Reactive and Concurrent Systems: Specification. Kluwer Academic, Norwell (1992)
38. McNaughton, R.: Infinite games played on finite graphs. Ann. Pure Appl. Logic **65**, 149–184 (1993)
39. Mitchell, T.M.: Machine Learning. McGraw Hill, Maidenhead (1997)
40. Neider, D.: Small strategies for safety games. In: ATVA, pp. 306–320 (2011)
41. Neider, D., Saha, S., Madhusudan, P.: Synthesizing piece-wise functions by learning classifiers. In: TACAS, pp. 186–203 (2016)
42. Neider, D., Topcu, U.: An automaton learning approach to solving safety games over infinite graphs. In: TACAS, pp. 204–221 (2016)
43. Pnueli, A.: The temporal logic of programs. In: FOCS, pp. 46–57 (1977)
44. Pnueli, A., Rosner, R.: On the synthesis of a reactive module. In: POPL, pp. 179–190 (1989)
45. Quinlan, J.R.: Induction of decision trees. Mach. Learn. **1**(1), 81–106 (1986)
46. Quinlan, J.R.: Learning with continuous classes. In: Australian Joint Conference on Artificial Intelligence, pp. 343–348 (1992)
47. Rabin, M.: Automata on infinite objects and Church's problem. In: Conference Series in Mathematics (1969)
48. Ramadge, P., Wonham, W.: Supervisory control of a class of discrete-event processes. SIAM J. Control Optim. **25**(1), 206–230 (1987)
49. Thomas, Wolfgang: Languages, automata, and logic. In: Rozenberg, Grzegorz, Salomaa, Arto (eds.) Handbook of Formal Languages, pp. 389–455. Springer, Heidelberg (1997). https://doi.org/10.1007/978-3-642-59126-6_7
50. R. Wimmer, et al.: Symblicit calculation of long-run averages for concurrent probabilistic systems. In: QEST, pp. 27–36 (2010)

Neural Network Precision Tuning

Arnault Ioualalen[1] and Matthieu Martel[1,2(✉)]

[1] Numalis, Cap Omega, Rond-Point Benjamin Franklin, 34960 Montpellier, France
ioualalen@numalis.com
[2] Laboratoire de Mathématiques et Physique (LAMPS),
Université de Perpignan Via Domitia, Perpignan, France
matthieu.martel@univ-perp.fr

Abstract. Minimizing the precision in which the neurons of a neural network compute is a desirable objective to limit the resources needed to execute it. This is specially important for neural networks used in embedded systems. Unfortunately, neural networks are very sensitive to the precision in which they have been trained and changing this precision generally degrades the quality of their answers. In this article, we introduce a new technique to tune the precision of neural networks in such a way that the optimized network computes in a lower precision without modifying the quality of the outputs of more than a percentage chosen by the user. From a technical point of view, we generate a system of linear constraints among integer variables that we can solve by linear programming. The solution to this system is the new precision of the neurons. We present experimental results obtained by using our method.

Keywords: Formal methods · Floating-point arithmetic ·
Static analysis · Dynamic analysis · Linear programming ·
Numerical accuracy

1 Introduction

Neural networks are more and more used in many domains, including critical embedded systems in aeronautics, space, defense, automotive, etc. These neural networks also become larger and larger while embedded systems still have limited resources, mainly in terms of computing power and memory. As a consequence, running large neural networks on embedded systems with limited resources introduces several new challenges. While recent work has focused on safety [7,9,12,21,22] and security properties [14,24], a different problem is addressed in this article which concerns the accuracy of the computations. It is well-known that neural networks are sensitive to the precision of the computations, or, in other terms to the computer arithmetic used during their training and execution. Indeed, a neural network working correctly in some computer

This work is supported by La Region Occitanie under Grant GRAINE - Syfi. https://
www.laregion.fr.

© Springer Nature Switzerland AG 2019
D. Parker and V. Wolf (Eds.): QEST 2019, LNCS 11785, pp. 129–143, 2019.
https://doi.org/10.1007/978-3-030-30281-8_8

arithmetic (e.g. IEEE754 single precision [1]) may behave poorly if we run it in lower or even in higher precision (e.g. in IEEE754 half or double precision).

We consider the problem of tuning the precision of an already trained neural network, assumed to behave correctly at some precision, in such a way that, after tuning, the network behaves *almost* like the original one while performing its computations in lower precision. In this article, we focus on interpolator networks, i.e. in networks computing mathematical functions. In this case, we will say that the original and optimized networks behave almost identically if they compute functions f and $\hat{f}$ respectively, such that, for any input x, the relative error between the numerical results computed by both networks is less than some user defined constant δ:

$$\left| \frac{f(x) - \hat{f}(x)}{f(x)} \right| \leq \delta. \tag{1}$$

This definition should be adapted for classifier networks without impacting the rest of the techniques presented here. More precisely, in this case, we should compare the original and optimized networks with respect to a performance metric (recall, precision, F1-score, etc.)

Recently, a lot of work has been done concerning precision tuning of general programs (without direct connection to neural networks), based on static analysis [2,5] or dynamic analysis [13,18,20]. In this article, we adapt the approach introduced in [15] to neural networks. We consider fully connected networks with `ReLU` or `tanh` functions (see Sect. 2.1). We always assume that these networks are already trained and work correctly in the sense that they have satisfying performances in terms of interpolation or classification. We assume that each neuron has its own precision for the computations. However, we assume that all the computations performed inside the same neuron (summation and activation function) use the same precision. Finally, we assume that the ranges of the inputs and outputs of each neuron are given. Several techniques have been developed recently to solve precisely this problem [7,9], which is orthogonal to our. Currently, in our implementation, we compute these ranges by dynamic analysis even if we aim at implementing static analysis techniques in (near) future work. We generate a set of constraints describing the propagation of the errors throughout the neural network. The strength of our approach is that we only generate linear constraints among integers (and only integers). These constraints are easy to solve by standard tools. Optimizing the precision of the network under the correctness constraint of Eq. (1) then becomes a linear programming problem. We demonstrate the efficiency of our technique by showing how the size of interpolator neural networks can be reduced in function of the parameter δ of Eq. (1).

The rest of this article is organized as follows. Preliminary notions and notations are introduced in Sect. 2. They concern neural networks and computer arithmetic. The propagation of the roundoff errors throughout a neural network is modeled in Sect. 3. The generation of constraints is introduced in Sect. 4 and experimental results are given in Sect. 5. Section 6 concludes.

$$W_1 = \begin{pmatrix} 0.9 & 0.0 & 2.3 \\ 1.1 & -0.7 & 0.0 \\ 0.1 & -2.1 & 0.4 \end{pmatrix} \quad b_1 = \begin{pmatrix} 0.1 \\ 0.2 \\ 0.3 \end{pmatrix}$$

$$W_2 = \begin{pmatrix} 0.0 & -0.3 & 1.1 \\ 1.0 & 0.2 & 0.0 \\ -0.4 & 0.4 & 1.1 \end{pmatrix} \quad b_2 = \begin{pmatrix} -0.1 \\ 0.0 \\ -0.1 \end{pmatrix}$$

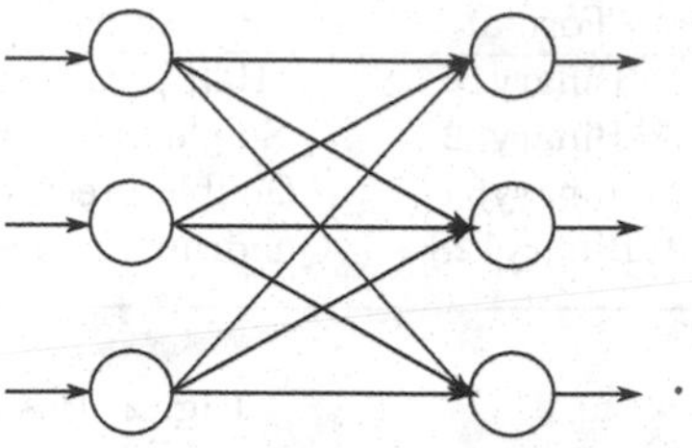

Fig. 1. Example of a fully-connected two-layer network with three neurons by layer.

2   Preliminary Definitions

In this section, we introduce preliminary notions and notations concerning neural networks and computer arithmetics. Section 2.1 is dedicated to neural networks while Sect. 2.2 focuses on the floating-point arithmetic. Finally, Sect. 2.3 introduces precision tuning.

2.1   Neural Networks

In this article, a neural network is defined by means of affine transformations defined by the grammar of Eq. (2) in function of an input vector $\overline{x} \in \mathbb{R}^m$.

$$f(\overline{x}) ::= \text{ReLU}(W \cdot \overline{x} + \overline{b}) \mid \tanh(W \cdot \overline{x} + \overline{b}) \mid f_1(f_2(\overline{x})) \tag{2}$$

The hyperbolic tangent is denoted **tanh** and a rectified linear unit (**ReLU**) activation function is defined by

$$\text{ReLU}(\overline{x}) = \big(\max(0, x_1), \ldots, \max(0, \overline{x}_m) \big)^T. \tag{3}$$

Following Eq. (2), an affine function is either an affine map $f : \mathbb{R}^m \to \mathbb{R}^n$ composed with a ReLU or tanh function or the composition of the former elements. In general, an affine function with ReLU or tanh $f : \mathbb{R}^m \to \mathbb{R}^n$ defines a fully connected layer of a neural network. The whole network is a sequence of ℓ layers, which corresponds to the composition of ℓ affine functions $f_1 \circ f_2 \ldots \circ f_\ell$.

An example of neural network is given in Fig. 1. This fully connected neural network is made of two layers, each layer containing three neurons. The matrices W_1 and W_2 correspond to the first and second layers respectively and b_1 and b_2 are the second members of each layer. For example, the first neuron of the first layer computes $0.9\overline{x}_1 + 2.3\overline{x}_3 + 0.1$ in function of the entry $\overline{x} \in \mathbb{R}^3$.

Note that other operations, different from affine transformations and usually performed by some layers of other kinds of neural networks, such as convolutional layers or max pooling layers, can be reduced to affine transformations [9]. We may then omit them in our work without loss of generality.

Format	Name	p	e bits	e_{min}	e_{max}
Binary16	Half precision	11	5	-14	$+15$
Binary32	Single precision	24	8	-126	$+127$
Binary64	Double precision	53	11	-1122	$+1223$
Binary128	Quadruple precision	113	15	-16382	$+16383$

Fig. 2. Basic binary IEEE754 formats.

2.2 Computer Arithmetics

We introduce here some elements of floating-point arithmetic [1,17]. First of all, a *floating-point number* x in base β is defined by

$$x = s \cdot (d_0.d_1 \ldots d_{p-1}) \cdot \beta^e = s \cdot m \cdot \beta^{e-p+1} \tag{4}$$

where $s \in \{-1, 1\}$ is the sign, $m = d_0 d_1 \ldots d_{p-1}$ is the *significand*, $0 \leq d_i < \beta$, $0 \leq i \leq p-1$, p is the *precision* and e is the exponent, $e_{min} \leq e \leq e_{max}$.

A floating-point number x is *normalized* whenever $d_0 \neq 0$. The IEEE754 Standard defines binary formats (with $\beta = 2$) and decimal formats (with $\beta = 10$). In this article, without loss of generality, we only consider normalized numbers and we always assume that $\beta = 2$ (which is the most common case in practice). The IEEE754 Standard also specifies a few values for p, e_{min} and e_{max} which are summarized in Fig. 2. Finally, special values are also defined: nan (Not a Number) resulting from an invalid operation, $\pm\infty$ corresponding to overflows, and $+0$ and -0 (signed zeros).

The IEEE754 Standard also defines five rounding modes for elementary operations over floating-point numbers. These modes are towards $-\infty$, towards $+\infty$, towards zero, to the nearest ties to even and to the nearest ties to away and we write them $\circ_{-\infty}$, $\circ_{+\infty}$, $\circ_0$, $\circ_{\sim_e}$ and $\circ_{\sim_a}$, respectively. The semantics of the elementary operations $\diamond \in \{+, -, \times, \div\}$ is then defined by

$$f_1 \diamond_\circ f_2 = \circ (f_1 \diamond f_2) \tag{5}$$

where $\circ \in \{\circ_{-\infty}, \circ_{+\infty}, \circ_0, \circ_{\sim_e}, \circ_{\sim_a}\}$ denotes the rounding mode. Equation (5) states that the result of a floating-point operation $\diamond_\circ$ done with the rounding mode $\circ$ returns what we would obtain by performing the exact operation $\diamond$ and next rounding the result using $\circ$. The IEEE754 Standard also specifies how the square root function must be rounded in a similar way to Eq. (5) but does not specify the roundoff of other functions like sin, log, etc.

We introduce hereafter two functions which compute the *unit in the first place* and the *unit in the last place* of a floating-point number. These functions are used further in this article to generate constraints encoding the way roundoff errors are propagated throughout computations. The ufp of a number x is

$$\mathsf{ufp}(x) = \min \{i \in \mathbb{N} : 2^{i+1} > x\} = \lfloor \log_2(x) \rfloor. \tag{6}$$

The ulp of a floating-point number which significant has size p is defined by

$$\mathsf{ulp}(x) = \mathsf{ufp}(x) - p + 1. \tag{7}$$

The ufp of a floating-point number corresponds to the binary exponent of its most significant digit. Conversely, the ulp of a floating-point number corresponds to the binary exponent of its least significant digit.

2.3 Precision Tuning

The method developed in this article aims at tuning the precision of neural networks. While this subject is new for neural networks, some work has been carried out recently in this domain for usual computer programs and, in this section, we introduce some background material about this domain. Precision tuning consists of finding the least floating-point formats enabling a program to compute some results with an accuracy requirement. Precision tuning allows compilers to select the most appropriate formats (for example IEEE754 [1] half, single, double or quadruple formats [1,17]) for each variable. It is then possible to save memory, reduce CPU usage and use less bandwidth for communications whenever distributed applications are concerned. So, the choice of the best floating-point formats is an important compile-time optimization in many contexts. Precision tuning is also of great interest for the fixed-point arithmetic [11] for which it is important to determine data formats, for example in FPGAs [8,16]. In mixed precision, i.e. when every variable or intermediary result may have its own format, possibly different from the format of the other variables, this problem leads to a combinatorial explosion.

Several approaches have been proposed to determine the best floating-point formats as a function of the expected accuracy on the results. Darulova and Kuncak use a forward static analysis to compute the propagation of errors [6]. If the computed bound on the accuracy satisfies the post-conditions then the analysis is run again with a smaller format until the best format is found. Note that in this approach, all the values have the same format (contrarily to our framework where each control-point has its own format). While Darulova and Kuncak develop their own static analysis, other static techniques [10,23] could be used to infer from the forward error propagation the suitable formats. This approach has also been improved in [5]. Chiang *et al.* [2] have proposed a method to allocate a precision to the terms of an arithmetic expression (only). They use a formal analysis via Symbolic Taylor Expansions and error analysis based on interval functions. In spite of our linear constraints, they solve a quadratically constrained quadratic program to obtain annotations.

Other approaches rely on dynamic analysis. For instance, the Precimonious tool tries to decrease the precision of variables and checks whether the accuracy requirements are still fulfilled [18,20]. Lam *et al.* instrument binary codes in order to modify their precision without modifying the source codes [13]. They also propose a dynamic search method to identify the pieces of code where the precision should be modified.

Another related research direction concerns the compile-time optimization of programs in order to improve the accuracy of the floating-point computation in function of given ranges for the inputs, without modifying the formats of the numbers [4,19].

3   Roundoff Error Modelling

In this section, we introduce some theoretical results concerning the numerical errors done inside a neural network. The error on the output of an affine transformation function can be decomposed in two parts, the propagation of the errors on the input vector and the roundoff errors arising in the computation of the affine function itself. We show in Proposition 2 that the numerical error on the output of an affine transformation function can be expressed by $\max(p+\mu, q+\nu)+1$ where p is related to the precision of the input vector, q is the precision in which the affine transformation is computed and where μ and ν are constants depending only on the neural networks, i.e. on W and b.

Following Eq. (2), a fully connected layer of a neural network computes an output vector $\overline{u} \in \mathbb{R}^n$ in function of an input vector $\overline{x} \in \mathbb{R}^m$ such that

$$\overline{u} = f(\overline{x}) = W \cdot \overline{x} + \overline{b}, \tag{8}$$

for some $n \times m$ matrix W and for some vector $\overline{b} \in \mathbb{R}^n$ (the case of ReLU and tanh functions will be discussed at the end of Sect. 4.) Proposition 1 states how to bound the numerical errors arising in Eq. (8).

Proposition 1. *Let us consider a fully connected layer of a neural network as defined in Eq. (8). Let p_i, $1 \le i \le n$, denote the precision of the ith neuron of the layer. Let $\hat{x}$ be an approximated input and $\overline{e}$ some absolute error bound on the input, i.e. the exact input $\overline{x}$ satisfies $|\overline{x} - \hat{x}| \le \overline{e}$. Then the networks computes the output $f(\hat{x})$ and, for all i, $1 \le i \le n$, the absolute error $\overline{err}_i$ on the ith component of the output $\overline{err} = |f(\overline{x}) - f(\hat{x})|$ on this output is bound by*

$$\overline{err}_i \le \sum_{j=1}^{m} W_{ij} \cdot \overline{e}_j + 2^{-p_i} \cdot \left(\overline{b}_i + (m+1) \cdot \sum_{j=1}^{m} |W_{ij} \cdot \hat{x}_j| \right). \tag{9}$$

Proof. Using the notations of Eq. (8), we have

$$\overline{u}_i = \sum_{j=1}^{m} W_{ij} \cdot \overline{x}_j + \overline{b}_i, \quad 1 \le i \le n. \tag{10}$$

Then the error $\overline{err}$ on the output is

$$\overline{err} = W \cdot \overline{e} + \overline{c} \tag{11}$$

where $W \cdot \overline{e}$ is the propagation of the initial error on the input and $\overline{c}$ the error introduced by the computation of $\overline{u} = f(\hat{x})$ in machine. We need to bound $\overline{c}$. Explicitly,

$$\overline{u}_i = \hat{f}_i(\hat{x}) = W_{i1} \cdot \hat{x}_1 + W_{i2} \cdot \hat{x}_2 + \ldots + W_{im} \cdot \hat{x}_m + \overline{b}_i. \tag{12}$$

First, the errors due to products in Eq. (12) are bound by

$$err_\times(\overline{u}_i) \le |W_{i1} \cdot \hat{x}_1| \cdot 2^{-p_i} + |W_{i2} \cdot \hat{x}_2| \cdot 2^{-p_i} + \ldots + |W_{im} \cdot \hat{x}_m| \cdot 2^{-p_i} \tag{13}$$

$$= \left(|W_{i1} \cdot \hat{x}_1| + |W_{i2} \cdot \hat{x}_2| + \ldots + |W_{im} \cdot \hat{x}_m| \right) \cdot 2^{-p_i}. \tag{14}$$

Then the errors due to additions are bound by

$$err_+(\overline{u}_i) \le 2^{-p_i} \cdot (m-1) \cdot \sum_{j=1}^{m} |W_{ij} \cdot \hat{x}_j| + 2^{-p_i} \cdot (\overline{b}_i + \sum_{j=1}^{m} |W_{ij} \cdot \hat{x}_j|) \quad (15)$$

and, consequently,

$$err(\overline{u}_i) = err_\times(\overline{u}_i) + err_+(\overline{u}_i) \quad (16)$$

$$\le 2^{-p_i} \cdot m \cdot \sum_{j=1}^{m} |W_{ij} \cdot \hat{x}_j| + 2^{-p_i} \cdot \left(\hat{x}_i + \sum_{j=1}^{m} |W_{ij} \cdot \hat{x}_j| \right). \quad (17)$$

Finally, by combining Eqs. (11) and (16), we bound $\overline{err}$ the global error vector on the output $\overline{u}$ by

$$\overline{err}_i \le \sum_{j=1}^{m} W_{ij} \cdot \overline{e}_j + 2^{-p_i} \cdot \left(\overline{b}_i + (m+1) \cdot \sum_{j=1}^{m} |W_{ij} \cdot \hat{x}_j| \right). \quad (18)$$

$\square$

The next step consists of linearizing the equations in order to make them easier to solve by a solver.

Proposition 2. *Let us consider a fully connected layer of a neural network as defined in Eq. (8). Let $\overline{p}_i$, $1 \le i \le n$, denote the precision of the ith neuron of the layer. Let $\hat{x}$ be an approximated input of precision $\overline{q}$, i.e. the absolute error $\overline{e}$ on the input, is bound by $|\overline{x}_i - \hat{x}_i| \le \overline{e}_i < 2^{-q_i}$, $\forall i$, $1 \le i \le n$. Then the accuracy of the ith output $\overline{u}_i = f_i(\hat{x})$ is $\max(\mu_i + p_i, \nu_i + q_i) + 1$ with μ_i and ν_i two constants defined by*

$$\mu_i = \mathbf{ufp}\left(\overline{b}_i + (m+1) \cdot \sum_{j=1}^{m} |W_{ij} \cdot \hat{x}_j| \right) \quad \nu_i = \mathbf{ufp}\left(\left| \sum_{j=1}^{m} W_{ij} \right| \right) \quad \forall i, \ 1 \le i \le n.$$

Proof. Let us write

$$\overline{\alpha}_i = \overline{b}_i + (m+1) \cdot \sum_{j=1}^{m} |W_{ij} \cdot \hat{x}_j| \quad (19)$$

Indeed, the vector $\overline{\alpha}$ is constant. Let $\overline{e}$ be the data error vector introduced in Eq. (11) and let q be the accuracy of the input, we have

$$q_i = \min \left\{ r \in \mathbb{N} \ \overline{e}_i \le 2^r \right\} \quad (20)$$

Let

$$\overline{\beta}_i = \left| \sum_{j=1}^{m} W_{ij} \right| \quad (21)$$

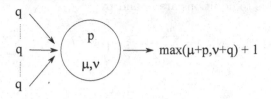

Fig. 3. Accuracy propagation throughout a neuron as defined in Proposition 2.

Again, the $\overline{\beta}_i$, $1 \leq i \leq n$ are constants. Then

$$\left| \sum_{j=1}^{m} W_{ij} \cdot \overline{e}_j \right| \leq \overline{\beta}_i \cdot 2^{q_i} \tag{22}$$

Consequently,

$$\overline{err} \leq \overline{\alpha} \cdot 2^p + \overline{\beta} \cdot 2^q \tag{23}$$

Let $\overline{\mu} = \mathbf{ufp}(\overline{\alpha})$ and $\overline{\nu} = \mathbf{ufp}(\overline{\beta})$. Equation (23) becomes

$$\forall i, \ 1 \leq i \leq n, \ \overline{err}_i \leq 2^{\overline{\mu}_i} \cdot 2^{\overline{p}_i} + 2^{\overline{\nu}_i} \cdot 2^{\overline{q}_i} = 2^{\overline{\mu}_i + \overline{p}_i} + 2^{\overline{\nu}_i + \overline{q}_i} \tag{24}$$

$$\leq 2^{\max(\overline{\mu}_i + \overline{p}_i, \overline{\nu}_i + \overline{q}_i) + 1} \tag{25}$$

$\square$

The way Proposition 2 defines the propagation of a neuron is summarized in Fig. 3.

4   Constraint Generation

In this section, we describe our algorithm to tune the precision of a neural network. We assume that the input network correctly computes a function $f(\overline{x})$. When we decrease this precision, the network computes a new function $\hat{f}(\overline{x})$. Then we aim at finding the smallest precision such that the relative error

$$\left| \frac{f(\overline{x}) - \hat{f}(\overline{x})}{f(\overline{x})} \right| < \delta \quad , \tag{26}$$

for a given tolerance δ specified by the user.

We generate the constraints of Fig. 4, explained hereafter. Let us consider a neural network made of ℓ layers of n fully connected neurons. The variables of the constraint system are $\mathtt{Prec}(W[k,i])$, for $0 \leq i < n$, $0 \leq k < \ell$ and $\mathtt{Prec}(X[k,i])$, for $0 \leq i < n$, $0 \leq k < \ell + 1$. They correspond respectively to the accuracy used to compute inside the neurons and the accuracy of the output of each neuron. Next the constraints depend on values computed *a priori* in

$$\forall 0 \leq i < n,\ \forall 0 \leq k < \ell,\ 0 < \text{Prec}(X[k,i]) \leq \text{ComputedPrec}(X[k,i]) \quad \text{(PB)}$$

$$\forall 0 \leq i < n,\ \forall 0 \leq k < \ell,\ 0 < \text{Prec}(W[k,i]) \leq \text{InitialPrec}(W[k,i]) \quad \text{(IP)}$$

$$\forall 0 \leq i < n,\ \text{Prec}(\text{Output}[i]) \leq \text{Prec}(X[\ell,i]) \quad \text{(PC)}$$

$$\forall 0 \leq i < n,\ \forall 0 \leq k < \ell,\ \text{Prec}(X[k+1,i]) - \text{Prec}(W[k,i]) \leq \mu - 1 \quad \text{(EW)}$$

$$\forall 0 \leq i < n,\ \forall 0 \leq k < \ell,\ \text{Prec}(X[k+1,i]) - \text{Prec}(X[k,i]) \leq \nu - 1 \quad \text{(EX)}$$

Fig. 4. Constraints generated for precision optimization of neural networks.

function of the network and its input datasets. First, $\forall 0 \leq i < n$, $\forall 0 \leq k < \ell$, $\text{InitialPrec}(W[k,i])$ is the initial precision of the ith neuron of the kth layer, i.e. the precision used for this neuron in the original network before optimization. Second, $\forall 0 \leq i < n$, $\text{Output}[i]$ is the precision wanted by the user for the ith neuron of the output of the last layer. This precision can be computed from the parameter δ. Finally, $\forall 0 \leq i < n$, $\forall 0 \leq k < \ell + 1$, $\text{ComputedPrec}(X[k,i])$ is the precision of the output of the ith neuron of the layer $k - 1$. This precision is computed by static or dynamic analysis by applying Proposition 2 to all the neurons of the original network (with its original precision). Note that for all $0 \leq i < n$, $X[0,i]$ corresponds to the input of the network. Our algorithm works as follows.

1. **Compute the forward accuracy of the network.** For each neuron $0 \leq i < n$ of the kth layer, $0 \leq k < \ell$, we compute the precision precX[k,i] of the output X[k,i] in the worst case, for all input vectors of the considered dataset D. We also compute at the same time, for each neuron, the minimum and maximum values Xmin[k,i] and Xmax[k,i] of its output for D. In our implementation, these computations are done by dynamic analysis but they can be done by static analysis using the techniques of [7,9].
2. **Generate constraints for precision bounds.** On one hand, the forward accuracy computed at Step 1 gives an upper bound on the accuracy of the output X[k,i] of each neuron such that $0 \leq i < n$, $0 \leq k < \ell$. On the other hand, the precision desired by the user (thanks to the parameter δ) gives a lower bound on the accuracy of the outputs of the last layer. For each neuron, we generate the constraints (PB), (IP) and (PC) of Fig. 4.
3. **Generate constraints for backward precision conditions.** Using Proposition 2, we generate the constraints (EW) and (EC) of Fig. 4. In function of the precision of the outputs of the neurons of some layer k, these constraints set conditions on the precision of the neurons of layer k and on the precision of the inputs of layer k. Hence, they propagate in a backward way the constraints set by the user on the final outputs of the network by means of the parameter δ.

The constraints of Fig. 4 are linear constraints among integers. We find an optimal solution the system by linear programming. This solution gives the accuracy needed for each neuron for the δ parameter chosen by the user.

As mentioned in Sect. 2.1 at Eq. (2), the dot product performed in each neuron can be composed with another mathematical function, typically a ReLU or tanh function. Indeed, the examples of Sect. 5 do use tanh functions. While a ReLU does not impact the accuracy in the worst case (it just keeps the input value or reset it to zero which does not make the accuracy decrease), this is not the case for the tanh function. However, $\forall x \in \mathbb{R}$, $|\tanh(x)| \leq x$ and the function only reduces the errors in absolute value. It is then possible to make the approximation $\tanh(x) \approx x$ without under-estimating the roundoff errors done in the computations of the neurons. In other word, we may get rid of the tanh function in the error analysis and constraint generation.

For neural networks combining dot products with other mathematical functions, or to improve the error bounds on the results of the tanh function, a fine error propagation can be computed by means of Taylor series developments. For example, for a value x approximated by a floating-point number f with an error e, i.e. $x = f + e$, we have

$$\tanh(f + e) \approx (f + e) - \frac{1}{3}(f + e) \tag{27}$$

$$= (f + e) - \frac{1}{3}\left(f^3 + 3f^2 e + 3fe^2 + e^3\right) \tag{28}$$

$$= \left(f - \frac{1}{3}f^3\right) + \left(e - f^2 e - fe^2 - \frac{1}{3}e^3\right) \tag{29}$$

$$\approx \tanh(f) + \left(\tanh(e) - f^2 e - fe^2\right) \tag{30}$$

Consequently, the error propagated by the tanh function can be approximated by $\tanh(e) - f^2 e - fe^2$. Similar reasonings can be done for other elementary functions.

5   Experimental Results

In this section, we show on two representative neural networks how our precision tuning method may optimize the precision. These networks originally work in IEEE754 double precision. The prototype used for these experiments has been implemented in Python 2.7 using the linprog function of the scipy library.

5.1   Neural Network Computing the Hyperbolic Sine

The first neural network we consider computes the hyperbolic sine of the point (x, y). This network, displayed in Fig. 5, is made of four layers containing 12, 8, 4 and 1 neurons respectively. The curve in the bottom left corner of Fig. 5 displays the percentage of bits that we can save by our method in function of the parameter δ which sets the relative error that we accept between the outputs of the original and transformed networks. On this curve, 100% corresponds to

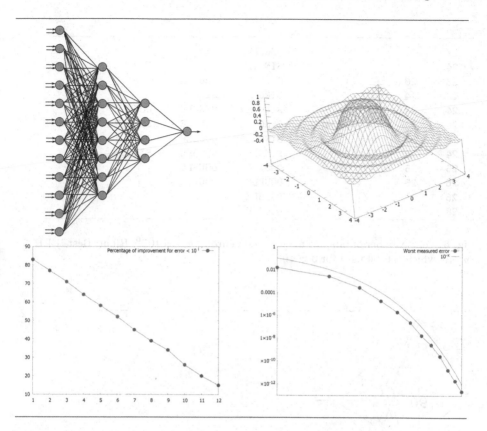

Fig. 5. Top left: Interpolator network used to compute the hyperbolic sine function. Top right: The hyperbolic sine function. Bottom left: Percentage of improvement in bits, compared to the initial network in double precision. Bottom right: Measured error between the original and optimized networks.

the case where all the computations are done in double precision. As we can observe, our technique makes it possible to save a significant amount of bits, depending on δ. The curve in the bottom right corner of Fig. 5 displays the measured distance between the results of the original and optimized networks. This error is compared to the worst accepted error defined by δ. We can see that the actual error is always less than δ.

In Fig. 6, we give more details on the results of our optimization for the case $\delta = 10^{-6}$. The left part of the figure shows the actual number of bits needed for each neuron in this case. Indeed, our method is able to save 54% of bits in this case. In addition, in the right part of Fig. 6, we display the best IEEE754 formats that we may choose according to the number of bits needed for each neuron. For this example, 56% of the neurons can be set in single precision while guaranteeing that the error between the neural network working fully in double precision and the optimized network will be less than $\delta = 10^{-6}$ for any input.

26				DOUBLE			
24				SINGLE			
23	26			SINGLE	DOUBLE		
22	24			SINGLE	SINGLE		
24	23	26		SINGLE	SINGLE	DOUBLE	
24	23	24	24	SINGLE	SINGLE	SINGLE	SINGLE
24	25	24		SINGLE	DOUBLE	SINGLE	
24	25	24		SINGLE	DOUBLE	SINGLE	
25	25			DOUBLE	DOUBLE		
26	25			DOUBLE	DOUBLE		
26				DOUBLE			
28				DOUBLE			

Fig. 6. Left: Number of bits needed for each neuron for $\delta = 10^{-6}$. Right: Best IEEE754 format which can be used for $\delta = 10^{-6}$.

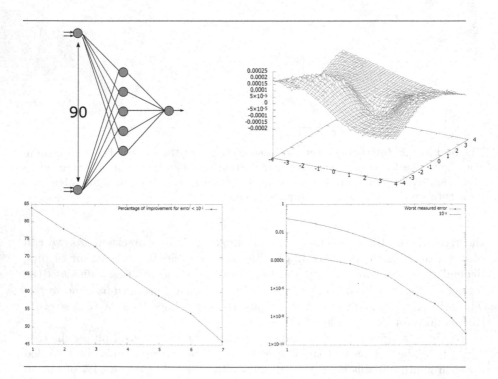

Fig. 7. Top left: An interpolator network used to compute the bump function displayed on the right of the top right corner of the figure. Bottom left: Percentage of improvement in bits, compared to the initial network in double precision. Bottom right: Measured error between the original and optimized networks.

The mean execution time to generate and solve the constraints is 0.9 s for $\delta = 10^{-6}$. This time does not change significantly is we take another value for δ.

5.2 Neural Network Computing a Bump Function

In this section, we introduce a second network computing a function of a point (x, y) displayed in Fig. 7. Again, we compute the percentage of bits that we can save by our method in function of the parameter δ. Again, our results, displayed in Fig. 7 show that our method makes it possible to save an important number of bits (curve at the bottom left corner of Fig. 7). As in Sect. 5.1, we also measure the error between the original and optimized networks and compare it to the theoretical error defined by the parameter δ (curve at the bottom right corner of Fig. 7).

The mean execution time for constraint generation and constraint solving is 25 s. Note that our implementation is not optimized and consider that all the layers have the same number of neurons (90 in this example).

6 Conclusion

In this article, we introduced a new method to tune the precision of the computations done inside the neurons of a network in order to save memory while ensuring that the network still answers correctly, compared to the original network. Our method models the propagation of the roundoff errors through a set of linear constraints among integers which can be solved by linear programming. Experimental results show the efficiency of our method.

A first perspective is to test our method on larger, real-size neural networks. This requires to improve our prototype to manage some implementation details. We believe that our method will scale up as long as the linear programming solver will scale up. If this is not enough, a solution would be to assign the same precision to a group of neurons in order to reduce the number of equations and variables in the constraint system. The choice of the best partition remains an open question currently and additional work should be carried out in this direction.

A second perspective is to extend our method to classifiers, i.e. to neural networks recognizing patterns given as inputs. While most our approach can be reused for classifier, it would be necessary to formally define what is an acceptable approximated output for the networks working with less precision. A possibility would be to check that the original and optimized networks almost always classify the inputs in the same way (in $\delta\%$ of the cases, δ being chosen by the user). In particular, we aim at testing our method on neural networks developed for standard recognition benchmarks such as CIFAR and MNIST.

A third perspective is to generate code for the fixed-point arithmetic [11]. Fixed-point arithmetic (or possibly integer arithmetic) are more and more used to run neural networks, specially in embedded systems. To cope with the fixed-point arithmetic, we need to adapt the error propagations equations of Sect. 3 without changing the general approach developed in this article.

A last perspective is to improve the method itself, by optimizing the equations of Sect. 3. For example, some errors are over-estimated. In addition, all the computations done inside the same neuron have the same accuracy and we could improve this point. The way the computations are done inside neurons could also be transformed by re-parsing of the computations in order to improve their accuracy [3, 4] and, consequently, to allow smaller formats.

References

1. ANSI/IEEE: IEEE Standard for Binary Floating-point Arithmetic (2008)
2. Chiang, W., Baranowski, M., Briggs, I., Solovyev, A., Gopalakrishnan, G., Rakamaric, Z.: Rigorous floating-point mixed-precision tuning. In: POPL, pp. 300–315. ACM (2017)
3. Damouche, N., Martel, M.: Mixed precision tuning with salsa. In: Proceedings of the 8th International Joint Conference on Pervasive and Embedded Computing and Communication Systems, PECCS 2018, pp. 185–194. SciTePress (2018)
4. Damouche, N., Martel, M., Chapoutot, A.: Improving the numerical accuracy of programs by automatic transformation. STTT **19**(4), 427–448 (2017). https://doi.org/10.1007/s10009-016-0435-0
5. Darulova, E., Horn, E., Sharma, S.: Sound mixed-precision optimization with rewriting. In: Proceedings of the 9th ACM/IEEE International Conference on Cyber-Physical Systems, ICCPS, pp. 208–219. IEEE Computer Society/ACM (2018)
6. Darulova, E., Kuncak, V.: Sound compilation of reals. In: POPL 2014, pp. 235–248. ACM (2014)
7. Dutta, S., Jha, S., Sankaranarayanan, S., Tiwari, A.: Output range analysis for deep feedforward neural networks. In: Dutle, A., Muñoz, C., Narkawicz, A. (eds.) NFM 2018. LNCS, vol. 10811, pp. 121–138. Springer, Cham (2018). https://doi.org/10.1007/978-3-319-77935-5_9
8. Gao, X., Bayliss, S., Constantinides, G.A.: SOAP: structural optimization of arithmetic expressions for high-level synthesis. In: International Conference on Field-Programmable Technology, pp. 112–119. IEEE (2013)
9. Gehr, T., Mirman, M., Drachsler-Cohen, D., Tsankov, P., Chaudhuri, S., Vechev, M.T.: AI2: safety and robustness certification of neural networks with abstract interpretation. In: 2018 IEEE Symposium on Security and Privacy, SP, pp. 3–18. IEEE (2018)
10. Goubault, E.: Static analysis by abstract interpretation of numerical programs and systems, and FLUCTUAT. In: Logozzo, F., Fähndrich, M. (eds.) SAS 2013. LNCS, vol. 7935, pp. 1–3. Springer, Heidelberg (2013). https://doi.org/10.1007/978-3-642-38856-9_1
11. Graphics, M.: Algorithmic C Datatypes, software version 2.6 edn. (2011). http://www.mentor.com/esl/catapult/algorithmic
12. Katz, G., Barrett, C., Dill, D.L., Julian, K., Kochenderfer, M.J.: Reluplex: an efficient SMT solver for verifying deep neural networks. In: Majumdar, R., Kunčak, V. (eds.) CAV 2017. LNCS, vol. 10426, pp. 97–117. Springer, Cham (2017). https://doi.org/10.1007/978-3-319-63387-9_5
13. Lam, M.O., Hollingsworth, J.K., de Supinski, B.R., LeGendre, M.P.: Automatically adapting programs for mixed-precision floating-point computation. In: Supercomputing, ICS 2013, pp. 369–378. ACM (2013)

14. Madry, A., Makelov, A., Schmidt, L., Tsipras, D., Vladu, A.: Towards deep learning models resistant to adversarial attacks. In: 6th International Conference on Learning Representations, ICLR 2018 (2018). OpenReview.net
15. Martel, M.: Floating-point format inference in mixed-precision. In: Barrett, C., Davies, M., Kahsai, T. (eds.) NFM 2017. LNCS, vol. 10227, pp. 230–246. Springer, Cham (2017). https://doi.org/10.1007/978-3-319-57288-8_16
16. Martel, M., Najahi, A., Revy, G.: Code size and accuracy-aware synthesis of fixed-point programs for matrix multiplication. In: Pervasive and Embedded Computing and Communication Systems, pp. 204–214. SciTePress (2014)
17. Muller, J.M., et al.: Handbook of Floating-Point Arithmetic. Birkhäuser, Boston (2010)
18. Nguyen, C., et al.: Floating-point precision tuning using blame analysis. In: International Conference on Software Engineering (ICSE). ACM (2016)
19. Panchekha, P., Sanchez-Stern, A., Wilcox, J.R., Tatlock, Z.: Automatically improving accuracy for floating point expressions. In: PLDI 2015, pp. 1–11. ACM (2015)
20. Rubio-Gonzalez, C., et al.: Precimonious: tuning assistant for floating-point precision. In: International Conference for High Performance Computing, Networking, Storage and Analysis, pp. 27:1–27:12. ACM (2013)
21. Singh, G., Gehr, T., Mirman, M., Püschel, M., Vechev, M.T.: Fast and effective robustness certification. In: Advances in Neural Information Processing Systems 31: Annual Conference on Neural Information Processing Systems 2018, NeurIPS 2018, pp. 10825–10836 (2018)
22. Singh, G., Gehr, T., Püschel, M., Vechev, M.T.: An abstract domain for certifying neural networks. PACMPL 3(POPL), 41:1–41:30 (2019)
23. Solovyev, A., Jacobsen, C., Rakamarić, Z., Gopalakrishnan, G.: Rigorous estimation of floating-point round-off errors with symbolic taylor expansions. In: Bjørner, N., de Boer, F. (eds.) FM 2015. LNCS, vol. 9109, pp. 532–550. Springer, Cham (2015). https://doi.org/10.1007/978-3-319-19249-9_33
24. Wang, S., Pei, K., Whitehouse, J., Yang, J., Jana, S.: Formal security analysis of neural networks using symbolic intervals. In: 27th USENIX Security Symposium, USENIX Security 2018, pp. 1599–1614. USENIX Association (2018)

Hybrid Systems

SOS: Safe, Optimal and Small Strategies
for Hybrid Markov Decision Processes

Pranav Ashok[1], Jan Křetínský[1], Kim Guldstrand Larsen[2], Adrien Le Coënt[2],
Jakob Haahr Taankvist[2], and Maximilian Weininger[1(✉)]

[1] Technical University of Munich, Munich, Germany
maxi.weininger@tum.de
[2] Aalborg University, Aalborg, Denmark

Abstract. For hybrid Markov decision processes, UPPAAL Stratego can compute strategies that are safe for a given safety property and (in the limit) optimal for a given cost function. Unfortunately, these strategies cannot be exported easily since they are computed as a very long list. In this paper, we demonstrate methods to learn compact representations of the strategies in the form of decision trees. These decision trees are much smaller, more understandable, and can easily be exported as code that can be loaded into embedded systems. Despite the size compression and actual differences to the original strategy, we provide guarantees on both safety and optimality of the decision-tree strategy. On the top, we show how to obtain yet smaller representations, which are still guaranteed safe, but achieve a desired trade-off between size and optimality.

1 Introduction

Cyber-physical systems often are safety-critical and hence strong guarantees on their safety are paramount. Furthermore, resource efficiency and the quality of the delivered service are strong requirements; the behaviour needs to be optimized with respect to these objectives, while of course staying within the bounds of what is still safe. In order to achieve this, controllers of such systems can be either implemented manually or automatically synthesized. In the former case, due to the complexity of the system, coming up with a controller that is safe is difficult, even more so with the additional optimization requirement. In the latter case, the synthesis may succeed with significantly less effort, though the requirement on both safety and optimality is still a challenge for current synthesis methods. However, due to the size of the systems, the produced controllers may be very complex, hard to understand, implement, modify, or even just output. Indeed, even for moderately sized systems, we can easily end up with gigabytes-long descriptions of their controllers (in the algorithmic context called strategies).

This research was funded in part by TUM IGSSE project 10.06 (PARSEC), the German Research Foundation (DFG) project KR 4890/2-1 "Statistical Unbounded Verification" and the ERC Advanced Grant LASSO.

ⓒ Springer Nature Switzerland AG 2019
D. Parker and V. Wolf (Eds.): QEST 2019, LNCS 11785, pp. 147–164, 2019.
https://doi.org/10.1007/978-3-030-30281-8_9

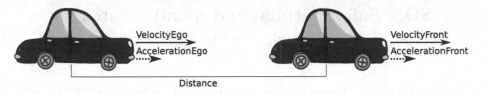

Fig. 1. The two cars, *Ego* and *Front*. We control *Ego* and the environment controls *Front*. Both cars have an acceleration and a velocity. In addition, we know the distance between the cars.

In this paper, we show how to provide a more compact representation, which can yield acceptably short and simple code for resource-limited embedded devices, and consequently can be more easily understood, maintained, modified, debugged, and the requirements are better traceable in the final controller. To this end, as the formalism for the compact representation we choose decision trees [41]. This representation is typically several orders of magnitude smaller than the classical explicit description and also is known for its interpretability and understandability [9,41,47]. The resulting encoded strategy may differ from the original one, but despite that and despite being smaller, it is still guaranteed to be safe and as nearly-optimal as the original one. Moreover, we can trade off additional decrease in size for decrease in performance (getting farther from optimum) to a desired degree, while maintaining safety.

Example 1. As a running example and one of the case studies, we use the following example introduced in [35] and expanded with stronger safety guarantees in [34].

We consider two cars *Ego* and *Front*, depicted in Fig. 1. We control *Ego*, whereas *Front* is controlled by the environment. *Ego* is driving behind *Front*, and both cars have a discrete input (the acceleration) and a continuous state (the velocity). The goal of the adaptive cruise control in *Ego* is, first, to stay safe (by keeping the distance between the cars greater than a given safe distance), and second, to drive as close to *Front* as possible, i.e. to optimize the aggregated distance between the cars.

We use UPPAAL TIGA [2] to get a safe strategy for *Ego* as in [34], and then UPPAAL STRATEGO [18] to learn a (near-)optimal strategy for a desired cost function, given the constraints from the safe strategy. The resulting strategy is output as a list with almost 6 million configurations. Using our new methods, we obtain a decision tree representing the strategy, that has only about 2713 nodes. Additionally, we can trade performance to reduce the size even further, e.g. by increasing the aggregated distance reasonably we can reduce the size to 1247 nodes.

Our Contribution:

– We design and implement a framework STRATEGO+ to transform safe and (near-)optimal strategies into their decision-tree representation, preserving safety and the same level of optimality, while being much smaller.

- We provide several transformations and ways to yet further decrease the size while preserving safety, but relaxing the optimality to a desired extent.
- We test our methods on three case studies, where we show size reductions of up to three orders of magnitude, and quantify the additional size-performance trade-off.

Our techniques can be used to represent (finite-memory non-stochastic) strategies for arbitrary systems exhibiting non-determinism (e.g. Markov decision processes, timed/concurrent/stochastic games). This paper demonstrates the technique on hybrid Markov decision processes, as that is the formalism used in UPPAAL STRATEGO.

Related Work: The problem of computing strategies for hybrid systems has been extensively studied in the past years. Most approaches rely on abstraction techniques: the continuous and infinite state space of the system is represented with a finite number of symbols, *e.g.* discrete points [24,50], sets of states [15], etc. However, it is still hard to deal with uncontrollable components, even though some approaches exist such as robust control [26], or contract-based design [51], but they usually consider the uncontrollable component as a bounded perturbation and do not tackle stochastic behaviour. The tool PESSOA [38,48] can synthesize controllers for cyber-physical systems represented by a set of smooth differential equations with a specification in a fragment of Linear Temporal Logic (LTL). Abstraction techniques are used in [27] for synthesizing strategies for a class of hybrid systems that involve random phenomena together with discrete and continuous behaviours. Discrete, stochastic dynamical systems are considered in [54], where the synthesis of strategies with respect to LTL objectives is made possible with an abstraction-refinement method. In [22] a number of benchmarks for hybrid system verification has been proposed, including a room heating benchmark. In [16] UPPAAL SMC was applied to the performance evaluation of several strategies proposed in the benchmark. However, there was no focus on safety in this approach. In our work, the safety strategy synthesis relies on a discretization of the continuous variables, leading to a decidable problem that can be handled by UPPAAL TIGA, but we furthermore provide safety guarantees for the original system with the use of a Timed Game abstraction based on a guaranteed Euler scheme [36].

In artificial intelligence, compact (factored) representations of Markov decision processes (MDPs) have been developed using dynamic Bayesian networks [7,31], probabilistic STRIPS [33], algebraic decision diagrams [30], and also decision trees [7]. For a detailed survey of compact representations see [5]. Formalisms used to represent MDPs can, in principle, be used to represent strategies as well. In particular, variants of decision trees are probably the most used [7,13,32]. Decision trees have been also used in connection with real-time dynamic programming and reinforcement learning [6,44].

In the context of verification, MDPs are often represented using variants of (MT) BDDs [19,28,39], and strategies by BDDs [55]. Learning a compact decision-tree representation of a strategy has been investigated in [37] for the case of body sensor networks, in [9] for finite (discrete) MDPs, and in [10] for finite games, but only with Boolean variables. Moreover, these decision trees can only predict a single action for a state configuration whereas in this work, we allow the trees to predict more than one action for a single configuration. In control theory, [56] proves that the problem of computing size-optimal determinisiation of controllers is NP-complete and hence discuss various heuristic-based determin- isation algorithms. None of these works consider the optimization aspect, which being a soft constraint enables the trade-offs.

Permissive strategies have been studied in e.g. [3,8,20].

2 Preliminaries

2.1 Hybrid Markov Decision Processes

We describe the mathematical modelling framework. The correspondence to the UPPAAL models is straightforward.

Definition 1 (HMDP). *A hybrid Markov decision process (HMDP) $\mathcal{M}$ is a tuple (C, U, X, F, δ) where:*

1. *the controller C is a finite set of (controllable) modes $C = \{c_1, \ldots, c_k\}$,*
2. *the uncontrollable environment U is a finite set of (uncontrollable) modes $U = \{u_1, \ldots, u_l\}$,*
3. *$X = (x_1, \ldots, x_n)$ is a finite tuple of continuous (real-valued) variables,*
4. *for each $c \in C$ and $u \in U$, $F_{c,u} : \mathbb{R}_{>0} \times \mathbb{R}^X \to \mathbb{R}^X$ is the flow-function that describes the evolution of the continuous variables over time in the combined mode (c, u), and*
5. *δ is a family of probability functions $\delta_\gamma : U \to [0, 1]$, where $\gamma = (c, u, \boldsymbol{x})$ is a global configuration. More precisely, $\delta_\gamma(u')$ is the probability that u in the global configuration $\gamma = (c, u, \boldsymbol{x})$ will change to the uncontrollable mode u'.*

In the following, we denote by $\mathbb{C}$ the set of global configurations $C \times U \times \mathbb{R}^X$ of the HMDP $\mathcal{M}$. The above notion of HMDP actually describes an infinite-state Markov Decision Process [43], where choices of mode for the controller is made periodically and choice of mode for the uncontrollable environment is made probabilistically according to δ. Note that abstracting δ_γ to the support $\hat{\delta}_\gamma = \{u \mid \delta_\gamma(u) > 0\}$, turns $\mathcal{M}$ into a (traditional) hybrid two-player game. The inclusion of δ allows for a probabilistic refinement of the uncontrolled environ- ment in this game. Such a refinement is irrelevant for the purposes of guaran- teeing safety; however, it will be useful for optimizing the cost of operating the system. Indeed, rather than optimizing only the worst-case performance, we wish to optimize the overall expected behaviour.

Strategies. A – memoryless and possibly non-deterministic – strategy σ for the controller C is a function $\sigma : \mathbb{C} \to 2^C$, i.e. given the current configuration $\gamma = (c, u, \boldsymbol{x})$, the expression $\sigma(\gamma)$ returns the set of allowed *actions* in that configuration; in our setting, the actions are the controllable modes to be used for the duration of the next period. Non-deterministic strategies are also called permissive since they permit many actions instead of prescribing one.

The evolution of system over time is defined as follows. Let $\gamma = (c, u, \boldsymbol{x})$ and $\gamma' = (c', u', \boldsymbol{x}')$. We write $\gamma \xrightarrow{\tau} \gamma'$ in case $c' = c, u' = u$ and $\boldsymbol{x}' = F_{(c,u)}(\tau, \boldsymbol{x})$.

A *run* is an interleaved sequence $\pi \in \mathbb{C} \times (\mathbb{R} \times \mathbb{C} \times \mathbb{C} \times \mathbb{C})^*$ of configurations and relative time-delays of some given period P:

$$\pi = \gamma_0 :: P :: \alpha_1 :: \beta_1 :: \gamma_1 :: P :: \alpha_2 :: \beta_2 :: \gamma_2 :: P :: \cdots$$

Then π is a *run according to the strategy* σ if after each period P the following sequence of discrete (instantaneous) changes are made:

1. the value of the continuous variables are updated according to the flow of the current mode, i.e. $\gamma_{i-1} = (c_{i-1}, u_{i-1}, \boldsymbol{x}_{i-1}) \xrightarrow{P} (c_{i-1}, u_{i-1}, \boldsymbol{x}_i) =: \alpha_i$;
2. the environment changes to any possible new mode, i.e. $\beta_i = (c_{i-1}, u_i, \boldsymbol{x}_i)$ where $\delta_{\alpha_i}(u_i) > 0$;
3. the controller changes mode according to the strategy σ, i.e. $\gamma_i = (c_i, u_i, \boldsymbol{x}_i)$ with $c_i \in \sigma(\beta_i)$.

Safety. A strategy σ is said to be *safe* with respect to a set of configuration $S \subseteq \mathbb{C}$, if for any run π according to σ all configurations encountered are within S, i.e. $\alpha_i, \beta_i, \gamma_i \in S$ for all i and also $\gamma'_i \in S$ whenever $\gamma_i \xrightarrow{\tau} \gamma'_i$ with $\tau \leq P$. Note that the notion of safety does not depend on the actual δ, only on its supports. Recall that almost-sure safety, i.e. with probability 1, coincides with sure safety.

We use a guaranteed set-based *Euler method* introduced in [34] to ensure safety of a strategy not only at the configurations where we make decisions, but also in the continuum in between them. We refer the reader to [1, Appendix A.2] for a brief reminder of this method.

Optimality. Under a given *deterministic* (i.e. permitting one action in each configuration) strategy σ the game $\mathcal{M}$ becomes a completely stochastic process $\mathcal{M} \restriction \sigma$, inducing a probability measure on sets of runs. In case σ is *non-deterministic* or *permissive*, the non-determinism in $\mathcal{M} \restriction \sigma$ is resolved uniformly at random. On such a process, we can evaluate a given optimization function. Let $H \in \mathbb{N}$ be a given time-horizon, and D a random variable on runs, then $\mathbb{E}^{\mathcal{M},\gamma}_{\sigma,H}(D) \in \mathbb{R}_{\geq 0}$ is the expected value of D on the space of runs of $\mathcal{M} \restriction \sigma$ of length[1] H starting in the configuration γ. As an example of D, consider the integrated deviation of a continuous variable, e.g. distance between *Ego* and *Front*, with respect to a given target value.

---

[1] Note that there is a bijection between length of the run and time, as the time between each step, P, is constant.

Consequently, given a (memoryless non-deterministic) safety strategy σ_{safe} with respect to a given safety set S, we want to find a deterministic sub-strategy[2] σ_{opt} that optimizes (minimizes or maximizes) $\mathbb{E}^{\mathcal{M},\gamma}_{\sigma_{safe},H}(D)$.

2.2 Decision Trees

From the perspective of machine learning, *decision trees* (DT) [41] are a classification tool assigning classes to data points. A data point is a d-dimensional vector $v = (v_1, v_2, \ldots, v_d)$ of features with each v_i drawing its value from some set D_i. If D_i is an ordered set, then the feature corresponding to it is called *ordered* or *numerical* (e.g. *velocity* $\in \mathbb{R}$) and otherwise, it is called *categorical* (e.g. *color* $\in \{red, green, blue\}$). A (multi-class) DT can represent a function $f \colon \prod_{i=1}^{d} D_i \to A$ where A is a finite set of classes.

A (single-label) DT over the domain $D = \prod_{i=1}^{d} D_i$ with labels A is a tuple $\mathcal{T} = (T, \rho, \theta)$, where T is a finite binary tree, ρ assigns to every inner node predicates of the form $x_i \sim c$ where $\sim \; \in \{\leq, =\}$, $c \in D_i$, and θ assigns to every leaf node a list of natural numbers $[m_1, m_2, \ldots, m_{|A|}]$. For every $v \in D$, there exists a *decision path* from the root node to some leaf ℓ_v. We say that v satisfies a predicate $\rho(t)$ if $\rho(t)$ evaluates to true when its variables are evaluated as given by v. Given v and an inner node t with a predicate $\rho(t)$, the decision path selects either the *left* or *right* child of t based on whether v satisfies $\rho(t)$ or not. For v from the training set, we say that the leaf node ℓ_v *contains* v. Then m_a of a leaf is the number of points contained in the leaf and classified a in the training set. Further, the classes assigned by a DT to a data point v (from or outside of the training set) are given by $\arg\max \theta(\ell_v) = \{i \mid \forall i, j \leq |A|. \; \theta(\ell_v)_i \geq \theta(\ell_v)_j\}$, i.e. the most frequent classes in the respective leaf.

Decision trees may also predict sets of classes instead of a single class. Such a generalization (representing functions of the type $\prod_{i=1}^{d} D_i \to 2^A$) is called a *multi-label* decision tree. In these trees, θ assigns to every leaf node a list of tuples $[(n_1, y_1), (n_2, y_2), \ldots, (n_{|A|}, y_{|A|})]$ where $n_a, y_a \in \mathbb{N}$ are the number of data points in the leaf *not* labelled by class a and labelled by class a, respectively. The (multi-label) classification of a data point is then typically given by the majority rule, i.e. it is classified as a if $n_a < y_a$.

A DT may be constructed using decision-tree learning algorithms such as ID3 [45], C4.5 [46] or CART [11]. These algorithms take as input a training set, i.e. a set of vectors whose classes are already known, and output a DT classifier. The tree constructions start with a single root node containing all the data points of the training set. The learning algorithms explore all possible predicates $p = x_i \sim c$, which split the data points of this node into two sets, X_p and $X_{\neg p}$. The predicate that minimizes the sum of entropies[3] of the two sets is selected. These sets are added as child nodes to the node being split and the whole process

---

[2] i.e. a strategy that for every configuration returns a (non-strict) subset of the actions allowed by the safe strategy.

[3] Entropy of a set X is $H(X) = \sum_{a \in A} p_a \log_2(p_a) + (1 - p_a) \log_2(1 - p_a)$, where p_a is the fraction of samples in X belonging to class a. See [14] for more details.

is repeated, splitting each node further and further until the entropy of the node becomes 0, i.e. all data points belong to the same class. Such nodes are called *pure* nodes. This construction is extended to the multi-label setting by some of the algorithms. A multi-label node is called *pure* if there is at least one class that is endorsed by all data points in that node, i.e. $\exists a \in A : n_a = 0$.

If the tree is grown until all leaves have zero entropy, then the classifier memorizes the training data exactly, leading to *overfitting* [41]. This might not be desirable if the classifier is trained on noisy data or if it needs to predict classes of unknown data. The learning algorithms hence provide some parameters, known as *hyperparameters*, which may be tuned to generalize the classifier and improve its accuracy. Overfitting is not an issue in our setup where we want to learn the strategy function (almost) precisely. However, we can use the hyperparameters to produce even smaller representations of the function, at the "expense" of not being entirely precise any more. One of the hyperparameters of interest in this paper is the minimum split size k. It can be used to stop splitting nodes once the number of data points in them become smaller than k. By setting larger k, the size of the tree decreases, usually at the expense of increasing the entropy of the leaves. There also exist several pruning techniques [21, 40], which remove either leaves or entire subtrees after the construction of the DT.

2.3 Standard UPPAAL STRATEGO Workflow

The process of obtaining an optimized safe strategy σ_{opt} using UPPAAL STRATEGO is depicted as the grey boxes in Fig. 2. First, the HMDP $\mathcal{M}$ is abstracted into a 2-player (non-stochastic) timed game $\mathcal{TG}$, ignoring any stochasticity of the behaviour. Next, UPPAAL TIGA is used to synthesize a safe strategy $\sigma_{safe} : \mathbb{C} \rightarrow 2^C$ for $\mathcal{TG}$ and the safety specification φ, which is specified using a simplified version of timed computation tree logic (TCTL) [2]. After that, the safe strategy is applied on $\mathcal{M}$ to obtain $\mathcal{M} \upharpoonright \sigma_{safe}$. It is now possible to perform reinforcement learning on $\mathcal{M} \upharpoonright \sigma_{safe}$ in order to learn a sub-strategy σ_{opt} that will optimize a given quantitative cost, given as any run-based expression containing e.g. discrete variables, locations, clocks, hybrid variables. For more details, see [17, 18].

3   Stratego$^+$

In this section, we discuss the new UPPAAL STRATEGO$^+$ framework following with each of its components are elucidated.

3.1   New Workflow

UPPAAL STRATEGO$^+$ extends the standard workflow in two ways: Firstly, in the top row, we generate the DT $\mathcal{T}_{opt}$ that exactly represents σ_{opt}, yielding a small representation of the strategy.

The DT learning algorithm can make use of two (hyper-)parameters k and p which may be used to prune the DT; this approach is described in Sect. 3.4.

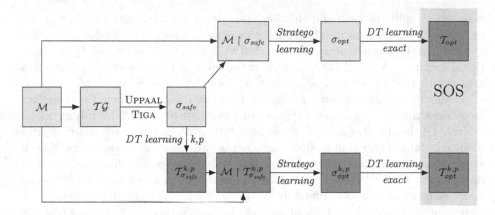

Fig. 2. UPPAAL STRATEGO$^+$ workflow. The dark orange nodes are the additions to the original workflow, which now involve DT learning, the yellow-shaded area delimits the desired safe, optimal, and small strategy representations. (Color figure online)

While pruning reduces the size of the DT, the resultant tree no longer represents the strategy exactly. Hence it is not possible to prune a DT representing deterministic strategies, like in the case of the σ_{opt} described in the first row of the workflow, as safety would be violated.

However, for our second extension we apply the DT learning algorithm to the non-deterministic, permissive strategy σ_{safe}, resulting in $\mathcal{T}^{k,p}_{\sigma_{safe}}$. This DT is less permissive, thereby smaller, since the pruning disallows certain actions; yet it still represents a safe strategy (details in Sect. 3.4). Next, as in the standard workflow, this less permissive safe strategy is applied to the game and STRATEGO is used to get a near-optimal strategy $\sigma^{k,p}_{opt}$ for the modified game $\mathcal{M} \restriction \mathcal{T}^{k,p}_{\sigma_{safe}}$. In the end, we again construct a DT exactly representing the optimal strategy, namely $\mathcal{T}^{k,p}_{opt}$. Note that in the game restricted to $T^{k,p}_{\sigma_{safe}}$ fewer actions are allowed than when it is restricted only to σ_{safe}, and hence the resulting strategy could perform worse. For example, let σ_{safe} allow decelerating or remaining neutral for some configuration, while $T^{k,p}_{\sigma_{safe}}$ pruned the possibility to remain neutral and only allows decelerating. Thus, σ_{opt} remains neutral, whereas $\sigma^{k,p}_{opt}$ has to decelerate and thereby increase the distance that we try to minimize.

In both cases, the resulting DT is safe by construction since we allow the DT to predict only pure actions (actions allowed by all configurations in a leaf, see next section for the formal definition). We convert these trees into a nested if-statements code, which can easily be loaded onto embedded systems.

3.2 Representing Strategies Using DT

A DT with domain $\mathbb{C}$ and labels C can learn a (non-deterministic) strategy $\sigma\colon \mathbb{C} \to 2^C$. The strategy is provided as a list of tuples of the form $(\gamma, \{a_1, \ldots, a_k\})$, where γ is a global configuration and $\{a_1, \ldots, a_k\}$ is the set of actions permitted by σ. The training data points are given by the integer configurations $\gamma \in \mathbb{C}$ (safety for non-integer points is guaranteed by the Euler method; see Sect. 2.1) and the set of

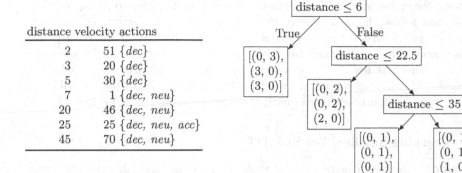

distance	velocity	actions
2	51	{dec}
3	20	{dec}
5	30	{dec}
7	1	{dec, neu}
20	46	{dec, neu}
25	25	{dec, neu, acc}
45	70	{dec, neu}

Fig. 3. A sample dataset (left); and a (multi-label) decision tree generated from the dataset (right). The leaf nodes contain the list of tuples assigned by θ, the inner nodes contain the predicates assigned by ρ

classes for each γ is given by $\sigma(\gamma)$. Consequently, a multi-label decision tree learning algorithm as described in Sect. 2.2 can be run on this dataset to obtain a tree $\mathcal{T}_\sigma$ representing the strategy σ.

Each node of the tree contains the set of configurations that satisfy the decision path traced from the root of the tree to the node. The leaf attribute θ gives, for each action a, the number of configurations in the leaf where the strategy disallows and allows a, respectively. For example, consider a node with 10 configurations with $\theta = [(0, 10), (2, 8), (9, 1)]$. This means that the first action is allowed by all 10 configurations in the node, the second action is disallowed by 2 configurations and allowed by 8, and the third action is disallowed by 9 configurations and allowed only by 1.

Since we want the DT to exactly represent the strategy, we need to run the learning algorithm until the entropy of all the leaves becomes 0, i.e. all configurations of the leaf agree on every action. More formally, given a leaf ℓ with n configurations we require $\theta(\ell) = (0, n)$ or $\theta(\ell) = (n, 0)$ for every action. We call an action that all configurations allow a *pure action*.

The table on the left of Fig. 3 shows a toy strategy. Based on values of distance d and velocity v, it permits a subset of the action set $\{dec, neu, acc\}$. A corresponding DT encoding is displayed on the right of Fig. 3.

3.3 Interpreting DT as Strategy

To extract a strategy from a DT, we proceed as follows: Given a configuration C, we pick the leaf ℓ_C associated with it by evaluating the predicates and following a path through the DT. Then we compute $\theta(\ell_C) = [(n_1, y_1), (n_2, y_2), \ldots, (n_{|A|}, y_{|A|})]$ where $n_a, y_a \in \mathbb{N}$ are the number of data points in the leaf *not* labelled by class a and labelled by class a, respectively. The classes assigned to ℓ_C are exactly its pure actions, i.e. $\{a \mid (0, y_a) \in \theta(\ell_C)\}$.

Note that allowing only pure actions is necessary in order to preserve safety. We do not follow the common (machine learning) way of assigning classes to the nodes based on the majority criterion, i.e. the majority of the data points in that node allow the action; because then the decision tree might prescribe unsafe actions just because they were allowed in most of the configurations in the node. This is also the reason why the DT-learning algorithm described in the previous section needs to run until the entropy of all leaves becomes 0.

3.4   Learning Smaller, Yet Safe DT

We now describe how to learn a DT for a safe strategy that is smaller than the exact representation, but still preserves safety. A tree obtained using off-the-shelf DT learning algorithms is unlikely to exactly represent the original strategy.[4] We use two different methods to achieve the goal: firstly, we use the standard hyperparameter named *minimum split size*, and secondly, we introduce a new post-processing algorithm called *safe pruning*. Both methods rely on the given strategy being non-deterministic/permissive, i.e. permitting several actions in a leaf.

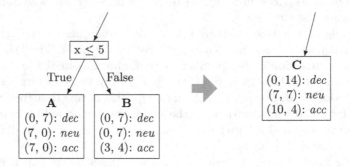

Fig. 4. Illustration of safe pruning applied to a node. The pure action of leaf **A** is just *dec*, for **B** it is both *dec* and *neu*. Safe pruning replaces the nodes with **C**, where only *dec* is a pure action.

(1) Using Minimum Split Size. The splitting process can be stopped before the entropy becomes 0. We do this by introducing a parameter k, which determines the minimum number of data points required in a node to consider splitting it further. During the construction of the tree, a node is usually split if its entropy is greater than 0. When k is set to an integer greater than 2, a node is split only if both the entropy is greater than 0 *and* the number of data points (configurations) in the node is at least k. The strategy given by such a tree is safe as long as it predicts only pure actions, i.e. a with $n_a = 0$. In order to obtain a fully expanded tree, k may be set to 2 (in nodes with <2 configurations,

---

[4] This is because DT learning algorithms are usually configured to avoid overfitting on the dataset.

there is nothing to split). For larger k, the number of pure actions in the leaves decreases. Ultimately, for too large k, we would obtain a tree that has some leaf nodes not containing *any* pure actions. In such a case, the strategy represented by the DT would not be well-defined, as for some data point no action could be picked. However, this can be detected immediately during the construction.

Algorithm 1. Safe Pruning

1: **procedure** SAFE-PRUNING(DT $\mathcal{T}_\sigma = (T, \rho, \theta), p \in \mathbb{N}$)
2:　　**for** $i \leftarrow 1..p$ **do**
3:　　　　$N \leftarrow \{n \in T \mid LEFT(n) \text{ and } RIGHT(n) \text{ are leaves}\}$
4:　　　　　　　　　　　　　　　　　　　▷ Candidate nodes for pruning
5:　　　　**for each** $n \in N$ **do**
6:　　　　　　$c_\ell \leftarrow LEFT(n), c_r \leftarrow RIGHT(n)$
7:　　　　　　**if** $\theta(c_\ell) \cap \theta(c_r) \neq \emptyset$ **then**
8:　　　　　　　　　　　　　　　　　▷ Prune and keep the common classification
9:　　　　　　　　Convert n to a leaf node
10:　　　　　　　$\theta(n) \leftarrow \theta(c_\ell) \cap \theta(c_r)$
11:　　　　　　　Remove c_ℓ and c_r from T

---

(2) Using Safe Pruning. Another way of obtaining a smaller tree is by using a procedure to prune the leaves of the produced tree by merging them while preserving safety. For example, consider the decision node on the left of Fig. 4 with two children that are leaves **A** and **B**. For **A**, only the action *dec* is pure (i.e. allowed by all configurations in the leaf), while for **B** both *dec* and *neu* are pure. Since the sets of pure actions of the two leaf nodes intersect, we can safely remove both **A** and **B** and replace the decision node with a new leaf node **C** that contains only those actions that are in the intersection, in this case only *dec*.

Algorithm 1 describes the pruning process formally. If θ returns only safe actions, then the tree obtained after pruning is guaranteed to represent a safe strategy, although a less-permissive one. The algorithm may be run for multiple (possibly 0) rounds, denoted by p, at most until we get a "fully pruned" tree representing a safe but deterministic strategy. We denote by $\mathcal{T}_{\sigma_{safe}}^{k,p}$ the decision tree for σ_{safe} constructed by only splitting nodes with k or more data points, followed by p rounds of safe pruning. Clearly, the more permissive the original strategy is, the more we can prune using safe pruning.

When generating $\mathcal{T}_{\sigma_{safe}}^{k,p}$, we use a modified implementation of the CART decision tree learning algorithm implemented in the `DecisionTreeClassifier` class of the Python-based machine learning library Scikit-learn [42]. Since we construct the DT from a safe strategy and as long as we let the DT-encoded strategy have at least one pure action in each leaf, the strategy will remain safe. With this in mind, we can freely change the parameters of the `DecisionTreeClassifier` class. However, in our experiments, we picked only the minimum split size k from the Scikit-parameters as a demonstrative example, as well as our newly introduced p. The methods described in this paper would work with other parameters as well.

3.5   Comparing DTs to Binary Decision Diagrams

A Binary Decision Diagram (BDD, e.g. [12]) is a popular data structure that can be used to represent boolean functions $f\colon \mathbb{B}^n \to \mathbb{B}$. It may also be used to represent strategies by encoding configurations and actions into a suitable form via bit-blasting, i.e. converting them into propositional formulae. For example, the configuration-action pair $((x = 6, y = 2), a_0)$ can be represented as $(x_2 \wedge x_1 \wedge \neg x_0 \wedge \neg y_2 \wedge y_1 \wedge \neg y_0 \wedge a_0)$, if it is known that the maximum value that x and y can take is less than 8 (3 bits). A strategy can be seen as a disjunction $\bigvee_{\gamma, a \in \sigma(\gamma)} (\gamma, a)$ of all configuration-action pairs (γ, a) permitted by the strategy σ. Such an encoding allows for an easy conversion into a BDD. Though theoretically straightforward, there are some practical concerns involved when constructing the BDD. Mainly, the ordering of the variables in the BDD can drastically change its size. While computing the optimal ordering so as to have the smallest BDD is an NP-complete problem [4], various heuristics exist that can be used to get better orderings. We use the CUDD package [53] to construct the BDD, along with Rudell's Sifting reordering technique [49].

The main disadvantage of DTs compared to BDDs is that isomorphic subgraphs are not merged (DTs are trees, BDDs are directed acyclic graphs); and even if merging was allowed, it would not save much. Indeed, since DT may choose different predicates on the same level (which is an advantage in contrast to BDD with a fixed variable ordering) isomorphic subgraphs occur rarely. There are further advantages of DT, related to learning, that make them more compact than BDD in some contexts, e.g. [9,10]. Firstly, they can be learnt fast, using the entropy-based heuristic, compared to the graph processing and variable reordering of BDDs. Secondly, a DT can ignore "don't-care inputs"; these inputs are encodings of things that are not valid configuration-action pairs, in the sense that either the action is not available in the configuration or that it is not a valid configuration at all. In contrast, a BDD has to explicitly either allow or disallow these inputs. Thirdly, DT learning can also be used to represent the strategy imprecisely using a smaller DT, which can be model checked for safety. For the modifications described in Sect. 3.4, we do not even need to re-verify safety, because this property is preserved by both our size reduction techniques. Fourthly, DT can use much wider class of predicates, compared to single bit tests for a bit representation in a BDD. This final point is also a reason (together with the smaller size) why DT is a more understandable representation than a BDD [9,10]. We also illustrate this point on a case-study in Remark 1.

4   Case Studies and Experimental Results

In this section, we evaluate the techniques discussed above on three different case studies: (1) the adaptive cruise control model introduced in the motivation; (2) a two tank case study introduced in [29]; and (3) the heating system of a two room apartment adapted from [25].

Table 1 compares representations for our case studies obtained in different ways. We discuss results for the three case studies, denoted cruise, twotanks,

Table 1. Sizes of the different representations: explicit list as output by UPPAAL STRATEGO, the relevant part of the list, BDD displaying [minimum/median/maximum] over the 40 trials, and DT according to the upper path in Fig. 2.

	#Variables	Stratego list	List	BDD[min/med/max]	DT $\mathcal{T}_{opt}$ Size
cruise$_{\text{non-Euler}}$	5	1,790,034	308,216	[3,718/5,066/5,890]	2,899
cruise	7	5,931,154	304,752	[3,470/4,728/4,742]	2,713
twotanks	9	23,182	23,182	[65/69/91]	1
tworooms	11	1,924,708	509,715	[16,370/20,214/25,909]	487

Table 2. Tables displaying the number $|\mathcal{T}_{opt}^{k,p}|$ of nodes of $\mathcal{T}_{opt}^{k,p}$ (left) and the expected performance $\mathbb{E}_{\sigma,H}^{\mathcal{M},\gamma}(D)$ (right) for various k and p, i.e. using the bottom path of Fig. 2, for the cruise model. Higher performance corresponds to a lower number. (Color table online)

Min split size (k)	Rounds of pruning (p)			Min split size (k)	Rounds of pruning (p)		
	0	1	2		0	1	2
2	2,713	1,725	1,267	2	2,627	3,618	4,240
10	2,705	1,733	1,249	10	2,696	3,596	4,210
20	2,667	1,733	1,131	20	2,778	3,625	14,039
30	2,657	1,695	993	30	2,778	3,589	14,108
40	2,627	1,669	1,015	40	2,778	3,600	14,096
50	2,557	1,695	1,003	50	2,825	3,614	14,037
60	2,635	1,489	963	60	2,905	3,673	14,074
70	2,613	1,441	955	70	2,898	3,714	14,095
80	2,519	1,537	915	80	2,907	3,717	14,092
90	2,455	1,323	923	90	3,006	3,741	14,077
100	1,929	1,023	877	100	3,030	14,061	14,292

and tworooms respectively. Additionally, the first line displays cruise without the integrated Euler method, to illustrate the effect of Euler method on the final size. All the representations are safe and as optimal as σ_{opt} produced by UPPAAL STRATEGO.

For each of the models we display the following information: the third column lists the number of items in the explicit list representation of σ_{opt} output by UPPAAL STRATEGO. The fourth column lists the number of those items that are actually relevant, i.e. sets of configurations where an actual decision is to be made. The fifth and sixth column list the sizes of BDD and DT representations learnt from σ_{opt}, i.e. the upper path in Fig. 2. For BDDs, since the initial ordering plays a role in the size of the final result despite applying the re-ordering heuristics, we ran 40 experiments for each model with random initial variable orderings. For creating BDDs, we used the free Python library *tulip-control/dd* as an interface to CUDD.

We conclude that both BDDs and DTs reduce the size by several order of magnitude. DTs are slightly better in all cases, and 2 orders of magnitude smaller in the tworooms model. Note that reliably achieving good results when

Table 3. Tables displaying the number $|\mathcal{T}_{opt}^{k,p}|$ of nodes of $\mathcal{T}_{opt}^{k,p}$ (left) and the expected performance $\mathbb{E}_{\sigma,H}^{\mathcal{M},\gamma}(D)$ (right) for various k and p, i.e. using the bottom path of Fig. 2, for the tworooms model. Higher performance corresponds to a lower number. (Color table online)

Min split	Rounds of pruning (p)				Min split	Rounds of pruning (p)			
size (k)	0	1	2	3	size (k)	0	1	2	3
2	543	403	283	191	2	2,096	2,353	2,821	3,156
10	525	387	271	185	10	2,156	2,460	3,285	3,283
50	497	365	251	171	50	1,989	2,778	3,287	3,281
125	445	317	219	151	125	2,374	2,053	3,280	3,284
250	387	265	179	123	250	2,283	2,071	3,288	3,282
500	323	211	139	97	500	2,563	2,155	3,280	3,282
750	277	175	111	77	750	2,333	2,210	3,279	3,286

constructing the BDD relies on repeating the construction several times; since already constructing a single BDD and applying the heuristics [49] already took roughly 10 times longer than DT learning, DT can be obtained one or two orders of magnitude faster than BDDs, depending on how many times one tries constructing the BDD. Further, for the two tanks, only DT realizes that the strategy is actually trivial. The main reason for BDD not to spot this is the point of ignoring "don't-care" inputs addressed in Sect. 3.5.

Table 2 shows how the size of the DT can be further reduced by the bottom path of Fig. 2, when the "exact representation" criterion is relaxed. It displays the performance, i.e. the aggregated distance to *Front* car, and size of $\mathcal{T}_{opt}^{k,p}$ for different combinations of the pruning parameters k and p. Recall that using no pruning ($k = 2, p = 0$) yields the same DT as the upper path of Fig. 2, i.e. $\mathcal{T}_{opt}^{2,0} = \mathcal{T}_{opt}$.

We observed that for cruise, increasing the values of k and p buys a reduction in size of the DT against a reduction in performance. For instance, using $k = 80, p = 0$, one can decrease the size to 2485 (by 8.4%) while deteriorating the performance to 2907 (by 10%). Allowing for half the performance (double the aggregated distance), one can make the DT even smaller than half of its original size, e.g. by setting $k = 10, p = 2$. The shading and colouring of the table display different "trade-off zones", each with comparable savings/losses. The same conclusions hold for cruise_non-Euler, see the similar Table in [1, Appendix A.3]. For tworooms (Table 3), the best performance is observed not with $k = 2, p = 0$, but with $k = 50, p = 0$. We conjecture that the less permissive safe strategy assists STRATEGO in performing the optimisation faster by reducing the size of the search space. As a result, here we get a both smaller and more performant strategy. In the case of twotanks, already $\mathcal{T}_{opt}$ has only a single node, hence no further reductions are possible.

Remark 1. Interestingly, domain knowledge can reduce the DT size further and make the representation more understandable. Indeed, for the cruise model we were able to construct a DT with only 25 nodes, designing our predicates based on the car kinematics. For example, the expected time until the front car reaches minimal velocity if it only decelerates from now on (1) plays an important role in the decision making and (2) can be easily expressed by solving the standard kinematics equation $v(t) = v_{\text{current}} - a_{\text{dec}} \cdot t$. The resulting DT (illustrated in [1, Appendix A.4] is thus very small and easy to interpret, as each of the few nodes has a clear kinematic interpretation. The DT thus open the possibility for strategy representation to profit from predicate/invariant synthesis.

5 Conclusion

We have provided a framework for producing small representations of safe and (near-)optimal strategies, without compromising safety. As to (near-)optimality, we can choose between two options: (i) not compromising it, or (ii) finding a suitable trade-off between compromising it (causing drops of performance) and additional size reductions. Compared to the original sizes, we achieve orders-of-magnitude reductions, allowing for efficient usage of the strategies in e.g. embedded devices. Compared to BDD representation, the size of the DT representation is smaller and can be computed faster; additionally trivial solutions are represented by trivial DTs. DTs are more readable as argued in [9,10].

A detailed examination of the latter point in the hybrid context remains future work. Further, candidates for more complex predicates could be automatically generated based on given domain knowledge or learnt from the data similarly to invariants from program runs [23,52]. As illustrated in Remark 1, this could lead to further reduction in size and improved understandability. Additionally, isomorphic/similar subtrees could be merged as in decision diagrams and further optimizations for algebraic decision diagrams [56] could be employed. Finally, we plan to visualize the DT representation of the strategies directly in UPPAAL STRATEGO$^+$ for convenience of the users.

References

1. Ashok, P. Křetínský, J., Larsen, K.G., Coënt, A.L., Taankvist, J.H., Weininger, M.: SOS: Safe, optimal and small strategies for hybrid Markov decision processes. Technical report (2019)
2. Behrmann, G., Cougnard, A., David, A., Fleury, E., Larsen, K.G., Lime, D.: UPPAAL-Tiga: time for playing games!. In: Damm, W., Hermanns, H. (eds.) CAV 2007. LNCS, vol. 4590, pp. 121–125. Springer, Heidelberg (2007). https://doi.org/10.1007/978-3-540-73368-3_14
3. Bernet, J., Janin, D., Walukiewicz, I.: Permissive strategies: from parity games to safety games. ITA 36, 261–275 (2002)
4. Bollig, B., Wegener, I.: Improving the variable ordering of OBDDs is NP-complete. IEEE Trans. Comput. 45(9), 993–1002 (1996)
5. Boutilier, C., Dean, T.L., Hanks, S.: Decision-theoretic planning: structural assumptions and computational leverage. J. Artif. Intell. Res. 11, 1–94 (1999)

6. Boutilier, C., Dearden, R.: Approximating value trees in structured dynamic programming. In: ICML (1996)
7. Boutilier, C., Dearden, R., Goldszmidt, M.: Exploiting structure in policy construction. In: IJCAI (1995)
8. Bouyer, P., Markey, N., Olschewski, J., Ummels, M.: Measuring permissiveness in parity games: mean-payoff parity games revisited. In: Bultan, T., Hsiung, P.-A. (eds.) ATVA 2011. LNCS, vol. 6996, pp. 135–149. Springer, Heidelberg (2011). https://doi.org/10.1007/978-3-642-24372-1_11
9. Brázdil, T., Chatterjee, K., Chmelík, M., Fellner, A., Křetínský, J.: Counterexample explanation by learning small strategies in Markov decision processes. In: Kroening, D., Păsăreanu, C.S. (eds.) CAV 2015. LNCS, vol. 9206, pp. 158–177. Springer, Cham (2015). https://doi.org/10.1007/978-3-319-21690-4_10
10. Brázdil, T., Chatterjee, K., Křetínský, J., Toman, V.: Strategy representation by decision trees in reactive synthesis. In: Beyer, D., Huisman, M. (eds.) TACAS 2018. LNCS, vol. 10805, pp. 385–407. Springer, Cham (2018). https://doi.org/10.1007/978-3-319-89960-2_21
11. Breiman, L.: Classification and Regression Trees. Routledge, Abingdon (2017)
12. Bryant, R.E.: Symbolic manipulation of boolean functions using a graphical representation. In: DAC (1985)
13. Chapman, D., Kaelbling, L.P.: Input generalization in delayed reinforcement learning: an algorithm and performance comparisons. In: IJCAI. Morgan Kaufmann (1991)
14. Clare, A., King, R.D.: Knowledge discovery in multi-label phenotype data. In: De Raedt, L., Siebes, A. (eds.) PKDD 2001. LNCS (LNAI), vol. 2168, pp. 42–53. Springer, Heidelberg (2001). https://doi.org/10.1007/3-540-44794-6_4
15. Coënt, A.L., Sandretto, J.A.D., Chapoutot, A., Fribourg, L.: An improved algorithm for the control synthesis of nonlinear sampled switched systems. Formal Methods Syst. Design 53(3), 363–383 (2018)
16. David, A., Du, D., Larsen, K.G., Mikucionis, M., Skou, A.: An evaluation framework for energy aware buildings using statistical model checking. Sci. China Inform. Sci. 55(12), 2694–2707 (2012)
17. David, A., et al.: On time with minimal expected cost!. In: Cassez, F., Raskin, J.-F. (eds.) ATVA 2014. LNCS, vol. 8837, pp. 129–145. Springer, Cham (2014). https://doi.org/10.1007/978-3-319-11936-6_10
18. David, A., Jensen, P.G., Larsen, K.G., Mikučionis, M., Taankvist, J.H.: Uppaal stratego. In: Baier, C., Tinelli, C. (eds.) TACAS 2015. LNCS, vol. 9035, pp. 206–211. Springer, Heidelberg (2015). https://doi.org/10.1007/978-3-662-46681-0_16
19. de Alfaro, L., Kwiatkowska, M., Norman, G., Parker, D., Segala, R.: Symbolic model checking of probabilistic processes using MTBDDs and the kronecker representation. In: Graf, S., Schwartzbach, M. (eds.) TACAS 2000. LNCS, vol. 1785, pp. 395–410. Springer, Heidelberg (2000). https://doi.org/10.1007/3-540-46419-0_27
20. Dräger, K., Forejt, V., Kwiatkowska, M., Parker, D., Ujma, M.: Permissive controller synthesis for probabilistic systems. In: Ábrahám, E., Havelund, K. (eds.) TACAS 2014. LNCS, vol. 8413, pp. 531–546. Springer, Heidelberg (2014). https://doi.org/10.1007/978-3-642-54862-8_44
21. Esposito, F., Malerba, D., Semeraro, G.: Decision tree pruning as a search in the state space. In: Brazdil, P.B. (ed.) ECML 1993. LNCS, vol. 667, pp. 165–184. Springer, Heidelberg (1993). https://doi.org/10.1007/3-540-56602-3_135
22. Fehnker, A., Ivančić, F.: Benchmarks for hybrid systems verification. In: Alur, R., Pappas, G.J. (eds.) HSCC 2004. LNCS, vol. 2993, pp. 326–341. Springer, Heidelberg (2004). https://doi.org/10.1007/978-3-540-24743-2_22

23. Garg, P., Löding, C., Madhusudan, P., Neider, D.: ICE: a robust framework for learning invariants. In: Biere, A., Bloem, R. (eds.) CAV 2014. LNCS, vol. 8559, pp. 69–87. Springer, Cham (2014). https://doi.org/10.1007/978-3-319-08867-9_5

24. Girard, A.: Controller synthesis for safety and reachability via approximate bisimulation. Automatica **48**(5), 947–953 (2012)

25. Girard, A.: Low-complexity quantized switching controllers using approximate bisimulation. Nonlinear Anal.: Hybrid Syst. **10**, 34–44 (2013)

26. Girard, A., Martin, S.: Synthesis for constrained nonlinear systems using hybridization and robust controllers on simplices. IEEE Trans. Automat. Control **57**(4), 1046–1051 (2012)

27. Hahn, E.M., Norman, G., Parker, D., Wachter, B., Zhang, L.: Game-based abstraction and controller synthesis for probabilistic hybrid systems. In: QEST (2011)

28. Hermanns, H., Kwiatkowska, M.Z., Norman, G., Parker, D., Siegle, M.: On the use of mtbdds for performability analysis and verification of stochastic systems. J. Log. Algebr. Program. **56**(1–2), 23–67 (2003)

29. Hiskens, I.A.: Stability of limit cycles in hybrid systems. In: HICSS (2001)

30. Hoey, J., St-Aubin, R., Hu, A., Boutilier, C.: SPUDD: stochastic planning using decision diagrams. In: UAI (1999)

31. Kearns, M., Koller, D.: Efficient reinforcement learning in factored MDPs. In: IJCAI (1999)

32. Koller, D., Parr, R.: Computing factored value functions for policies in structured MDPs. In: IJCAI (1999)

33. Kushmerick, N., Hanks, S., Weld, D.: An algorithm for probabilistic least-commitment planning. In: AAAI (1994)

34. Larsen, K.G., Le Coënt, A., Mikučionis, M., Taankvist, J.H.: Guaranteed control synthesis for continuous systems in Uppaal Tiga. In: Chamberlain, R., Taha, W., Törngren, M. (eds.) CyPhy/WESE -2018. LNCS, vol. 11615, pp. 113–133. Springer, Cham (2019). https://doi.org/10.1007/978-3-030-23703-5_6

35. Larsen, K.G., Mikučionis, M., Taankvist, J.H.: Safe and optimal adaptive cruise control. In: Meyer, R., Platzer, A., Wehrheim, H. (eds.) Correct System Design. LNCS, vol. 9360, pp. 260–277. Springer, Cham (2015). https://doi.org/10.1007/978-3-319-23506-6_17

36. Coënt, A.L., De Vuyst, F., Chamoin, L., Fribourg, L.: Control synthesis of nonlinear sampled switched systems using Euler's method. In: SNR (2017)

37. Liu, S., Panangadan, A., Talukder, A., Raghavendra, C.S.: Compact representation of coordinated sampling policies for body sensor networks. In: 2010 IEEE Globecom Workshops (2010)

38. Majumdar, R., Render, E., Tabuada, P.: Robust discrete synthesis against unspecified disturbances. In: HSCC (2011)

39. Miner, A., Parker, D.: Symbolic representations and analysis of large probabilistic systems. In: Baier, C., Haverkort, B.R., Hermanns, H., Katoen, J.-P., Siegle, M. (eds.) Validation of Stochastic Systems. LNCS, vol. 2925, pp. 296–338. Springer, Heidelberg (2004). https://doi.org/10.1007/978-3-540-24611-4_9

40. Mingers, J.: An empirical comparison of pruning methods for decision tree induction. Mach. Learn. **4**, 227–243 (1989)

41. Mitchell, T.M.: Machine Learning. McGraw-Hill, Inc., New York (1997)

42. Pedregosa, F., Varoquaux, G., Gramfort, A., Michel, V., Thirion, B., Grisel, O., Blondel, M., Prettenhofer, P., Weiss, R., Dubourg, V., VanderPlas, J., Passos, A., Cournapeau, D., Brucher, M., Perrot, M., Duchesnay, E.: Scikit-learn: machine learning in Python. J. Mach. Learn. Res. **12**, 2825–2830 (2011)

43. Puterman, M.L.: Markov Decision Processes. Wiley, Hoboken (1994)
44. Pyeatt, L.D.: Reinforcement learning with decision trees. Appl. Inform. 26–31 (2003)
45. Quinlan, J.R.: Induction of decision trees. Mach. Learn. **1**, 81–106 (1986)
46. Quinlan, J.R.: C4.5: Programs for Machine Learning. Elsevier, Amsterdam (2014)
47. Riddle, P.J., Segal, R., Etzioni, O.: Representation design and brut-force induction in a boeingmanufacturing domain. Appl. Artif. Intell. **8**, 125–147 (1994)
48. Roy, P., Tabuada, P., Majumdar, R.: Pessoa 2.0: a controller synthesis tool for cyber-physical systems. In: HSCC (2011)
49. Rudell, R.: Dynamic variable ordering for ordered binary decision diagrams. In: CAD (1993)
50. Rungger, M., Zamani, M.: Scots: a tool for the synthesis of symbolic controllers. In: HSCC (2016)
51. Saoud, A., Girard, A., Fribourg, L.: On the composition of discrete and continuous-time assume-guarantee contracts for invariance. In: ECC (2018)
52. Sharma, R., Gupta, S., Hariharan, B., Aiken, A., Nori, A.V.: Verification as learning geometric concepts. In: Logozzo, F., Fähndrich, M. (eds.) SAS 2013. LNCS, vol. 7935, pp. 388–411. Springer, Heidelberg (2013). https://doi.org/10.1007/978-3-642-38856-9_21
53. Somenzi, F.: CUDD: CU decision diagram package-release 2.4. 2 (2009). http://vlsi.colorado.edu/~fabio/CUDD
54. Svoreňová, M., Křetínský, J., Chmelík, M., Chatterjee, K., Černá, I., Belta, C.: Temporal logic control for stochastic linear systems using abstraction refinement of probabilistic games. Nonlinear Anal.: Hybrid Syst. **23**, 230–253 (2017)
55. Wimmer, R., et al.: Symblicit calculation of long-run averages for concurrent probabilistic systems. In: QEST (2010)
56. Zapreev, I.S., Verdier, C., Mazo, M.: Optimal symbolic controllers determinization for BDD storage. In: ADHS (2018)

Fast Falsification of Hybrid Systems
Using Probabilistically Adaptive Input

Gidon Ernst[1](✉), Sean Sedwards[2], Zhenya Zhang[3], and Ichiro Hasuo[3]

[1] Ludwig-Maximilians-University, Munich, Germany
gidon.ernst@lmu.de
[2] University of Waterloo, Waterloo, Canada
sean.sedwards@uwaterloo.ca
[3] National Institute of Informatics, Tokyo, Japan
{zhangzy,hasuo}@nii.ac.jp

Abstract. We present an algorithm that quickly finds falsifying inputs for hybrid systems, i.e., inputs that steer the system towards violation of a given temporal logic requirement. Our method is based on a probabilistically directed search of an increasingly fine grained spatial and temporal discretization of the input space. A key feature is that it adapts to the difficulty of a problem at hand, specifically to the local complexity of each input segment, as needed for falsification. In experiments with standard benchmarks, our approach shows comparable or better performance to existing techniques, while at the same time being relatively simple.

Keywords: Cyber-physical system · Falsification ·
Stochastic optimization · Temporal logic · Quantitative semantics ·
Las Vegas Tree Search

1 Introduction

The falsification problem we consider seeks a (time-bounded) input signal that causes a hybrid system model to violate a given temporal logic specification. A popular way to address this is to first construct a "score function" that quantifies how much of the specification has been satisfied during the course of an execution. The falsification can then be treated as an optimization problem, which can be solved using standard algorithms. This approach, especially using a quantitative "robustness" semantics [15] of requirements as the score function, has been successfully applied, resulting in a number of now mature tools [4,10] with practical applications as well as and friendly competitions [8,9,28].

Despite its apparent success, robustness is in general not a perfect optimization function, but only a heuristic score function [20] with respect to the falsification problem, as greedy hill climbing may lead to local optima. In practice, standard optimization algorithms overcome this limitation by including stochastic exploration. The most sophisticated of these can also model the dynamics of the system (e.g., [2]), in order to estimate the most productive direction of input signal space to explore. There is, however, "no free lunch" [27], and high performance general purpose optimization algorithms are not necessarily the best choice. For example, such algorithms often optimize with respect to the entire

© Springer Nature Switzerland AG 2019
D. Parker and V. Wolf (Eds.): QEST 2019, LNCS 11785, pp. 165–181, 2019.
https://doi.org/10.1007/978-3-030-30281-8_10

input trace, without exploiting the time causality of the problem, i.e., the fact that a good trace (one that eventually falsifies the property) may be dependent on a good trace prefix.

The contribution of this paper is a randomized falsification algorithm (Sect. 3.1) that exploits the time-causal structure of the problem and that adapts to local complexity. In common with alternative approaches, our algorithm searches a discretized space of input signals, but in our case the search space also includes multiple levels of spatial and temporal granularity (Sect. 3.2). The additional complexity is mitigated by an efficient tree search that probabilistically balances exploration and exploitation (Sect. 3.3).

The performance of our algorithm benefits from the heuristic idea to explore simple (coarse granularity) inputs first, then gradually switch to more complex inputs that include finer granularity. Importantly, the finer granularity tends only to be added where it is needed, thus avoiding the exponential penalty of searching the entire input space at the finer granularity. While it is always possible to construct pathological problem instances, we find that despite its simplicity, our approach is effective on benchmarks from the literature. Our experimental results (Sect. 4) demonstrate that our algorithm can achieve comparable or better performance than other methods, in terms of speed and reliability of finding a falsifying input.

2 Preliminaries

In this work we represent a deterministic black-box system model as an input/output function $\mathcal{M}\colon ([0, T] \to \mathbb{R}^n) \to ([0, T] \to \mathbb{R}^m)$. In general, $\mathcal{M}$ comprises continuous dynamics with discontinuities. $\mathcal{M}$ takes a time-bounded, real-valued input signal $\boldsymbol{u}\colon [0, T] \to \mathbb{R}^n$ of length $|\boldsymbol{u}| = T$ and transforms it to a time bounded output signal $\boldsymbol{y}\colon [0, T] \to \mathbb{R}^m$ of the same length, but potentially different dimensionality. The dimension n of the input indicates that at each moment $t \leq T$ within the time horizon T, the value $\boldsymbol{u}(t) \in \mathbb{R}^n$ of the input is an n-dimensional real vector (analogously for the output).

We denote by $\boldsymbol{u}_1 \boldsymbol{u}_2\colon [0, T_1 + T_2] \to \mathbb{R}^n$ the concatenation of signals $\boldsymbol{u}_1$ and $\boldsymbol{u}_2$ that have the same dimensions. Concatenation of more than two signals follows naturally and is denoted $\boldsymbol{u}_1 \boldsymbol{u}_2 \boldsymbol{u}_3 \cdots$. A constant input signal segment is written (t, v), where t is a time duration and $v \in \mathbb{R}^n$ is a vector of input values. A piecewise constant input signal is the concatenation of such segments.

In this work we adopt the syntax and robustness semantics of STL defined in [12]. The syntax of an STL formula is thus given by

$$\varphi ::= \neg\varphi \mid \varphi \vee \varphi \mid \varphi \wedge \varphi \mid \varphi \, \mathsf{U}_I \, \varphi \mid \Box_I \varphi \mid \Diamond_I \varphi \mid \mu, \tag{1}$$

where the logical connectives and temporal operators have their usual Boolean interpretations and equivalences, I is the interval of time over which the temporal operators range, and atomic formulas $\mu \equiv f(x_1, \ldots, x_m) > 0$ are predicates over the spatial dimensions of a trace. The robustness of trace $\boldsymbol{y}$ with respect to formula φ, denoted $\rho(\varphi, \boldsymbol{y})$, is calculated inductively according to the following robustness semantics, using the equivalence $\rho(\varphi, \boldsymbol{y}) \equiv \rho(\varphi, \boldsymbol{y}, 0)$.

$$\rho(\mu, \boldsymbol{y}, t) = f(x_1[t], \dots, x_m[t]), \qquad \text{for } \mu \equiv f(x_1, \dots, x_m) > 0$$

$$\rho(\neg\varphi, \boldsymbol{y}, t) = -\rho(\varphi, \boldsymbol{y}, t)$$

$$\rho(\varphi_1 \vee \varphi_2, \boldsymbol{y}, t) = \max(\rho(\varphi_1, \boldsymbol{y}, t), \rho(\varphi_2, \boldsymbol{y}, t)) \qquad \rho(\Diamond_I \varphi, \boldsymbol{y}, t) = \max_{t' \in t+I}(\rho(\varphi, \boldsymbol{y}, t'))$$

$$\rho(\varphi_1 \wedge \varphi_2, \boldsymbol{y}, t) = \min(\rho(\varphi_1, \boldsymbol{y}, t), \rho(\varphi_2, \boldsymbol{y}, t)) \qquad \rho(\Box_I \varphi, \boldsymbol{y}, t) = \min_{t' \in t+I}(\rho(\varphi, \boldsymbol{y}, t'))$$

$$\rho(\varphi_1 \mathsf{U}_I \varphi_2, \boldsymbol{y}, t) = \max_{t' \in t+I}\left(\min_{t'' \in [t,t']}(\rho(\varphi_1, \boldsymbol{y}, t'')), \min(\rho(\varphi_2, \boldsymbol{y}, t'))\right)$$

An important characteristic of the robustness semantics is that it is faithful to standard boolean satisfaction, such that

$$\rho(\varphi, \boldsymbol{y}) > 0 \implies \boldsymbol{y} \models \varphi \quad \text{and} \quad \rho(\varphi, \boldsymbol{y}) < 0 \implies \boldsymbol{y} \not\models \varphi. \tag{2}$$

Together, these equations justify using the robustness semantics $\rho(\varphi, \mathcal{M}(\boldsymbol{u}))$ to detect whether an input $\boldsymbol{u}$ corresponds to the violation of a requirement φ. This correspondence is exploited to find such falsifying inputs through global hill-climbing optimization:

$$\text{Find } \boldsymbol{u}^* = \operatorname*{arg\,min}_{\boldsymbol{u} \in ([0,T] \to \mathbb{R}^n)} \rho(\varphi, \mathcal{M}(\boldsymbol{u})) \text{ such that } \rho(\varphi, \mathcal{M}(\boldsymbol{u}^*)) < 0. \tag{3}$$

Of course, finding an adequate falsifying input $\boldsymbol{u}^*$ is generally hard and subject to the limitations of the specific optimization algorithm used.

Sound approximations of the lower and upper bounds of the robustness of a prefix $\boldsymbol{y}$ can sometimes be used to short-cut the search. We thus define lower and upper bounds in the following way.

$$\text{Lower: } \underline{\rho}(\varphi, \boldsymbol{y}) = \min_{\boldsymbol{y}'} \rho(\varphi, \boldsymbol{y}\boldsymbol{y}') \qquad \text{Upper: } \overline{\rho}(\varphi, \boldsymbol{y}) = \max_{\boldsymbol{y}'} \rho(\varphi, \boldsymbol{y}\boldsymbol{y}') \tag{4}$$

A lower bound $\underline{\rho}(\varphi, \mathcal{M}(\boldsymbol{u}))) > 0$ can be used to detect that a prefix cannot be extended to a falsifying trace (e.g., after the deadline for a harmful event has passed). An upper bound $\overline{\rho}(\varphi, \mathcal{M}(\boldsymbol{u})) < 0$ similarly implies $\mathcal{M}(\boldsymbol{u}\boldsymbol{u}') \not\models \varphi$ for all $\boldsymbol{u}'$, concluding that input $\boldsymbol{u}$ is already a witness for falsification (e.g., a limit is already exceeded). Robustness can be computed efficiently [11], as well as the respective upper and lower bounds [13].

3 Approach

We wish to solve the following falsification problem efficiently:

$$\text{Find } \boldsymbol{u}^* \text{ such that } \overline{\rho}(\varphi, \mathcal{M}(\boldsymbol{u}^*)) < 0. \tag{5}$$

Our approach is to repeatedly construct input signals $\boldsymbol{u} = \boldsymbol{u}_1 \boldsymbol{u}_2 \boldsymbol{u}_3 \cdots$, where $\boldsymbol{u}_i$ is drawn from a predetermined search space of candidate input segments, $\mathcal{A}$. The choice is probabilistic, according to a distribution $\mathcal{D}$ that determines the search strategy, i.e., which inputs are likely to be tried next given a partially explored search space. The construction of each input is done incrementally, to take advantage of the potential short cuts described at the end of Sect. 2.

Algorithm 1 "adaptive Las Vegas Tree Search" (aLVTS) codifies the high level functionality of this probabilistic approach, described in detail in Sect. 3.1.

The effectiveness of our algorithm in practice comes from the particular choices of $\mathcal{A}$ and $\mathcal{D}$, which let the search gradually *adapt* to the difficulty of the problem. The set $\mathcal{A}$ (defined in Sect. 3.2) contains input segments of diverse granularity, which intuitively corresponds to how precise the input must be in order to find a falsifying trace. The distribution $\mathcal{D}$ (defined in Sect. 3.3) initially assigns high probabilities to the "coarsest" input segments in $\mathcal{A}$. Coarse here means that the segments tend to be long in relation to the time horizon T and large in relation to the extrema of the input space. The algorithm probabilistically balances exploration and exploitation of segments, but as the coarser segments become fully explored, and the property has not been falsified, the algorithm gradually switches to finer-grained segments.

3.1   Algorithm

Algorithm 1 searches the space of input signals constructed from piecewise constant (over time) segments, which are chosen at random according to the distribution defined by $\mathcal{D}$ in line 6. This distribution is a function of the numbers of unexplored and explored edges at different levels of granularity, and thus defines the probabilities of exploration, exploitation and adaptation. The precise calculation made by $\mathcal{D}$ is described in Sect. 3.3.

As the search proceeds, the algorithm constructs a tree whose nodes each correspond to a unique input signal prefix. The edges of the tree correspond to the constant segments that make up the input signal. The root node corresponds to time 0 and the empty input signal (line 4).

To each node identified by an input signal prefix u is associated a set of unexplored edges, $unexplored(u) \subseteq \mathcal{A}$, that correspond to unexplored input signal segments, and a set of explored edges, $explored(u) \subseteq \mathcal{A}$, that remain inconclusive with respect to falsification. Initially, all edges are unexplored (line 1 and line 2). Once an edge has been chosen (line 6), the unique signal segment associated to the edge may be appended to the signal prefix associated to the node, to form an extended input signal. If the chosen edge is unexplored, it is removed from the set of unexplored edges (line 8) and the extended input signal uu' is transformed by the system into an extended output signal (line 9). If the requirement is falsified by the output signal (y in line 10), the algorithm immediately quits and returns the falsifying input signal (line 11). If the requirement is satisfied, with no possibility of it being falsified by further extensions of the signal (12), the algorithm quits the current signal (line 13) and starts a new signal from the root node (line 4). This is the case, in particular, when the length of the signal exceeds the time horizon of the formula as a consequence of the definition of ρ in (4). If the requirement is neither falsified nor satisfied, the edge is added to the node's set of explored edges (line 14). Regardless of whether the chosen edge was previously explored or unexplored, if the signal remains inconclusive, the extended input signal becomes the focus of the next iterative step (line 15).

Algorithm 1: Adaptive Las Vegas Tree Search (aLVTS)

Input:
system model $\mathcal{M} : \boldsymbol{u} \to \boldsymbol{y}$, with $\boldsymbol{u} : [0, t] \to \mathbb{R}^n$ and $\boldsymbol{y} : [0, t] \to \mathbb{R}^m$,
time-bounded specification ϕ, set of all possible input trace segments $\mathcal{A}$

Output:
$\boldsymbol{u}$ such that $\mathcal{M}(\boldsymbol{u}\boldsymbol{u}') \not\models \phi$ for all $\boldsymbol{u}'$, or $\bot$ after timeout or maximum iterations

1 $unexplored(\boldsymbol{u}) \leftarrow \mathcal{A}$ for all $\boldsymbol{u}$
2 $explored(\boldsymbol{u}) \leftarrow \varnothing$ for all $\boldsymbol{u}$
3 **repeat**
4      $\boldsymbol{u} \leftarrow \langle\rangle$
5   **while** $unexplored(\boldsymbol{u}) \neq \varnothing \vee explored(\boldsymbol{u}) \neq \varnothing$ **do**
6   sample $\boldsymbol{u}' \sim \mathcal{D}(\boldsymbol{u})$
7   **if** $\boldsymbol{u}' \in unexplored(\boldsymbol{u})$ **then**
8              $unexplored(\boldsymbol{u}) \leftarrow unexplored(\boldsymbol{u}) \setminus \{\boldsymbol{u}'\}$
9              $\boldsymbol{y} \leftarrow \mathcal{M}(\boldsymbol{u}\boldsymbol{u}')$
10   **if** $\overline{\rho}(\phi, \boldsymbol{y}) < 0$ **then**
11   **return** $\boldsymbol{u}\boldsymbol{u}'$
12   **if** $\rho(\phi, \boldsymbol{y}) > 0$ **then**
13   **continue** line 3
14             $explored(\boldsymbol{u}) \leftarrow explored(\boldsymbol{u}) \cup \{\boldsymbol{u}'\}$
15          $\boldsymbol{u} \leftarrow \boldsymbol{u}\boldsymbol{u}'$
16 **until** *timeout or maximum number of iterations*;
17 **return** $\bot$

While not explicit in the presentation of Algorithm 1, our approach is deliberately incremental in the evaluation of the system model. In particular, we can re-use partial simulations to take advantage of the fact that traces share common prefixes. Hence, for example, one can associate to every visited $\boldsymbol{u}$ the terminal state of the simulation that reached it, using this state to initialize a new simulation when subsequently exploring $\boldsymbol{u}\boldsymbol{u}'$. This idea also works for the calculation of robustness. We note, however, that incremental simulations may be impractical. For example, suspending and re-starting Simulink can be more expensive than performing an entire simulation from the start.

3.2 Definition of $\mathcal{A}$

The set $\mathcal{A}$ contains constant, n-dimensional input signal segments $\boldsymbol{u}'$ with values $(v_1, \ldots, v_n) \in \mathbb{R}^n$ and time duration t. Let $\underline{v}_i$ and $\overline{v}_i$ denote the minimum and maximum possible values, respectively, of dimension $i \in \{1, \ldots, n\}$. For each integer level $l \in \{0, \ldots, l_{\max}\}$, we define the set of possible proportions of the interval $[\underline{v}_i, \overline{v}_i]$ as

$$\mathbf{p}_l = \{(2j + 1)/2^l \mid j \in \mathbb{N}_0 \leq (2^l - 1)/2\}.$$

The numerators of all elements are coprime with the denominator, 2^l, hence $\mathbf{p}_i \cap \mathbf{p}_j = \varnothing$ for all $i \neq j$. By definition, $\mathbf{p}_0$ also includes 0. Hence, $\mathbf{p}_0 = \{0, 1\}$, $\mathbf{p}_1 = \{\frac{1}{2}\}$ and $\mathbf{p}_2 = \{\frac{1}{4}, \frac{3}{4}\}$, etc. The set of possible values of dimension i at level l is thus given by

$$\mathbf{v}_{i,l} = \underline{v}_i + \mathbf{p}_l \times (\overline{v}_i - \underline{v}_i).$$

Rather than making the granularity of each dimension independent, we interpret the value of l as a granularity "budget" that must be distributed among the (non-temporal) dimensions of the input signal. The set of possible per-dimension budget allocations for level l is given by

$$\mathbf{b}_l = \{(b_1, ..., b_n) \in \mathbb{N}_0^n \mid b_1 + \cdots + b_n = l\}.$$

For example, with $n = 2$, $\mathbf{b}_3 = \{(0, 3), (1, 2), (2, 1), (3, 0)\}$. If we denote the set of possible time durations at level l by $\mathbf{t}_l$, then the set of possible input segments at level l is given by

$$\mathcal{A}_l = \bigcup_{(b_1, ..., b_n) \in \mathbf{b}_l} \mathbf{t}_l \times \mathbf{v}_{1, b_1} \times \cdots \times \mathbf{v}_{n, b_n}.$$

Note that while $\mathbf{t}_l$ is not required here to share the granularity budget, this remains a possibility. Our implementation actually specifies $\mathbf{t}_l$ by defining a fixed number of control points per level, $(k_0, \ldots, k_{l_{\max}})$, such that the $\mathbf{t}_l = \{T/k_l\}$ are singleton sets. The sizes of various $\mathcal{A}_l$ for different choices of n and l, assuming $|\mathbf{t}_l| = 1$, are given in Table 1.

Table 1. Size of $\mathcal{A}_l$ for input dimensionality n and level l given that $|\mathbf{t}_l| = 1$.

n	$l = 0$	1	2	3	4	5	6	7	8	9	10
2	4	4	9	20	44	96	208	448	960	2048	4352
3	8	12	30	73	174	408	944	2160	4896	11008	24576

In summary, an input segment $\boldsymbol{u}' = (t, v_1, \ldots, v_n) \in \mathcal{A}_l$ has $t \in \mathbf{t}_l$ and corresponding budget allocation $b_1 + \cdots + b_n = l$, with the value v_i for each dimension given by $v_i = \underline{v}_i + p_i(\overline{v}_i - \underline{v}_i)$, where $p_i = (2j_i + 1)/2^{b_i}$, for some j_i, defines the proportion between minimum $\underline{v}_i$ and maximum $\overline{v}_i$.

By construction, $\mathcal{A}_i \cap \mathcal{A}_j = \varnothing$, for all $i \neq j$. Hence, we define

$$unexplored_l(\boldsymbol{u}) = unexplored(\boldsymbol{u}) \cap \mathcal{A}_l \text{ and}$$
$$explored_l(\boldsymbol{u}) = explored(\boldsymbol{u}) \cap \mathcal{A}_l.$$

The set of all possible input signal segments is given by $\mathcal{A} = \mathcal{A}_0 \cup \mathcal{A}_1 \cup \cdots \cup \mathcal{A}_{l_{\max}}$. Figure 1 depicts the construction of $\mathcal{A}$ for two dimensions. The majority of candidate input points is concentrated on the outer contour, corresponding to an extreme choice for one dimension and a fine-grained choice for the other dimension. While this bias appears extreme, as layers are exhausted, finer-grained choices become more likely, e.g., after two points from $\mathcal{A}_0$ have been tried, all remaining points from both levels in the second panel would be equally probable.

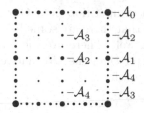

Fig. 1. Construction of $\mathcal{A}$ for $n = 2$ and $l_{\max} = 4$, with $\mathcal{A}_0$ representing the extremes of the two spatial dimensions. Showing $\mathcal{A}_0 \cup \mathcal{A}_1 \cup \mathcal{A}_2 \cup \mathcal{A}_3 \cup \mathcal{A}_4$. Larger points correspond to more likely values. Many points lie on the contour: the algorithm tends to combine finer choices in one dimension with coarse choices in the other dimension.

3.3 Definition of $\mathcal{D}$

The distribution $\mathcal{D}(\boldsymbol{u})$ is constructed implicitly. First, a granularity level $l \in \{0, \ldots, l_{\max}\}$ is chosen at random, with probability in proportion to the fraction of edges remaining at the level, multiplied by an exponentially decreasing scaling factor. Defining the overall weight of level l as

$$w_l = \frac{|unexplored_l(\boldsymbol{u})| + |explored_l(\boldsymbol{u})|}{2^l \cdot |\mathcal{A}_l|},$$

level l is chosen with probability $w_l / \sum_{i=0}^{l_{\max}} w_i$.

Having chosen l, one of the following strategies is chosen uniformly at random:

1. select $\boldsymbol{u}' \in unexplored_l(\boldsymbol{u})$, uniformly at random;
2. select $\boldsymbol{u}' \in explored_l(\boldsymbol{u})$, uniformly at random;
3. select $\boldsymbol{u}' \in explored_l(\boldsymbol{u})$, uniformly at random from those that minimise $\rho(\varphi, \mathcal{M}(\boldsymbol{u}\boldsymbol{u}'))$;
4. select $\boldsymbol{u}' \in explored_l(\boldsymbol{u})$, uniformly at random from those that minimise $\rho(\varphi, \mathcal{M}(\boldsymbol{u}\boldsymbol{u}'\boldsymbol{u}^*))$, where $\boldsymbol{u}^*$ denotes any, arbitrary length input signal suffix that has already been explored from $\boldsymbol{u}\boldsymbol{u}'$.

Strategy 1 can be considered pure exploration, while strategies 3–4 are three different sorts of exploitation. In the case that $unexplored_l(\boldsymbol{u})$ or $explored_l(\boldsymbol{u})$ are empty, their corresponding strategies are infeasible and a strategy is chosen uniformly at random from the feasible strategies. If for all $\boldsymbol{u}' \in explored(\boldsymbol{u})$, $explored(\boldsymbol{u}\boldsymbol{u}') = \varnothing$, then strategy 4 is equivalent to strategy 3, but it *is* feasible.

4 Evaluation

We evaluate our approach on a selection of standard driving-related automotive benchmarks, which are of particular interest to us and common in the falsification literature. Each benchmark has at least one time-varying input to discover (the motivation of aLVTS), in contrast to simply searching for an initial configuration. In addition to experiments conducted by us, we include some recent results from the ARCH 2019 falsification competition [28]. Our selection is representative, but not exhaustive, and note that there are some benchmarks used in [28] where our algorithm does not work well. There are, however, benchmarks where our algorithm is significantly better than the alternatives, motivating its potential inclusion in an ensemble approach.

4.1   Benchmarks

Automatic Transmission. This benchmark was proposed in [16] and consists of a Simulink model and its related requirements. The model has two inputs: *throttle* and *brake*. Depending on the speed and engine load, an appropriate gear is automatically selected. Besides the current gear g, the model outputs the car's speed v and engine rotations ω. We consider the following requirements:

$$\text{AT1}_t = \Box_{[0,t]}\, v < 120 \qquad \text{AT2} = \Box_{[0,10]}\omega < 4750$$
$$\text{AT5}_i = \Box_{[0,30]}((g \neq i \wedge \circ g = i) \implies \circ \Box_{[0,2.5]}g = i)$$
$$\text{AT6}_{t,\overline{v},\underline{\omega}} = (\Box_{[0,10]}\, v < \overline{v}) \vee (\Diamond_{[0,t]}\, \underline{\omega} < \omega)$$
$$\text{ATX1}_i = \Box_{[0,30]}\, (g = i \implies v > 10 \cdot i),\ i \in \{3,4\}$$
$$\text{ATX2} = \neg(\Box_{[10,30]}v \in [50,60])$$

AT* are from [16], with AT5 here subsuming AT3 and AT4 of [16]. ATX* are additional requirements with interesting characteristics. The syntax $\circ\, \phi$ denotes $\Diamond_{[0.001,0.1]}\,\phi$. Note that for falsification, AT1 and AT2 require extreme inputs, whereas AT6 and ATX2 require fine-grained inputs. The robustness scores of AT5 and ATX1 can be $\pm\infty$ and are discontinuous at gear changes.

The input signal for the benchmarks is piecewise constant with 4 control points for random sampling/Breach/S-TaLiRo, which is sufficient to falsify all requirements. We choose 6 levels, with 2, 2, 3, 3, 3, 4 control points, respectively, corresponding to a time granularity of input segment durations between 15 ($= \frac{30}{2}$, coarsest) to 7.5 ($= \frac{30}{4}$, finest).

Powertrain Control. The benchmark was proposed for hybrid systems verification and falsification in [21]. Falsification tries to detect amplitude and duration of spikes in the air-to-fuel AF ratio with respect to a reference value AF_{ref}. Those occur as a response to changes in the throttle θ. The input $\theta \in [0, 62.1)$ varies throughout the trace, whereas $\omega \in [900, 1100]$ is constant. The abbreviation $\mu = |AF - AF_{\text{ref}}|/AF_{\text{ref}}$ denotes the normalized deviation from the reference.

We consider requirements 27 from [21], which states that after a falling or rising edge of the throttle, μ should return to the reference value within 1 time unit and stay close to it for some time. We also consider requirement 29, which expresses an absolute error bound.

$$\text{AFC27} = \Box_{[11,50]}\, (rise \vee fall) \implies \Box_{[1,5]}\, |\mu| < \beta \qquad \text{AFC29} = \Box_{[11,50]}|\mu| < \gamma$$

where rising and falling edges of θ are detected by $rise = \theta < 8.8 \wedge \Diamond_{[0,\epsilon]}\, 40.0 < \theta$ and $fall = 40.0 < \theta \wedge \Diamond_{[0,\epsilon]}\, \theta < 8.8$ for $\epsilon = 0.1$. The concrete bounds $\beta = 0.008$

and $\gamma = 0.007$ are chosen from [8,9,28] as a balance between difficulty of the problem and ability to find falsifying traces.

The input signal is piecewise constant with 10 control points for all falsification methods, specifically 5 levels with 10 control points each for aLVTS.

Chasing Cars. In the model from [17] five cars drive in sequence. The time varying input controls the throttle and brake of the front car (each in $[0,1]$), while the other cars simply react to maintain a certain distance. We consider five artificial requirements from [28], where y_i is the position of the ith car. FALSTAR uses the same input parameterization as for the AT benchmarks.

$$CC1 = \Box_{[0,100]} y_5 - y_4 \le 40 \qquad CC2 = \Box_{[0,70]} \Diamond_{[0,30]} y_5 - y_4 \ge 15$$
$$CC3 = \Box_{[0,80]} ((\Box_{[0,20]} y_2 - y_1 \le 20) \vee (\Diamond_{[0,20]} y_5 - y_4 \ge 40))$$
$$CC4 = \Box_{[0,65]} \Diamond_{[0,30]} \Box_{[0,20]} y_5 - y_4 \ge 8$$
$$CC5 = \Box_{[0,72]} \Diamond_{[0,8]} ((\Box_{[0,5]} y_2 - y_1 \ge 9) \to (\Box_{[5,20]} y_5 - y_4 \ge 9))$$

4.2   Experimental Results

We have implemented Algorithm 1 in the prototype tool FALSTAR, which is publicly available on github, including repeatability instructions.[1] In our own experiments, we compare the performance of aLVTS with uniform random sampling (both implemented in FALSTAR) and with the state-of-the-art stochastic global optimization algorithm CMA-ES [18] implemented in the falsification tool Breach.[2] The machine and software configuration was: CPU Intel i7-3770, 3.40 GHz, 8 cores, 8 Gb RAM, 64-bit Ubuntu 16.04 kernel 4.4.0, MATLAB R2018a, Scala 2.12.6, Java 1.8.

We compare two performance metrics: *success rate* (how many falsification trials were successful in finding a falsifying input) and the *number of iterations* made, which corresponds to the number of simulations required and thus indicates time. To account for the stochastic nature of the algorithms, the experiments were repeated for 50 trials. For a meaningful comparison of the number of iterations until falsification, we tried to maximize the falsification rate for a limit of 300 iterations per trial.

The number of iterations of the top-level loop in Algorithm 1 in our implementation corresponds exactly to one complete Simulink simulation up to the time horizon. For random sampling and CMA-ES, the number of iterations likewise corresponds to samples taken by running exactly one simulations each. Hence the comparison is fair and, as the overhead is dominated by simulation time, the numbers are roughly proportional to wall-clock times.

---

[1] https://github.com/ERATOMMSD/falstar.

[2] https://github.com/decyphir/breach release version 1.2.9.

Table 2. Successes in 50 trials ("succ.", higher is better) and number of iterations averaged over successful trials ("iter.", lower is better) of uniform random sampling, Breach/CMA-ES, and FALSTAR/ Algorithm 1 for a maximum of 300 iterations per trial. est results for each requirement are highlighted. $AT6_a$: $\overline{v} = 80, \underline{\omega} = 4500$. $AT6_b$: $\overline{v} = 50, \underline{\omega} = 2700$. $t = 30$ in both cases.

Req.	Random		Breach: CMA-ES		FalStar: aLVTS	
	succ. /50	iter. mean	succ. /50	iter. mean	succ. /50	iter. mean
$AT1_{30}$	43	106.6	50	39.7	50	8.5
$ATX1_3$	50	41.0	50	13.2	50	33.4
$ATX1_4$	49	67.0	6	17.8	50	23.4
$ATX2$	19	151.1	50	145.2	50	86.3
$AT6_a$	36	117.3	50	97.0	50	22.8
$AT6_b$	2	117.7	49	46.7	50	47.6
$AFC27$	15	129.1	41	121.0	50	3.9

Table 3. Results from the ARCH competition. $AT6_a$: $t = 35, \overline{v} = 80$,. $AT6_b$: $t = 50, \overline{v} = 50$. $AT6_c$: $t = 65, \overline{v} = 50$. $\underline{\omega} = 3000$ in all three cases.

Req.	S-TaLiRo: SOAR		Breach: GNM		FalStar: aLVTS	
	succ. /50	iter. mean	succ. /50	iter. mean	succ. /50	iter. mean
$AT1_{20}$	50	118.8	50	11.0	50	33.0
$AT2$	50	23.9	50	2.0	50	4.3
$AT5_1$	50	26.7	41	74.6	50	69.5
$AT5_2$	50	4.1	49	72.0	26	125.3
$AT5_3$	50	3.4	49	74.5	50	70.8
$AT5_4$	50	10.5	21	84.9	50	71.1
$AT6_a$	49	78.4	50	97.9	50	76.1
$AT6_b$	33	132.6	49	112.9	50	82.4
$AT6_c$	47	61.3	50	94.1	0	–
$AFC27$	50	70.3	50	3.0	50	3.9
$AFC29$	50	13.5	50	3.0	50	1.2
$CC1$	50	9.4	50	3.0	50	4.1
$CC2$	50	6.0	50	1.0	50	4.0
$CC3$	50	19.9	50	3.0	50	6.9
$CC4$	20	188.0	0	–	2	52.0
$CC5$	50	42.9	49	26.1	46	91.2

Table 2 summarizes our results in terms of success rate, and mean number of iterations of successful trials. The unambiguously (possibly equal) best results are highlighted in blue. Where the lowest average number of iterations was achieved without finding a falsifying input for every trial, we highlight in grey the lowest average number of iterations for 100% success. We thus observe that aLVTS achieves the best performance in all but one case, $ATX1_3$. Importantly, within the budget of 300 iterations per trial, aLVTS achieves a perfect success rate. CMA-ES is successful in 296 trials out of the total 350, with sub maximal success for ATX_4 and $AFC27$. In comparison, random sampling succeeds in only 214 trials, with sub maximal success in all but ATX1.

The number of iterations required for falsification varies significantly between the algorithms and between the benchmarks. For the automatic transmission benchmarks, as an approximate indication of relative performance, CMA-ES requires about 50% more iterations than aLVTS, and random sampling requires again twice as many as CMA-ES. For the powertrain model (AFC27), the performance of aLVTS is more than an order of magnitude better: 3.9 iterations on average, compared to 121 for CMA-ES.

Figure 2 compares all trial runs for AFC27, ordered by the number of iterations required for falsification. Similar plots for the automatic transmission benchmarks are shown in Fig. 3. The shape of each curve gives an intuition of the performance and consistency of its corresponding algorithm. In general, fewer iterations and more successful results are better, so it is clear that aLVTS performs better than random sampling and CMA-ES.

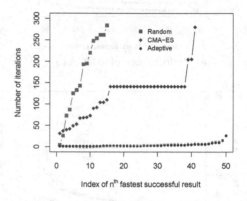

Fig. 2. Relative performance of falsification algorithms with AFC27 powertrain requirement: fewer iterations and more successful results are better.

To reinforce that the performance of aLVTS is at least comparable to other approaches, Table 3 presents some results of the recent ARCH competition [28]. The values for S-TaLiro and Breach (in different configurations to our experiments) were provided by the respective participants.

4.3   Discussion

For AT1, aLVTS quickly finds the falsifying input signal, as the required throttle of 100 and brake of 0 are contained in level 0 and are very likely to be tried early on. In contrast, even though this is a problem that is well-suited to hill-climbing, CMA-ES has some overhead to sample its initial population, cf. Fig. 3(a).

While CMA-ES deals very well with ATX1 for $i = 3$, it struggles to find falsifying inputs for $i = 4$ (cf. Figs. 3(c) and (d)). We attribute this to the fact that reaching gear 4 by chance occurs rarely in the exploration of CMA-ES when the robustness score is uninformative. aLVTS not only explores the spatial dimensions, but takes opportunistic jumps to later time points, which increases the probability of discovering a trace (prefix) where the gear is reached.

A priori, one would expect CMA-ES to perform well with ATX2 and AT6, exploiting its continuous optimization to fine tune inputs between conflicting requirements. E.g., ATX2 requires that v is both above 50 and below 60; AT4 requires that v is high while maintaining low ω, which is proportional to v. One would similarly expect the limited discrete choices made by aLVTS to hinder its ability to find falsifying inputs. Despite these expectations, our results

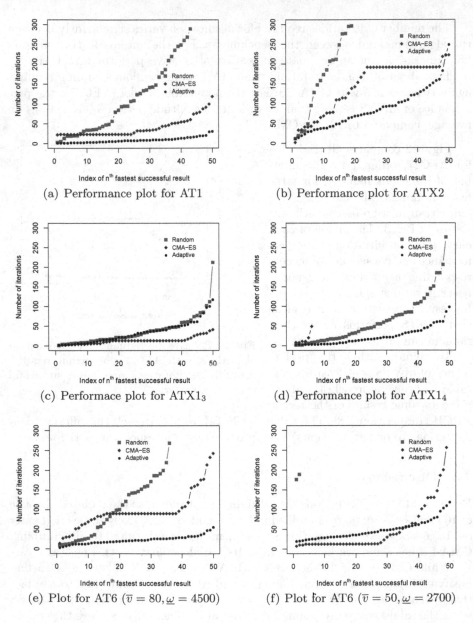

(a) Performance plot for AT1

(b) Performance plot for ATX2

(c) Performace plot for ATX1$_3$

(d) Performance plot for ATX1$_4$

(e) Plot for AT6 ($\overline{v} = 80, \underline{\omega} = 4500$)

(f) Plot for AT6 ($\overline{v} = 50, \underline{\omega} = 2700$)

Fig. 3. Performance comparison for the automatic transmission benchmark.

demonstrate that in most situations aLVTS converges to a falsifying input more consistently and with fewer iterations than CMA-ES. We speculate that this is because CMA-ES is too slow to reach the "sweet spots" in the input space, where its optimization is efficient.

For ATX2, there are a few instances where aLVTS does not quickly find a good prefix towards the corridor $v \in [50, 60]$ at time 10 (the rightmost points in Fig. 3(b)), which can be explained by the probabilistic nature of the search.

Regarding two instances of AT6 in Figs. 3(e) and (f), the graph for aLVTS is generally smoother and shallower, whereas CMA-ES shows consistent performance for only some of the trials but takes significantly more time on the worst 10 trials. We remark that the two parameter settings seem to pose opposite difficulty for the two algorithms, as CMA-ES is significantly quicker for two thirds of the trials for the second instance. It is unclear what causes this variation in performance of CMA-ES.

The plateaux apparent in some of the results using CMA-ES are difficult to explain, but suggest some kind of procedural logic or counter to decide termination. In contrast, the curves for random sampling and aLVTS are relatively smooth, reflecting their purely probabilistic natures.

For the results in Table 3, Breach was configured to use the GNM algorithm, which has a phase of sampling extreme and random values, thus sharing some of the characeristics of the aLVTS algorithm. As a consequence, many results of these two approaches are quite similar. A take-away is that algorithms that use random sampling in a disciplined way but are otherwise fairly simple work well on many falsification problem. There is no best overall tool: S-TaLiRo is quickest on the AT5 requirements involving discrete gear changes, whereas GNM yields the best results for the chasing cars model.

The AT and CC results for aLVTS show its capability to adapt to the (informal) difficulty of the problem, where the number of iterations increases but the falsification rate stays high. For $AT5_2$ and $AT6_b$ in Table 3 we conjecture that the available granularities of the search space are misaligned with the actual values to find a violation. Precisely, the required values are not contained in a set $\mathcal{A}_i$ that is sampled within the budget of 300; i is too large.

5 Related Work

The idea to find falsifying inputs using robustness as an optimization function originates from [15] and has since been extended to the parameter synthesis problem (e.g., [22]). Approaches to make the robustness semantics more informative include [3, 14], which use integrals instead of min/max in the semantics of temporal operators. Two mature implementations in MATLAB are S-Taliro [4] and Breach [10], which have come to define the benchmark in this field. Users of S-Taliro and Breach can select from a range of optimization algorithms, including Uniform Random, Nelder-Mead, Simulated Annealing, Cross-Entropy and CMA-ES. These cover a variety of trade-offs between exploration of the search space and exploitation of known good intermediate results.

Underminer [5] is a recent falsification tool that learns the (non-) convergence of a system to direct falsification and parameter mining. It supports STL formulas, SVMs, neural nets, and Lyapunov-like functions as classifiers. Other global approaches include [1], which partitions the input space into sub-regions from which falsification trials are run selectively. This method uses coverage metrics to balance exploration and exploitation. Comprehensive surveys of simulation based methods for the analysis of hybrid systems are given in [6,23].

The characteristic of our approach to explore the search space incrementally is shared with *rapidly-exploring random trees* (RRTs). The so-called star discrepancy metric guides the search towards unexplored regions and a local planner extends the tree at an existing node with a trajectory segment that closely reaches the target point. RRTs have been used successfully in robotics [24] and also in falsification [13]. On the other hand, the characteristic of our approach taking opportunistic coarse jumps in time is reminiscent of stochastic local search [7] and multiple-shooting [31].

Monte Carlo tree search (MCTS) has been applied to a model of aircraft collisions in [25], and more recently in a falsification context to guide global optimization [30], building on the previous idea of time-staging [29]. That work noted the strong similarities between falsification using MCTS and statistical model checking (SMC) using *importance splitting* [19]. The robustness semantics of STL, used in [29,30] and the present approach to guide exploration, can be seen as a "heuristic score function" [20] in the context of importance splitting. All these approaches construct trees from traces that share common prefixes deemed good according to some heuristic. The principal difference is that importance splitting aims to construct a diverse set of randomly-generated traces that all satisfy a property (equivalently, falsify a negated property), while falsification seeks a single falsifying input. The current work can be distinguished from standard MCTS and reinforcement learning [26] for similar reasons. These techniques tend to seek optimal policies that make good decisions in *all* situations, unnecessarily (in the present context) covering the entire search space.

6   Conclusion

The falsification problem is inherently hard (no theoretically best solution for all examples can exist), but our simple approach can provide useful results in isolation or as part of an ensemble. We have demonstrated this by matching and outperforming existing state-of-the-art methods on a representative selection of standard benchmarks. We hypothesize the reason our approach works well stems from the fact that there tends to be a significant mass of simple falsifying inputs for common benchmarks. As future work we will test this hypothesis (and the limits of our approach) by applying our algorithm to a wider range of benchmarks. In addition, we propose to fine-tune the probabilities of exploration vs. exploitation, and find better inputs by interpolating from previously seen traces, in a manner reminiscent of the linear combinations computed by the Nelder-Mead algorithm.

Acknowledgement. This work is supported by the ERATO HASUO Metamathematics for Systems Design Project (No. JPMJER1603), JST; and Grants-in-Aid No. 15KT0012, JSPS.

References

1. Adimoolam, A., Dang, T., Donzé, A., Kapinski, J., Jin, X.: Classification and coverage-based falsification for embedded control systems. In: Majumdar, R., Kunčak, V. (eds.) CAV 2017. LNCS, vol. 10426, pp. 483–503. Springer, Cham (2017). https://doi.org/10.1007/978-3-319-63387-9_24
2. Akazaki, T.: falsification of conditional safety properties for cyber-physical systems with gaussian process regression. In: Falcone, Y., Sánchez, C. (eds.) RV 2016. LNCS, vol. 10012, pp. 439–446. Springer, Cham (2016). https://doi.org/10.1007/978-3-319-46982-9_27
3. Akazaki, T., Hasuo, I.: Time robustness in MTL and expressivity in hybrid system falsification. In: Kroening, D., Păsăreanu, C.S. (eds.) CAV 2015. LNCS, vol. 9207, pp. 356–374. Springer, Cham (2015). https://doi.org/10.1007/978-3-319-21668-3_21
4. Annpureddy, Y., Liu, C., Fainekos, G., Sankaranarayanan, S.: S-TALiRo: a tool for temporal logic falsification for hybrid systems. In: Abdulla, P.A., Leino, K.R.M. (eds.) TACAS 2011. LNCS, vol. 6605, pp. 254–257. Springer, Heidelberg (2011). https://doi.org/10.1007/978-3-642-19835-9_21
5. Balkan, A., Tabuada, P., Deshmukh, J.V., Jin, X., Kapinski, J.: Underminer: a framework for automatically identifying nonconverging behaviors in black-box system models. ACM Trans. Embed. Comput. Syst. **17**(1), 1–28 (2017)
6. Bartocci, E., et al.: Specification-based monitoring of cyber-physical systems: a survey on theory, tools and applications. In: Bartocci, E., Falcone, Y. (eds.) Lectures on Runtime Verification. LNCS, vol. 10457, pp. 135–175. Springer, Cham (2018). https://doi.org/10.1007/978-3-319-75632-5_5
7. Deshmukh, J., Jin, X., Kapinski, J., Maler, O.: Stochastic local search for falsification of hybrid systems. In: Finkbeiner, B., Pu, G., Zhang, L. (eds.) ATVA 2015. LNCS, vol. 9364, pp. 500–517. Springer, Cham (2015). https://doi.org/10.1007/978-3-319-24953-7_35
8. Dokhanchi, A., Yaghoubi, S., Hoxha, B., Fainekos, G.E.: ARCH-COMP17 category report: preliminary results on the falsification benchmarks. In: Frehse, G., Althoff, M. (eds.) Applied Verification of Continuous and Hybrid Systems (ARCH). EPiC Series in Computing, vol. 48, pp. 170–174. EasyChair (2017)
9. Dokhanchi, A., et al.: ARCH-COMP18 category report: results on the falsification benchmarks. In: Frehse, G. (ed.) Applied Verification of Continuous and Hybrid Systems (ARCH). EPiC Series in Computing, vol. 54, pp. 104–109. EasyChair (2019)
10. Donzé, A.: Breach, a toolbox for verification and parameter synthesis of hybrid systems. In: Touili, T., Cook, B., Jackson, P. (eds.) CAV 2010. LNCS, vol. 6174, pp. 167–170. Springer, Heidelberg (2010). https://doi.org/10.1007/978-3-642-14295-6_17
11. Donzé, A., Ferrère, T., Maler, O.: Efficient robust monitoring for STL. In: Sharygina, N., Veith, H. (eds.) CAV 2013. LNCS, vol. 8044, pp. 264–279. Springer, Heidelberg (2013). https://doi.org/10.1007/978-3-642-39799-8_19

12. Donzé, A., Maler, O.: Robust satisfaction of temporal logic over real-valued signals. In: Chatterjee, K., Henzinger, T.A. (eds.) FORMATS 2010. LNCS, vol. 6246, pp. 92–106. Springer, Heidelberg (2010). https://doi.org/10.1007/978-3-642-15297-9_9

13. Dreossi, T., Dang, T., Donzé, A., Kapinski, J., Jin, X., Deshmukh, J.V.: Efficient guiding strategies for testing of temporal properties of hybrid systems. In: Havelund, K., Holzmann, G., Joshi, R. (eds.) NFM 2015. LNCS, vol. 9058, pp. 127–142. Springer, Cham (2015). https://doi.org/10.1007/978-3-319-17524-9_10

14. Eddeland, J., Miremadi, S., Fabian, M., Åkesson, K.: Objective functions for falsification of signal temporal logic properties in cyber-physical systems. In: Conference on Automation Science and Engineering (CASE), pp. 1326–1331. IEEE (2017)

15. Fainekos, G.E., Pappas, G.J.: Robustness of temporal logic specifications for continuous-time signals. Theor. Comp. Sci. 410(42), 4262–4291 (2009)

16. Hoxha, B., Abbas, H., Fainekos, G.E.: Benchmarks for temporal logic requirements for automotive systems. In: Frehse, G., Althoff, M. (eds.) Applied veRification for Continuous and Hybrid Systems (ARCH). EPiC Series in Computing, vol. 34, pp. 25–30. EasyChair (2014)

17. Hu, J., Lygeros, J., Sastry, S.: Towards a theory of stochastic hybrid systems. In: Lynch, N., Krogh, B.H. (eds.) HSCC 2000. LNCS, vol. 1790, pp. 160–173. Springer, Heidelberg (2000). https://doi.org/10.1007/3-540-46430-1_16

18. Igel, C., Hansen, N., Roth, S.: Covariance matrix adaptation for multi-objective optimization. Evol. Comput. 15(1), 1–28 (2007)

19. Jegourel, C., Legay, A., Sedwards, S.: Importance splitting for statistical model checking rare properties. In: Sharygina, N., Veith, H. (eds.) CAV 2013. LNCS, vol. 8044, pp. 576–591. Springer, Heidelberg (2013). https://doi.org/10.1007/978-3-642-39799-8_38

20. Jegourel, C., Legay, A., Sedwards, S.: An effective heuristic for adaptive importance splitting in statistical model checking. In: Margaria, T., Steffen, B. (eds.) ISoLA 2014. LNCS, vol. 8803, pp. 143–159. Springer, Heidelberg (2014). https://doi.org/10.1007/978-3-662-45231-8_11

21. Jin, X., Deshmukh, J.V., Kapinski, J., Ueda, K., Butts, K.R.: Powertrain control verification benchmark. In: Fränzle, M., Lygeros, J. (eds.) Hybrid Systems: Computation and Control (HSCC), pp. 253–262. ACM (2014)

22. Jin, X., Donzé, A., Deshmukh, J.V., Seshia, S.A.: Mining requirements from closed-loop control models. IEEE Trans. Comput. Aided Des. Integr. Circuits Syst. 34(11), 1704–1717 (2015)

23. Kapinski, J., Deshmukh, J.V., Jin, X., Ito, H., Butts, K.: Simulation-based approaches for verification of embedded control systems: an overview of traditional and advanced modeling, testing, and verification techniques. IEEE Control Syst. Mag. 36(6), 45–64 (2016)

24. LaValle, S.M., Kuffner Jr., J.J.: Randomized kinodynamic planning. Int. J. Robot. Res. (IJRR) 20(5), 378–400 (2001)

25. Lee, R., Kochenderfer, M.J., Mengshoel, O.J., Brat, G.P., Owen, M.P.: Adaptive stress testing of airborne collision avoidance systems. In: IEEE/AIAA 34th Digital Avionics Systems Conference (DASC 2015), pp. 6C2:1–6C2:13 (2015)

26. Sutton, R.S., Barto, A.G.: Reinforcement Learning: An Introduction, 2nd edn. MIT press, Cambridge (2018)

27. Wolpert, D., Macready, W.G.: No free lunch theorems for optimization. IEEE Trans. Evol. Comput. 1(1), 67–82 (1997)

28. Yaghoubi, S., et al.: ARCH-COMP19 category report: results on the falsification benchmarks. In: Frehse, G. (ed.) Applied Verification of Continuous and Hybrid Systems (ARCH). EPiC Series in Computing. EasyChair (2019)

29. Zhang, Z., Ernst, G., Hasuo, I., Sedwards, S.: Time-staging enhancement of hybrid system falsification. In: 2018 IEEE Workshop on Monitoring and Testing of Cyber-Physical Systems (MT-CPS 2018), pp. 3–4. IEEE, April 2018

30. Zhang, Z., Ernst, G., Sedwards, S., Arcaini, P., Hasuo, I.: Two-layered falsification of hybrid systems guided by Monte Carlo tree search. In: IEEE Transactions on Computer-Aided Design of Integrated Circuits and Systems (TCAD 2018) (2018)

31. Zutshi, A., Deshmukh, J.V., Sankaranarayanan, S., Kapinski, J.: Multiple shooting, CEGAR-based falsification for hybrid systems. In: Embedded Software (EMSOFT), pp. 5:1–5:10 (2014)

State-Space Construction of Hybrid Petri Nets with Multiple Stochastic Firings

Jannik Hüls, Carina Pilch, Patricia Schinke, Joanna Delicaris, and Anne Remke[✉]

Westfälische Wilhelms-Universität, Einsteinstr. 62, 48149 Münster, Germany
{jannik.huels,carina.pilch,p_schi16,joanna.delicaris,
anne.remke}@uni-muenster.de

Abstract. Hybrid Petri nets have been extended to include general transitions that fire after a randomly distributed amount of time. With a single general one-shot transition the state space and evolution over time can be represented either as a *Parametric Location Tree* or as a *Stochastic Time Diagram*. Recent work has shown that both representations can be combined and then allow multiple stochastic firings. This work presents an algorithm for building the *Parametric Location Tree* with multiple general transition firings and shows how its transient probability distribution can be computed. A case study on a battery-backup system shows the feasibility of the approach.

Keywords: Petri nets · Stochastic hybrid model · Transient probability

1 Introduction

Hybrid Petri nets with general transitions (HPnG) [9] extend Hybrid Petri nets [1] by adding stochastic behaviour through general transitions with a randomly distributed delay. HPnGs provide a high-level formalism for a class of stochastic hybrid systems with piece-wise linear continuous behaviour without resets and a probabilistic resolution of non-determinism. Hybrid Petri nets have been shown useful for the evaluation of critical infrastructures [7,15], even though they have previously been restricted to a single random variable. This paper shows how the state-space of a Hybrid Petri net with multiple general transition firings can be constructed as a *Parametric Location Tree* (PLT) [9] and analyzed over time. As such, this paper provides the missing link to recent work [13,14] which provides model checking capabilities for HPnGs with multiple random variables, assuming the existence of the Parametric Location Tree.

The approach of this paper is twofold: Firstly, a purely numerical iterative algorithm for the construction of a PLT in the presence of a finite but arbitrary number of stochastic firings before a certain maximum time is presented. This algorithm is based on introducing an order on the random variables that occur

© Springer Nature Switzerland AG 2019
D. Parker and V. Wolf (Eds.): QEST 2019, LNCS 11785, pp. 182–199, 2019.
https://doi.org/10.1007/978-3-030-30281-8_11

due to the firing of general transitions. Each of these firings influences the evolution of the Petri net. The idea of parametric analysis, as presented in [9], is to collect values of random variables into intervals, which lead to a similar system evolution. These intervals are called potential domains.

Secondly, to facilitate a transient analysis, we identify those parametric locations the HPnG can be in at a specific point in time (the so-called candidates) and compute those subsets of the domains of the random variables present, for which the HPnG is guaranteed to be in a specific candidate. A multi-dimensional integration over these so-called time-restricted domains yields a transient distribution over locations.

Previous work had to restrict the number of general transition firings, since (i) identifying the potential domains of child locations requires linear optimization, (ii) transient analysis relies on multi-dimensional integration over polytopes. For one random variable, this was solved using the simplex method and discretization [9]. For two random variables, [8] used hyperplane arrangement and triangulation to compute numerically exact results. Vertex enumeration was proposed in [14] to construct a PLT for an arbitrary but finite number of stochastic firings, but did not provide a general algorithm. The strict order on the firings of the general transitions, as in this paper, simplifies the computation of the child locations to solving linear inequalities. This allows the implementation of (i) an efficient iterative algorithm for constructing the PLT up to a certain maximum time and (ii) a transient analysis based on *Fourier-Motzkin* variable elimination.

A scalable case study on a battery backed-up system shows the feasibility of the approach and discusses its performance.

Related Work. Hybrid Petri nets without general transitions form a subclass of non-initialized singular automata [1]. Hence, un-bounded reachability is not decidable, and [8,9] resort to computing time-bounded reachability, however only in the presence of at most two random variables. Approaches for Hybrid automata extended with discrete probability distributions [16,21,22,24] compute time-bounded reachability using abstraction.

Another variant of Petri nets with hybrid and stochastic behaviour are FSPNs [10,12], which only allow exponentially distributed and immediate discrete transitions. Related Petri net approaches are all restricted w.r.t. the number of continuous variables [11] or to Markovian jumps [5]. Semi-Markov processes have been used to evaluate the dependability of uninterrupted power supply (UPS) systems [23] and Stochastic Activity networks have been used to study the resilence of smart grid distribution networks [2]. They are however restricted to negative exponential stochastic behaviour.

Organisation. Section 2 recalls the modeling formalism, Sect. 3 repeats the Parametric Location Tree. Section 4 introduces its construction for an arbitrary but finite number of random variables. Transient analysis is explained in Sects. 5 and 6 presents a feasibility study. Section 7 concludes the paper.

2   The Model Formalism of HPnGs

Syntax. Hybrid Petri nets with multiple general transition firings (HPnG) consist of the three components *places, transitions* and *arcs*. HPnGs extend hybrid Petri nets [1] with so called general transitions [9]. Places are either *discrete* or *continuous*. Any discrete place $P_i^d \in \mathcal{P}^d$ holds a marking of $m_i \in \mathbb{N}_0$ tokens and any continuous place $P_j^c \in \mathcal{P}^c$ holds a continuous marking described by a fluid level $x_j \in \mathbb{R}_0^+$. A continuous place has a predefined upper bound. The lower bound is always zero. The initial marking $\mathbf{M}_0$ is given by the initial number of tokens $\mathbf{m}_0$ and fluid levels $\mathbf{x}_0$ of all places. Transitions change the marking upon firing, i.e. they change the content of discrete and continuous places. How to change the content is defined by assigning weights and priorities to discrete and continuous arcs, i.e., $A^d \in \mathcal{A}^d$ and $A^f \in \mathcal{A}^f$ respectively. They connect places and transitions and depending on the direction of connection, corresponding places are called input or output places. General, deterministic and immediate transitions, together also called discrete transitions, change the discrete marking. A discrete transition is enabled when its input places match the weight of their connecting arcs. A continuous transition is enabled if all connected input places hold fluid. Guard arcs $A^t \in \mathcal{A}^t$ may further influence the enabling of transitions. They carry a comparison operator and a weight. A discrete guard arc connects a discrete place to any transition and conditions its enabling on the comparison of the discrete marking and its weight. Correspondingly, continuous places may be connected to (only) discrete transitions via (continuous) guard arcs. Discrete transitions are each associated with a clock c_i, which if enabled evolves with $dc_i/dt = 1$, otherwise $dc_i/dt = 0$. Note that upon disabling, the clock value is preserved. For general transitions this corresponds to the preemptive resume strategy. A deterministic transition $T_k^D \in \mathcal{T}^D$ fires when c_i reaches the predefined transitions firing time. For an immediate transition $T_k^I \in \mathcal{T}^I$ the predefined firing time is always zero. The firing time of a general transition $T_m^G \in \mathcal{T}^G$ is modelled by a cummulative distribution function (CDF), which is assumed to be absolutely continuous. Each *stochastic firing* results in a random variable that follows the CDF of the general transition.

Rate adaptation. Every static continuous transition $T_n^F \in \mathcal{T}^F$ has a constant nominal flow rate. Dynamic continuous transitions $T_o^{Dyn} \in \mathcal{T}^{Dyn}$ represent a set $D \subset \mathcal{T}^F$ of static continuous transitions. Hence their nominal flow rate is a function of the actual flow rates of all static continuous transitions in D (c.f. [7]). Continuous transitions change the fluid level of connected input and output places with a constant rate. The rates of transitions that are connected to a continuous place that is at either of its boundaries require *rate adaptation* (c.f. [9]), which changes the actual flow rate. At the upper boundary the inflow is decreased to match the outflow and at the lower boundary, the outflow is reduced accordingly. A continuous place then evolves with a drift, which equals the sum of the actual inflow rates minus the sum of the actual outflow rates. If multiple immediate or deterministic transitions fire at the same time, this conflict is resolved using priorities and weights. For details on *conflict resolution* and *the concept of enabling* c.f. [9]. The probability that a general transition fires at the same time as a deterministic one is zero. We exclude zeno behaviour by banning cycles of immediate and general transitions. Section 3 further describes the model evolution and the interplay between stochastic and deterministic transitions.

Model Evolution. A state of an HPnG is a tuple $\Gamma = (\mathbf{m}, \mathbf{x}, \mathbf{c}, \mathbf{d}, \mathbf{g}, \mathbf{e})$, where $\mathbf{m}$ is the discrete marking, $\mathbf{x}$ the continuous marking, $\mathbf{c}$ contains for each deterministic transition the enabling time. The drift $\mathbf{d}$ describes the current change of fluid level per time unit for each continuous place, $\mathbf{g}$ contains the enabling time for the general transitions, and $\mathbf{e}$ describes the enabling status of all transitions. Events trigger changes in states as introduced in [9,19]:

Definition 1. *An event* $\Upsilon(\Gamma_i, \Gamma_{i+1}) = (\Delta\tau_{i+1}, \varepsilon_{i+1})$ *describes the change from one state* Γ_i *to another state* Γ_{i+1}, *with* $\varepsilon_{i+1} \in \mathcal{P}^c \cup \mathcal{T}^I \cup \mathcal{T}^D \cup \mathcal{T}^G \cup \mathcal{A}^t$ *specifying the model element that caused the event. Note that* $\Delta\tau_{i+1} \in \mathbb{R}_0^+$ *is a relative time between two events, such that one of the following conditions is fulfilled:*

1. *An immediate, deterministic or general transition fires, such that* $\mathbf{m}_i \neq \mathbf{m}_{i+1} \wedge \varepsilon_{i+1} \in \mathcal{T}^I \cup \mathcal{T}^D \cup \mathcal{T}^G$.
2. *A continuous place reaches its lower or upper boundary, such that* $\mathbf{d}_i \neq \mathbf{d}_{i+1} \wedge \varepsilon_{i+1} \in \mathcal{P}^c$.
3. *A guard arc condition is fulfilled or stops being fulfilled, such that* $\mathbf{e}_i \neq \mathbf{e}_{i+1} \wedge \varepsilon_{i+1} \in \mathcal{A}^t$.

The set of all possible events which can occur in state Γ_i is finite and its size depends on the number of continuous places, the number of guard arcs and the number of enabled discrete transitions. It is denoted $\mathcal{E}(\Gamma_i)$ and the set of events with minimum remaining time for that state is defined as follows:

$$\mathcal{E}^{\min}(\Gamma_i) = \{\Upsilon_j(\Gamma_i, \Gamma_j) \in \mathcal{E}(\Gamma_i) \,|\, \nexists \Upsilon_k(\Gamma_i, \Gamma_k) \in \mathcal{E}(\Gamma_i) : \Delta\tau_k < \Delta\tau_j\}. \quad (1)$$

Note that multiple events can happen at the same point in time, e.g. due to conflicts between deterministic transitions or due to the scheduling of stochastic transitions. We split the set $\mathcal{E}^{\min}(\Gamma_i)$ into two subsets, representing the next random events $\mathcal{E}_{\mathrm{ran}}^{\min}(\Gamma_i)$ and the set of all next deterministic events $\mathcal{E}_{\mathrm{det}}^{\min}(\Gamma_i) = \mathcal{E}^{\min} \backslash \mathcal{E}_{\mathrm{ran}}^{\min}(\Gamma_i)$. The next minimum event time is unique before the first stochastic firing, it simply is the minimum of the remaining times to fire of all enabled deterministic transitions. After at least one stochastic firing, the entry time of locations, the clocks and the continuous marking may linearly depend on random variable(s). The computation of the next minimum event then is based on polynomials and not scalars. Hence, after at least one stochastic firing the set of deterministic events with minimum remaining time may consist of more than one element. Their remaining time to fire is derived by minimizing the elements of $\Delta\mathbf{T}_{\mathrm{det}}^{\min}(\Gamma_i)$ over the support of the random variables that already have fired, where

$$\Delta\mathbf{T}_{\mathrm{det}}^{\min}(\Gamma_i) = \{\Delta\tau_m \in \mathbb{R}_0^+ \,|\, \exists \Upsilon(\Gamma_i, \Gamma_m) \in \mathcal{E}_{\mathrm{det}}^{\min}(\Gamma_i)\}. \quad (2)$$

3   The Parametric Location Tree

The evolution over time until a fixed maximum time $\tau_{\max}$ of an HPnG can be described as a so-called Parametric Location Tree [9,14]. Each node in the tree is called a parametric location and represents a set of states. The edges of a PLT represent events. The event corresponding to the edge is called *source event* w.r.t. the child node location. Extending previous notation, a parametric location is now described as a tuple $\Lambda = (ID, t, p, \Gamma, \mathbf{S}, \mathbf{o})$, where the entry time is $\Lambda.t$, the probability of choosing that specific location with identifier $\Lambda.ID$ is $\Lambda.p$ w.r.t. the parent node. $\Lambda.\Gamma$ denotes the unique state of the HPnG when entering the location and $\Lambda.\mathbf{S}$ denotes the potential domain of the random variables as described in the following. There can be multiple general transitions present in the system and each can possibly fire multiple times. The firing order is unique for each parametric location and hence is stored as vector $\Lambda.\mathbf{o}$.

Each firing of a general transition corresponds to a random variable which equals the enabling time between two consecutive firings. The random variable corresponding to the j-th firing of T_i^G is denoted s_i^j and is added to $\mathbf{o}$ upon the firing of the transition. The r-th stochastic firing is then stored in $\mathbf{o}[r]$. A new random variable is instantiated each time a general transition becomes enabled. Hence, the number of random variables n equals the number of stochastic firings plus the number of general transitions that are currently enabled. Concurrently enabled general transitions yield competing random variables of potentially different distributions, whereas consecutive firings of a single general transition result in a series of identically distributed random variables.

As the number of stochastic firings differ per location, the size of the potential domains, as indicated by $|\mathbf{S}|$, also differs. The potential domain for a random variable s_i^j in location Λ collects all values of the domain of s_i^j which are possible in that location. This is stored in $\mathbf{S}_i^j$ as a lower and an upper boundary $S_i^j.l \leq S_i^j.u$, such that $s_i^j \in [S_i^j.l, S_i^j.u]$. This extends the definition presented in [14] by separately collecting the possible intervals for each general transition.

The PLT of an HPnG is defined as a tree $(\mathbf{V}, \mathbf{E}, v_{\Lambda_0})$, where $\mathbf{V}$ is the set of nodes representing the parametric locations of the HPnG. $\mathbf{E}$ is the set of edges with $e_i = (v_{\Lambda_j}, v_{\Lambda_k}) \in \mathbf{E}$ for $v_{\Lambda_j}, v_{\Lambda_k} \in \mathbf{V}$ if an event $\Upsilon(\Lambda_j.\Gamma, \Lambda_k.\Gamma)$ exists which leads from the parametric location Λ_j to its child location Λ_k. The root node v_{Λ_0} represents the initial location. The PLT is iteratively constructed by adding a child location Λ_c for each possible event that can take place from a given parametric location. The finite set of possible next events (c.f. Eq. 1) together with the exclusion of cycles of immediate and general transitions in the model definition yields a finite PLT if computed up to time τ_{max}.

The absolute point in time at which the event takes place is then stored as the locations entry time $\Lambda_c.t$. The other parameters of the location are derived by executing the event in the HPnG. The random variable domain for a specific child location is derived from the parent location, by taking into account the event that leads to the specific location. Since the location entry times may depend on previously expired random variables, the set which consists of the remaining minimum times for all deterministic events $\mathbf{\Delta T}_{\text{det}}^{\min}(\Lambda_p.\Gamma)$ may contain more than one element and the order of the timed events depends on the values of the random variables that have expired. Since different child nodes symbolically describe sets of values leading to different system evolutions, the resulting potential domains are always disjoint for non-conflicting successors.

4   Random Variables Support

When allowing more than one stochastic firing the construction of the PLT becomes more complex [14]. For each enabled general transition, at least one location is added as child node, where the edge represents its next firing before every other event. Additionally, each general transition needs to be scheduled after each minimum deterministic event. Hence, if there are n deterministic events, for each enabled general transition, first $n + 1$ child nodes are generated, with one edge for the general transition firing and n for all possible deterministic next events $\mathcal{E}_{\text{det}}^{\min}(\Gamma_i)$. The deterministic next events and the enabled general transitions can then be arranged in two different ways: (i) either deterministic events are scheduled first (ii) or a general transitions fires before the next deterministic events. We need to consider all possible combinations of deterministic events and general transitions firings and we need to adjust the boundaries in all cases accordingly, which is discussed in the following.

4.1   Adjust Boundaries for Expired Random Variables

Each parametric location provides an interval $\mathbf{S}_i^j$ of possible values for all random variables as previously described. The bounds of those intervals may depend on firings with a lower order. When scheduling the deterministic successors of a location, we first need to compute the set of minimum events. In case this set contains more than one element, their order depends on the values of other random variables. This is handled by adjusting the potential domains of the corresponding random variables per event, such that each event takes place before all other minimum events for all values in the potential domain of the corresponding successor location.

The computation of the correct potential domains corresponds to minimizing the time to the minimum event over the domain of all random variables. In earlier work this has been solved, e.g., using the simplex method [9] in case of one random variable, hyperplane arrangement in the case of two [8] or as suggested by [14] using vertex enumeration. Instead, this paper defines a strict total order $\prec$ on the random variables that is based on their firing order: Let s_i^j and s_k^l be two random variables, then $s_i^j \prec s_k^l$ holds iff the i-th general transition fires for the j-th time before the k-th general transition fires for the l-th time: $\exists u, v \in \mathbb{N} : \mathbf{o}[u] = s_i^j \wedge \mathbf{o}[v] = s_k^l \wedge u < v \Rightarrow s_i^j \prec s_k^l$. This ensures that the interval of a random variable s_i^j may only depend on earlier expired random variables s_k^l. The interval bounds of a random variable s_i^j can in general be described by two linear expressions:

$$s_i^j \in S_i^j = [a_0 + \sum_{s_i^j \prec s_k^l} a_i^j \cdot s_i^j, \; b_0 + \sum_{s_i^j \prec s_k^l} b_i^j \cdot s_i^j], \text{ with } a_0, a_i^j, b_0, b_i^j \in \mathbb{R}. \qquad (3)$$

Each interval bound is given by the sum of previously expired random variables, each potentially multiplied with a real constant, which relates to the firing of deterministic transitions. In the following we describe how these bounds can be computed.

Scheduling the deterministic successors of a parametric location Λ_p, the set of minimum events $\mathcal{E}_{\det}^{\min}(\Lambda_p.\Gamma)$ is required. When generating a child location for a deterministic successor, the potential domains of the expired random variables are determined and stored for all deterministic events $\Upsilon(\Lambda_p.\Gamma, \Lambda_c.\Gamma) \in \mathcal{E}_{\det}^{\min}(\Lambda_p.\Gamma)$. If the set $\mathbf{\Delta T}_{\det}^{\min}(\Lambda_p.\Gamma)$ contains more than one value, the intervals of the potential domains for the (expired) random variables have to be limited for the corresponding deterministic successor, such that its source event $\Upsilon(\Lambda_p.\Gamma, \Lambda_c.\Gamma) = (\Delta\tau_c, \varepsilon_c)$, takes place before any other event. Hence $\Delta\tau_c$ is smaller than or equal to the remaining time for any other event:

$$\forall \Delta\tau^* \in \mathbf{\Delta T}_{\det}^{\min}(\Lambda_p.\Gamma) : \Delta\tau_c \leq \Delta\tau^*. \qquad (4)$$

Hence, the potential domain of the random variables in Λ_c is computed by comparing $\Delta\tau_c$ in a pairwise fashion to each other value $\Delta\tau^* \in \mathbf{\Delta T}_{\det}^{\min}(\Lambda_p.\Gamma)$ and limiting the intervals of the corresponding random variables such that Eq. 4 holds for the included potential domains. Having $\Delta\tau_c = \Delta\tau^*$ means that both linear expressions intersect. Note that this results in closed intervals for the potential domains with overlapping interval bounds, which however does not make a difference, since the probability for such case equals zero. The computation can be done by considering only random variables which correspond to firings in the past, i.e. with a lower order, since only past firings can affect clocks and markings and the remaining time to fire, as stored in $\Delta\mathbf{T}_{\det}^{\min}(\Lambda_p.\Gamma)$. Since $\Delta\tau_c$ is compared to the other remaining times one-by-one, we first present the pairwise comparison in the following and then explain how the results are brought together to obtain the potential domains for the child location Λ_c. The remaining times to fire, $\Delta\tau_c$ and each $\Delta\tau^* \in \mathbf{\Delta T}_{\det}^{\min}(\Lambda_p.\Gamma)$ for $\alpha_0, \alpha_z, \beta_0, \beta_z \in \mathbb{R}$ and $n = |\mathbf{S}|$, can be written as:

$$\Delta\tau_c = \alpha_0 + \sum_{z=1}^{n} \alpha_z \cdot \mathbf{o}[z], \text{ and } \Delta\tau^* = \beta_0 + \sum_{z=1}^{n} \beta_z \cdot \mathbf{o}[z]. \qquad (5)$$

According to Eq. 3 also the expressions above are dependent on the random variables in firing order. Each variable may again be multiplied by a real constant, which calculation is presented below. The intervals for any random variable $\mathbf{o}[z]$ with the same multiplier in both linear expressions, i.e. $\alpha_z = \beta_z$, do not need to be adapted since the validity of $\Delta\tau_c \leq \Delta\tau^*$ is independent of this random variables.

Using these linear expressions, the inequality $\Delta\tau_c \leq \Delta\tau^*$ can be solved for a specific random variable, i.e., the random variable with the highest order for which the multipliers differ in both inequalities. Hence, to ensure that the condition $\Delta\tau_c \leq \Delta\tau^*$ is fulfilled, it is sufficient to limit the intervals of the potential domain of this specific random variable, such that the rearranged inequality (that is solved for this random variable) is fulfilled. Precisely, we determine the maximum index k with respect to the order $\prec$, for which $\alpha_k \neq \beta_k$ holds. If no such index k exists, it follows that $\Delta\tau_c = \Delta\tau^*$. This results in two events with the same remaining occurrence time. In this case only the event with the higher order (c.f. Table 1 in [9]) is considered. If both events have the same order, the conflict resolution (c.f. Section 3.4 in [9]) is used to compute probabilities and $\Lambda_c.p$ is updated accordingly. In this case the potential domains of the random variables do not need to be adjusted. Due to the definition of $\mathcal{E}_{\det}^{\min}(\Gamma_i)$, if such a maximum index k exists, it needs to be larger than zero. Otherwise, the linear expressions of the remaining times $\Delta\tau_c$ and $\Delta\tau^*$ would be parallel functions and one of the remaining event times is always larger and hence the corresponding event is not included in $\mathcal{E}_{\det}^{\min}(\Gamma_i)$. For any $k > 0$, the potential domain of the random variable $\mathbf{o}[k]$ is adjusted by rearranging the inequality $\Delta\tau_c \leq \Delta\tau^*$ as:

$$\alpha_0 + \sum_{z=1}^{n} \alpha_z \cdot \mathbf{o}[z] \leq \beta_0 + \sum_{z=1}^{n} \beta_z \cdot \mathbf{o}[z] \Leftrightarrow (\alpha_k - \beta_k) \cdot \mathbf{o}[k] \leq (\beta_0 - \alpha_0) + \sum_{z=1}^{k-1} (\beta_z - \alpha_z) \cdot \mathbf{o}[z].$$

$$(6)$$

First $\Delta\tau_c$ as well as $\Delta\tau^*$ are replaced by the expressions presented in Eq. 5. Then both summations are combined on the right side of the inequality and the largest differentiating term, i.e., the one with index k, is pushed to the left side of the inequality. The sum $\sum_{z=k+1}^{n} (\beta_z - \alpha_z) \cdot \mathbf{o}[z]$ can be omitted since $\forall z > k :$ $\beta_z - \alpha_z = 0$. Rearranging the inequality and dividing it by $\alpha_k - \beta_k$, which is always possible, as $\alpha_k - \beta_k \neq 0$ by definition of the index k. We need to distinguish two cases: *Case $\alpha_k > \beta_k$*:

$$(\alpha_k - \beta_k) \cdot \mathbf{o}[k] \leq (\beta_0 - \alpha_0) + \sum_{z=1}^{k-1} (\beta_z - \alpha_z) \cdot \mathbf{o}[z]$$

$$(7)$$

$$\Rightarrow_{\alpha_k > \beta_k} \qquad \mathbf{o}[k] \leq \frac{\beta_0 - \alpha_0}{\alpha_k - \beta_k} + \sum_{z=1}^{k-1} \frac{\beta_z - \alpha_z}{\alpha_k - \beta_k} \cdot \mathbf{o}[z].$$

The upper bound for the random variable s_i^j stored in $\mathbf{o}[k]$ is adjusted as follows:

$$\Lambda_c.S_i^j = \left[\Lambda_c.S_i^j.l, \frac{\beta_0 - \alpha_0}{\alpha_k - \beta_k} + \sum_{z=1}^{k-1} \frac{\beta_z - \alpha_z}{\alpha_k - \beta_k} \cdot \mathbf{o}[z] \right].$$

$$(8)$$

Algorithm 1. scheduling child locations

```
1: for all (deterministicEvent Υ : Ε_det^min(Λ_p.Γ)) do
2:     Λ_c := generateChild();
3:     Λ_c.setExpiredRVBounds(Υ);
4:     for all (s: enabled RV) do
5:         Λ_c.s.l = Λ_p.s.l + Δτ_c; {Eq 11}
6:         Λ_c' := generateChild();
7:         Λ_c'.copyExpiredRVBounds(Λ_c);
8:         Λ_c'.s.u = Λ_p.s.l + Δτ_c; {Eq. 12}
9:         for all (s' : enabled RV) do
10:            if (s != s') then
11:                Λ_c'.s'.l = Λ_p.s'.l + s - Λ_p.s.l; {Eq. 13}
```

Case $\alpha_k < \beta_k$:

$$(\alpha_k - \beta_k) \cdot \mathbf{o}[k] \leq (\beta_0 - \alpha_0) + \sum_{z=1}^{k-1} (\beta_z - \alpha_z) \cdot \mathbf{o}[z]$$

$$\Rightarrow_{\alpha_k < \beta_k} \qquad \mathbf{o}[k] \geq \frac{\beta_0 - \alpha_0}{\alpha_k - \beta_k} + \sum_{z=1}^{k-1} \frac{\beta_z - \alpha_z}{\alpha_k - \beta_k} \cdot \mathbf{o}[z]. \tag{9}$$

The lower bound for the random variable s_i^j stored in $\mathbf{o}[k]$ is adjusted as follows:

$$\Lambda_c.S_i^j = \left[\frac{\beta_0 - \alpha_0}{\alpha_k - \beta_k} + \sum_{z=1}^{k-1} \frac{\beta_z - \alpha_z}{\alpha_k - \beta_k} \cdot \mathbf{o}[z], \Lambda_c.S_i^j.u \right]. \tag{10}$$

Repeating the pairwise comparison for every $\Delta\tau^* \in \mathbf{\Delta T}_{det}^{min}(\Lambda_p.\Gamma)$, we store the resulting upper and lower bounds and set the potential domain for child Λ_c by taking the maximum of the pair-wisely computed lower bounds and the minimum of the corresponding upper bounds. This ensures that the resulting potential domain for each random variable equals the intersection of all intervals obtained by the pairwise comparison of $\Delta\tau_c$ and any $\Delta\tau^* \in \mathbf{\Delta T}_{det}^{min}(\Lambda_p.\Gamma)$. In the following, we assume that for the child location Λ_c with source event Υ the above explained procedure is summarized in the function $\Lambda_c.\texttt{setExpiredRVBounds}(\Upsilon)$.

For each scheduled child location Λ_c, the potential domains further need to be adjusted in case one or more general transitions are currently enabled in the parent location. This is discussed in the following section, where we assume that for all expired random variables, the potential domains in $\Lambda_{c'}.\mathbf{S}$ are set to the corresponding potential domains in $\Lambda_c.\mathbf{S}$, that have been determined in the procedure described above, by calling a function $\Lambda_{c'}.\texttt{copyExpiredRVBounds}(\Lambda_c)$.

4.2   Scheduling the Child Locations

After computing the potential domains for each minimum next event, the corresponding child location needs to be scheduled for each combination of possible next deterministic event and each enabled stochastic firing. This requires iterating over all elements in the set of minimum deterministic events, generating the corresponding child locations and computing the potential domains for already expired random variables, as explained in the previous section. Then the firings of all enabled general transitions need to be scheduled before each minimum next event and the corresponding child locations are generated. To ensure that all random variables are scheduled once before and once after each minimum next event, the potential domains of the currently enabled random variables need to be further restricted.

Algorithm 1 presents pseudo code for the scheduling of all child locations for a given parametric location Λ_p. In Line 2 `generateChild()` creates a location for each child Λ_c for every event in the set of minimum deterministic events $\mathcal{E}_{\text{det}}^{\min}$. The entry time of each child location $\Lambda_c.t$ is set to $\Lambda_p.t + \Delta\tau_c$ and the boundaries for expired random variables are adjusted as explained in the previous section by calling the function $\Lambda_c.\texttt{setExpiredRVBounds}(\Upsilon)$ (Line 3). In order to schedule all competing general transition firings after the event $\Upsilon(\Lambda_p.\Gamma, \Lambda_c.\Gamma) = (\Delta\tau_c, \varepsilon_c)$, the lower interval bound for each of the still enabled random variables is increased by $\Delta\tau_c$ in the location Λ_c (Lines 5–6). Formally, the lower interval bound of the potential domain $\Lambda_c.S_i^j$ is adapted by increasing the lower bound of the corresponding potential domain $\Lambda_p.S_i^j$:

$$\Lambda_c.S_i^j.l = \Lambda_p.S_i^j.l + \Delta\tau_c. \tag{11}$$

After adapting the already scheduled successors which correspond to deterministic events, we need to schedule each enabled stochastic firing before each of the deterministic events. This requires the construction of new parametric locations.

Hence, for each enabled stochastic firing, i.e., for each event $\Upsilon(\Lambda_p.\Gamma, \Lambda_{c'}.\Gamma) = (\Delta\tau_{c'}, \varepsilon_{c'}) \in \mathcal{E}_{\text{ran}}^{\min}(\Lambda_p.\Gamma)$ one or more new parametric locations $\Lambda_{c'}$ need to be constructed (Line 7) and the potential domains of the expired random variables are inherited from the corresponding deterministic location Λ_c by calling the function $\Lambda_{c'}.\texttt{copyExpiredRVBounds}(\Lambda_c)$ (Line 8).

Furthermore, the upper interval bound for the potential domain of the random variable s_i^j is decreased in the new child location $\Lambda_{c'}$, such that the general transition fires before any other competing event occurs (Lines 9–10). Hence, s_i^j cannot take any value that is larger than the sum of its previous upper bound in the parent location Λ_p and the time to the next competing event.

The upper interval bound in the domain $\Lambda_{c'}.S_i^j$ of the child location is decreased to ensure that the enabled random variable s_i^j fires first:

$$\Lambda_{c'}.S_i^j.u = \Lambda_p.S_i^j.l + \Delta\tau_c. \tag{12}$$

Next, the potential domains for all other enabled random variables have to be adjusted, such that the corresponding firings do not occur before the firing related to s_i^j takes place. Hence, for all enabled random variables (except s_i^j), the potential domain is adjusted (Lines 11–16). For every random variable $s_u^v \neq s_i^j$ that corresponds to the v-th firing of transition $T_u^G \in \mathcal{T}^G$, which is enabled in $\Lambda_p.\Gamma$, the lower interval bound is increased in the potential domain $\Lambda_{c'}.S_u^v$:

$$\Lambda_{c'}.S_u^v.l = \Lambda_p.S_u^v.l + s_i^j - \Lambda_p.S_i^j.l. \tag{13}$$

Furthermore, the support of the random variable s_i^{j+1} that corresponds to the next firing of T_i^G has to be initialized if T_i^G is still enabled in $\Lambda_{c'}.\Gamma$. The interval is then always set to $S_i^{j+1} = [0, \infty)$. All of the new child locations Λ_c and $\Lambda_{c'}$ together create the set of children of location Λ_p, i.e. children(Λ_p).

The computational complexity of Algorithm 1 follows the nested for loops and is in $\mathcal{O}(n^2 \times |\mathcal{E}_{\det}^{\min}(\Lambda_p.\Gamma)|)$, where $n = |\mathbf{S}|$ forms an upper bound on the number of enabled random variables. The size of $\mathcal{E}_{\det}^{\min}(\Lambda_p.\Gamma)$ depends on the number of continuous places, guard arcs and immediate and deterministic transitions, which are present in the model.

5   Transient Analysis

Transient analysis computes the probability to be in a certain state of the model at time t'. This can easily be extended to computing the probability that the model fulfills a certain atomic property, w.r.t. the discrete or continuous marking. As the PLT is a symbolic representation of the state space evolution over time, the system can be in different parametric locations at time t', depending on the values of the random variables. For each of these so-called *candidate locations*, the subset of the potential domains need to be computed for which the system actually is in that location at time t'. The transient probabilities can then be computed by integrating over the computed subsets.

5.1   Candidate Locations and Restricting Their Potential Domains

Since the entry time $\Lambda.t$ of a location Λ depends on the potential domain of the random variables $\Lambda.\mathbf{S}$, the entry time point is a linear function of the random variables. However, we can specify the minimum entry time, i.e., the earliest possible entry time $\Lambda.t_{\min}^{\text{entry}}$ of a location Λ. Accordingly, $\Lambda.t_{\max}^{\text{entry}}$ specifies the latest possible entry time into a specific location. To identify the candidate locations for t', also the latest possible exit time of a location needs to be considered, which is obtained as the latest maximum entry time of all child locations:

$$\Lambda.t_{\max}^{\text{exit}} = \max\{\Lambda_c.t_{\max}^{\text{entry}} \mid \Lambda_c \in \text{children}(\Lambda)\}. \tag{14}$$

Note, that the number of locations in the tree for a maximum considered time $\tau_{\max}$ is finite, according to [14]. To obtain candidate locations for a specific time t' we check for each location of the PLT, whether the potential domains are such that values for random variables exist, that bring the system in that location at time t'. This is done by checking whether $\Lambda.t_{\min}^{\text{entry}} \leq t'$ and $\Lambda.t_{\max}^{\text{exit}} \geq t'$. If the first condition is violated, the entry time of the location Λ is guaranteed to be later than the considered time point, and hence its children do also not need to be considered as candidates. The second condition ensures that it is possible to still be in location Λ at time t'. In case this condition is violated, the child locations of Λ still need to be considered as candidates. Hence, the set of candidate locations $\mathcal{C}_{t'}$ at time t' equals $\mathcal{C}_{t'} = \{\Lambda \in \mathbf{V} \mid \Lambda.t_{\min}^{\mathrm{entry}} \le t' \le \Lambda.t_{\max}^{\mathrm{exit}}\}$.

The earliest and latest possible entry time of a parametric location Λ are computed iteratively by replacing random variables by their lower or upper boundaries. The boundaries of a random variable may depend on random variables with a lower order, i.e. that fired earlier. These dependencies need to be resolved when minimizing the entry time of a location or maximizing the entry time of its children. Using the previously defined order, the linear equation defining $\Lambda.t$ can be rearranged such that the random variables together with their factors occur according to their firing order: $\Lambda.t = f_0 + \sum_{k=1}^{n} f_k \cdot \mathbf{o}[k]$.

By iterating over the random variables it is possible to resolve their dependencies and to minimize the entry time at the same time. In case random variable s_i^j is stored in $\mathbf{o}[l]$ the random variable s_i^j needs to be replaced within $\Lambda.t$ by the corresponding interval bound $S_i^j.l$ or $S_i^j.u$, depending on the sign of f_l. This step is repeated for all $\mathbf{o}[l]$ in descending order and the resulting $\Lambda.t$ after each step then only depends on all $\mathbf{o}[k] \prec \mathbf{o}[l]$:

$$\Lambda.t = f_0 + \sum_{k=1}^{l-1} f_k \cdot \mathbf{o}[k] + \begin{cases} f_l \cdot S_i^j.l, & \text{if } f_l > 0, \\ f_l \cdot S_i^j.u, & \text{if } f_l < 0. \end{cases} \tag{15}$$

Once all random variables have been resolved by the corresponding lower or upper boundary, $\Lambda.t$ has been minimized within the potential domain of location Λ. When computing the maximum of the entry time for all children, i.e. the latest exit time point, the conditions to replace the random variables are both reversed.

In all candidate locations the potential domain of all random variables present up to t' need to be restricted, such that only those values remain for which the system certainly is in that location at time t'. The computation of the restricted potential domains per location corresponds to a *cylindrical algebraic decomposition* [3], which returns the restricted potential domains $\mathbf{S}'$ as a set of multi-dimensional intervals. For linear inequalities this can be computed in $O(n^2)$ using a variant of the *Fourier-Motzkin elimination* [4].

Under the assumption that no conflicts have occurred, the restricted potential domain of a location and the restricted potential domains of its children form disjoint subsets. In case a conflict occurs, it needs to be resolved by adapting the value $\Lambda_c.p$, for all children that participate in the conflict. In that case the restricted domains of all conflicting child locations may overlap. Note that conflicting child locations have the same minimum entry time.

5.2 Computing Transient Probabilities

The probability to be in a specific candidate location Λ at time t' is computed by first integrating over the joint probability density function, and then multiplying the result with the accumulated *conflict probability* towards that location. The latter is computed by multiplying the conflict probability of each location visited when traversing the tree from the root location to Λ.

Recall that the potential domain $\mathbf{S}$ consists of one multi-dimensional interval per location. The restricted potential domain $\mathbf{S}'$, however, may consist of several multi-dimensional intervals per location, as the Fourier-Motzkin elimination potentially splits the potential domain in several dimensions. Hence, the restricted domain $\mathbf{S}'$ is a set of multi-dimensional intervals. Each element $\mathbf{r} \in \mathbf{S}'$ is stored as a vector of intervals defining an upper and a lower bound for each random variable present in the order specified by $\mathbf{o}$.

Let function $g_i(s_i^j)$ denote the probability density function of all random variables s_i^j for $j \geq 0$, which is uniquely determined by the CDF assigned to the general transition T_i^G. All random variables are accessed in their firing order, as specified in $\mathbf{o}$. The probability that the values of all random variables in $\mathbf{s}$ lie within $\mathbf{S}'$ is given by Eq. 16, whereas the product over the $|\mathbf{S}| = n$ probability density functions yields the joint probability density function $\prod_{i=1}^{n} g_i(\mathbf{o}[i])$, due to the independence of the random variables. The resulting probability then is the sum of the integration over each restricted potential domain, as

$$Prob(\mathbf{s}, \mathbf{S}') = \sum_{\mathbf{r} \in \mathbf{S}'} \int_{\mathbf{r}[1].l}^{\mathbf{r}[1].u} \cdots \int_{\mathbf{r}[n].l}^{\mathbf{r}[n].u} \prod_{i=1}^{n} g_i(\mathbf{o}[i]) \, d\mathbf{o}[n] \cdots d\mathbf{o}[1]. \qquad (16)$$

Since no exact integration techniques exist, we use approximate methods together with an estimation of the (statistical) error. Existing libraries for numerical integration approaches only support integration over (bounded) rectangular regions, i.e. with constant limits. To use such libraries, a transformation of variables from S^n onto a rectangular region S_{rect}^n is needed, for which we refer to [18]. Monte Carlo methods [17] provide algorithms for the approximation of such integrals including a reliable error estimate, which are well-suited to compute integrals of higher dimensions since their convergence rates are independent of the dimension. The main idea for estimating an integral as in [6,20] is to select N random points from a distribution of points in the (rectangular) integration region S_{rect}^n. Our implementation uses the adaptive Monte Carlo scheme VEGAS [17] as in *GNU Scientific Library (GSL)*[1] [6], based on *importance sampling*.

6   Case Study: Battery Back-Up

As feasibility study, we model a factory whose power intake from the grid is limited by a service level agreement (SLA) with the energy provider. The factory relies on battery back-up which is discharged in the case of a peak-demand or a power outage. Figure 1(a) shows the core of the HPnG model: The continuous place P_0^C models the battery with capacity B, inflow (T_0^{Dy}) from the power grid and an outflow (T_1^{Dy}) which corresponds to the power that exceeds the constant peak demand P_A, specified by the SLA. The rates of the dynamic inflow and outflow are specified as $f_{T_0^{Dy}}(\mathbf{d}) = \max((P_A - \sum_{d_i \in \mathbf{d}} d_i), 0)$ and $f_{T_1^{Dy}}(\mathbf{d}) = \max((\sum_{d_i \in \mathbf{d}} d_i - P_A), 0)$, where $\mathbf{d} \in \mathbb{R}^{|\mathcal{T}^{St}|}$ is the vector of currently

---

[1] https://www.gnu.org/software/gsl/.

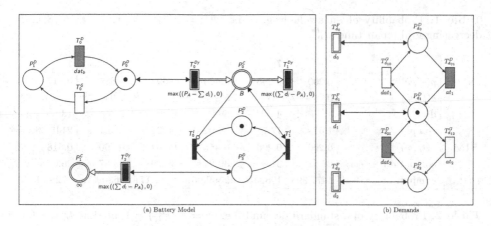

(a) Battery Model   (b) Demands

Fig. 1. Battery model and demands.

active demands, as specified in Fig. 1(b). The upper left part of Fig. 1(a) models the status of the grid and enables the inflow of the battery.

The failure of the grid (T_0^G) is modeled by a general transition, the recovery T_O^D is deterministic. Place P_3^D indicates an empty battery, which in turn enables transition T_2^{Dy}, which models the cost that occur due to service level violations when the accumulated demand exceeds the amount specified in the SLA. Its flow rate is $f_{T_2^{Dy}}(\mathbf{d}) = \max((\sum_{d_i \in \mathbf{d}} d_i - P_A), 0)$ and the extra cost is collected in the continuous place P_1^C (with infinite capacity).

Figure 1(b) models three levels of demand via static continuous transitions, i.e. standard $(T_{d_1}^F)$, reduced $(T_{d_0}^F)$, and extended $(T_{d_2}^F)$, that are enabled via a guard arcs. General transitions switch from standard to reduced demand $(T_{d_{10}}^G)$ and from standard to extended demand $(T_{d_{12}}^G)$. The number of random variables in the system depends on the maximum time τ_{max} of the analysis and the repair and switching times. At least three random variables are present, since all three general transitions are initially enabled concurrently.

Results. Table 1 shows the probability that P_0^D contains a token at time $t_c = 8$, computed numerically via the PLT[2] and Monte Carlo integration, for which error estimates are provided. For comparison, we include results obtained from the statistical model checking tool HYPEG [19]. The general transition T_0^G follows a uniform distribution $\mathcal{U}(0, 10)$ and the firing time of T_0^D takes values between 8 and 1, resulting in an increasing availability of the grid. The number of random variables present in the system increases, since T_0^G can fire more often before time $t_c = 8$. Due to the increasing dimensionality, the computation times of the analysis grow exponentially, mainly for the integration. The numerical results are well supported by the simulation and the Table 1 illustrates the trade-off between the computation times of both methods w.r.t. the number of dimensions.

---

[2] Tool available at: https://github.com/jannikhuels/hpnmg/tree/dynamicTransitions.

Table 1. Probability of an available grid, i.e. $\Phi := m(P_0^D) = 1$, for $t_c = 8$ and decreasing grid repair times T_0^D.

T_0^D		8	7	5	3	2	1
NUM	$p(t_c, \Phi)$	0.200	0.295	0.454	0.593	0.689	0.817
	$e(t_c, \Phi)$	0	$5.5 * e^{-6}$	$5.8 * e^{-6}$	$1.2 * e^{-5}$	$1.0 * e^{-5}$	$1.9 * e^{-5}$
	dim	4	4	5	6	8	13
	comp. t	1.67 s	17.74 s	38.92 s	90.22 s	250.15 s	2145.66 s
SIM	$p(t_c, \Phi)$	0.200	0.296	0.455	0.594	0.690	0.818
	comp. t.(CI: ±0.001)	27.61 s	31.30 s	39.87 s	39.29 s	33.57 s	26.04 s
	comp. t.(CI: ±0.0001)	1477.39 s	1983.97 s	2599.05 s	2761.17 s	2630.43 s	1728.41 s

Table 2. Probability of a standard demand, i.e. $\Phi := m(P_{d_1}^D) = 1$, at time $t_c = 8$ for randomly distributed demand changing times with different distributions.

$T_{d_{10}}^G, T_{d_{12}}^G$		$\mathcal{U}(0,10)$	$\mathcal{U}(6,10)$	$\mathcal{N}(\mu = 8, \sigma = 1)$	$\mathcal{N}(7,1)$	$\mathcal{N}(7,2)$
NUM	$p(t_c, \Phi)$	0.040	0.241	0.249	0.023	0.070
	$e(t_c, \Phi)$	$1.3 * e^{-5}$	$8.5 * e^{-5}$	$5.5 * e^{-6}$	$9.8 * e^{-6}$	$2.4 * e^{-5}$
	dim	4	4	4	4	4
	comp. t	4.16 s	4.04 s	8.17 s	8.21 s	7.62 s
SIM	$p(t_c, \Phi)$	0.041	0.240	0.250	0.025	0.072
	comp. t	8.89 s	19.54 s	32.91 s	7.09 s	16.87 s

The number of dimensions influences the run time of the numerical analysis, whereas the width of the confidence interval significantly influences the run time of the simulation. A confidence interval of ±0.00002 would be comparable to the largest error made by the numerical analysis. This is however not feasible, since the simulation does not terminate.

Simulation results are presented for two confidence interval widths, i.e. ±0.001 and ±0.0001. With such intervals the statistical simulation needs (significantly) more time than the numerical analysis for four or five dimensions. For 13 dimensions, the computation time of numerical analysis and simulation with a confidence interval of ±0.0001 are of the same magnitude. However note that the integration error of the numerical analysis is still approx. five times smaller than the confidence level.

Table 2 summarizes the probability that the standard demand is enabled at time $t_c = 8$ for varying CDFs assigned to the competing general transitions $T_{d_{10}}^G$ and $T_{d_{12}}^G$. The results are well supported by simulation and in most cases the computation time for the numerical approach is considerably slower.

As a feasibility study, we analyzed the model for an increasing number of random variables. In this setup the transition T_0^G follows a folded normal distribution with mean $\mu = 14$ and variance $\sigma = 4$. Once the grid failed, it needs 22 time units to be repaired. $T_{d_{01}}^D$ and $T_{d_{21}}^D$ follow a uniform distribution $\mathcal{U}(0,2)$ and

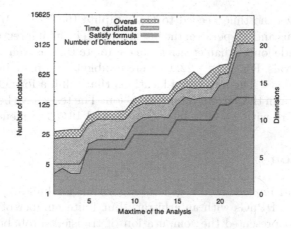

Fig. 2. Number of dimensions and locations created.

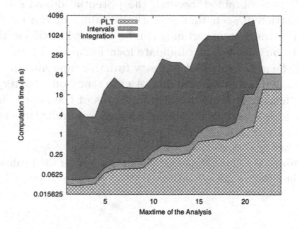

Fig. 3. Time to create the different location sets.

the firing times of both $T^D_{d_{10}}$ and $T^D_{d_{20}}$ equals 5. The checktime t_c always equals τ_{max}. Hence, a larger τ_{max} results in more random variables present in the system. We compute the probability that costs have occurred, i.e. $\Phi := x_{I(P^C_1)} \geq 0$.

Figure 2 shows the number of dimensions (right y-axis) for different values of τ_{max}. Values of $\tau_{max} \leq 5$ results in 4 dimensions, as all three general transitions can fire only once and the time also adds a dimension. In general, two dimensions are added every 5 time units, since the two general transitions that change the demand each can fire once more. At $\tau_{max} = 22$ the number of dimensions increases to 13, because then the grid can fail two times. On the other hand the left y-axis in Fig. 2 shows the total number of locations and the number in subsets thereof. Analogous to the dimensions, the number of locations also increases periodically. At $\tau_{max} \geq 22$ the number of locations rises drastically.

Figure 3 shows the time needed to (1) compute the PLT, (2) compute the restricted domains and (3) perform the integration. The final integration takes an order of magnitude longer than creating the complete PLT as well as creating the integration intervals. For $\tau_{max} \geq 22$, we were unable to complete integration due to memory overflows. The number of locations that fulfill a formula is therefore the limiting factor in the current implementation. The tests have been performed on a MacBook Pro with 2.5 GHz Inter Core i7 and 16 GB of memory.

7   Conclusion

We proposed and implemented a general algorithm for building a Parametric Location Tree for HPnGs with an arbitrary but finite number of general transition firings and presented the computation of transient probabilities in three stages. First the candidate locations, i.e. parametric locations the system can be in at time t', were obtained. Second, the potential domain of all candidate locations was restricted, such that only those values of the random variables remained for which the system certainly can be in that location at time t'. Third, the probability to be in a specific candidate location at time t' was computed by integrating over the joint probability density function, and multiplying the result with the accumulated conflict probability. A case study on a battery-backup system showed the feasibility and current limitations of the approach. We plan to conduct a large-scale case study to allow a more comprehensive analysis of the efficiency of the current implementation.

Acknowledgement. We thank the anonymous reviewers for their valuable comments. Especially regarding the computation times in the original version of Table 1.

References

1. Alla, H., David, R.: Continuous and hybrid Petri nets. J. Circ. Syst. Comput. **8**(01), 159–188 (1998)
2. Amare, T., Helvik, B., Heegaard, P.: A modeling approach for dependability analysis of smart distribution grids. In: 21st Conference on Innovation in Clouds, Internet and Networks and Workshops (ICIN), pp. 1–8. IEEE (2018)
3. Collins, G.E., Hong, H.: Partial cylindrical algebraic decomposition for quantifier elimination. J. Symbolic Comput. **12**(3), 299–328 (1991)
4. Dantzig, G.B.: Fourier-Motzkin elimination and its dual. The Basic George B. Dantzig, p. 255 (2003)
5. Everdij, M., Blom, H.: Piecewise deterministic Markov processes represented by dynamically coloured petri nets. Stochastics **77**(1), 1–29 (2005)
6. Galassi, M., et al.: GNU scientific library. Release 2.5. Network Theory Ltd, June 2018
7. Ghasemieh, H., Remke, A., Haverkort, B.: Survivability analysis of a sewage treatment facility using hybrid Petri nets. In: Performance Evaluation, vol. 97, pp. 36–56. Elsevier (2016)

8. Ghasemieh, H., Remke, A., Haverkort, B., Gribaudo, M.: Region-based analysis of hybrid petri nets with a single general one-shot transition. In: Jurdziński, M., Ničković, D. (eds.) FORMATS 2012. LNCS, vol. 7595, pp. 139–154. Springer, Heidelberg (2012). https://doi.org/10.1007/978-3-642-33365-1_11
9. Gribaudo, M., Remke, A.: Hybrid petri nets with general one-shot transitions. Perform. Eval. **105**, 22–50 (2016)
10. Gribaudo, M., Sereno, M., Horváth, A., Bobbio, A.: Fluid stochastic petri nets augmented with flush-out arcs: modelling and analysis. Discrete Event Dyn. Syst. **11**(1–2), 97–117 (2001)
11. Horton, G., Kulkarni, V., Nicol, D., Trivedi, K.: Fluid stochastic petri nets: theory, applications, and solution techniques. J. Oper. Res. **105**(1), 184–201 (1998)
12. Horton, G., Kulkarni, V.G., Nicol, D.M., Trivedi, K.S.: Fluid stochastic petri nets: theory, applications, and solution techniques. Eur. J. Oper. Res. **105**(1), 184–201 (1998)
13. Hüls, J., Remke, A.: Model checking HPnGs in multiple dimensions: representing state sets as convex polytopes. In: Pérez, J.A., Yoshida, N. (eds.) FORTE 2019. LNCS, vol. 11535, pp. 148–166. Springer, Cham (2019). https://doi.org/10.1007/978-3-030-21759-4_9. https://uni-muenster.sciebo.dc/s/xCJ8oGhrltYrKp4
14. Hüls, J., Schupp, S., Remke, A., Ábrahám, E.: Analyzing hybrid petri nets with multiple stochastic firings using HyPro. In: 11th EAI International Conference on Performance Evaluation Methodologies and Tools, VALUETOOLS 2017, pp. 178–185 (2017)
15. Jongerden, M.R., Hüls, J., Remke, A., Haverkort, B.R.: Does your domestic photovoltaic energy system survive grid outages? Energies **9**(9), 736 (2016)
16. Kwiatkowska, M., Norman, G., Segala, R., Sproston, J.: Automatic verification of real-time systems with discrete probability distributions. Theoret. Comput. Sci. **282**(1), 101–150 (2002)
17. Lepage, G.P.: A new algorithm for adaptive multidimensional integration. J. Comput. Phys. **27**(2), 192–203 (1978)
18. McNamee, J., Stenger, F.: Construction of fully symmetric numerical integration formulas. Numer. Math. **10**(4), 327–344 (1967)
19. Pilch, C., Remke, A.: Statistical model checking for hybrid petri nets with multiple general transitions. In: 47th IEEE/IFIP International Conference on Dependable Systems and Networks, DSN 2017, pp. 475–486. IEEE (2017)
20. Press, W., Teukolsky, S., Vetterling, W., Flannery, B.: Numerical Recipes in C: The Art of Scientific Computing, 2nd edn. Cambridge University Press, Cambridge (1992)
21. Sproston, J.: Decidable model checking of probabilistic hybrid automata. In: Joseph, M. (ed.) FTRTFT 2000. LNCS, vol. 1926, pp. 31–45. Springer, Heidelberg (2000). https://doi.org/10.1007/3-540-45352-0_5
22. Teige, T., Fränzle, M.: Constraint-based analysis of probabilistic hybrid systems. IFAC Proc. Volumes **42**(17), 162–167 (2009)
23. Yin, L., Fricks, R., Trivedi, K.: Application of semi-markov process and CTMC to evaluation of ups system availability. In: Annual Reliability and Maintainability Symposium, pp. 584–591. IEEE (2002)
24. Zhang, L., She, Z., Ratschan, S., Hermanns, H., Hahn, E.M.: Safety verification for probabilistic hybrid systems. Eur. J. Control **18**(6), 572–587 (2012)

Security

Expected Cost Analysis
of Attack-Defense Trees

Julia Eisentraut[(⊠)] and Jan Křetínský

Technical University of Munich, Munich, Germany
`julia.kraemer@in.tum.de`

Abstract. Attack-defense trees (ADT) are an established formalism for assessing system security. We extend ADT with costs and success probabilities of basic events. We design a framework to analyze the probability of a successful attack/defense, its expected cost, and its probability for a given maximum cost. On the conceptual level, we show that a proper analysis requires to model the problem using sequential decision making and non-tree structures, in contrast to classical ADT analysis. On the technical level, we provide three algorithms: (i) reduction to PRISM-games, (ii) dedicated game solution utilizing the structure of the problem, and (iii) direct analysis of ADT for certain settings. We demonstrate the framework and compare the solutions on several examples.

1 Introduction

Attack trees and their extension *attack-defense trees* (ADT) are established formalisms for security assessment [15–17,27,28]. Essentially, attack trees are labelled trees, where the root represents the goal an attacker wants to reach. This goal is refined into subgoals with the help of logic operations such as AND and OR. The leaves of the tree are called *basic events* and the inner nodes *gates*. On the one hand, they allow for formal reasoning and analysis of the system design. On the other hand, the formalism is at the same time appropriate for human interaction since ADT are easy to understand also for non-computer scientists. Together, this facilitates the model-driven development of security-critical systems.

Cost and Probability. To reflect the practical (in)feasibility of attacks one can decorate the basic events with *costs* [3,4,11,19], reflecting, for instance, time durations. This enables us to distinguish in the analysis between purely theoretical risks, such as brute-forcing a password, and more realistic ones. Further, modelling the *success probabilities* of attacks provides a complementary way to identify more likely scenarios. Accordingly, the analysis algorithms for models supporting both costs and success probabilities become more involved [3–5].

This research was funded in part by the Studienstiftung des deutschen Volkes project "Formal methods for analysis of attack-defence diagrams" and the German Research Foundation (DFG) project KR 4890/2-1 "Statistical Unbounded Verification".

© Springer Nature Switzerland AG 2019
D. Parker and V. Wolf (Eds.): QEST 2019, LNCS 11785, pp. 203–221, 2019.
https://doi.org/10.1007/978-3-030-30281-8_12

The OWASP CISO AppSec Guide[1] states that attack trees are a useful measure to identify the point of weakest resistance and explicitly mentions the importance of costs and probabilities there. NATO's *Improving Common Security Risk Analysis* report[2] says that attack trees might help to understand attacks in retrospective, i.e., by so-called *red teams*. Fortunately, in contrast to other model checking domains, probabilities and costs can typically be estimated using numerous approaches to retrospective system analysis, as done in [9]. For instance, it is known that in 2014 78% of all phishing attacks were spoofed to look like they were sent from a company's IT department or AV vendors[3].

Our goal in this paper is to provide a framework for efficient analysis of *ADT* in the presence of both sources of information: costs and success probabilities. In particular, we algorithmically answer the following questions:

– What is a game-theoretically optimal strategy for a successful attack?
– What is a game-theoretically optimal strategy for a successful attack for a given cost budget?
– What is the minimum expected cost among all optimally succeeding attacks?

This in turn allows us to answer practical questions such as

– Who can afford to try to attack the system (with high success probability)?
– How secure can a system become with a limited amount of resources and how to efficiently attack/defend it?
– What are the bottlenecks in the system where defense should be strengthened?

Conceptual Contribution. Due to the presence of probabilities, an analysis should take into account several fundamental issues, which are also the distinguishing points compared to the existing literature. We briefly mention these points here. Detailed justification of our approach follows in Section 3 and the comparison to the literature in Sect. 9.

– We shall argue that decision making should be modelled in a *sequential way*.
– Consequently, instead of classical Boolean valuation of successful/failed events a *3-valued* logic captures that some events have not been attempted (yet).
– Due to the sequential nature of the problem, the classical bottom-up techniques do not work any more since the problem is no more modular and cannot be solved for the subtrees in isolation.
– Besides, if an event is relevant for different subtrees, its occurrences cannot be treated in the analysis as probabilistically independent and we have to transform the tree into a *directed acyclic graph* instead.
– Once the sequential reasoning is necessarily present anyway, we can use a *richer class of gates* that require ordering of events, e.g., SAND, at no additional cost.

---

[1] https://www.owasp.org/index.php/CISO_AppSec_Guide:_Criteria_for_Managing_Application_Security_Risks.
[2] https://www.sto.nato.int/publications/STO%20Technical%20Reports/RTO-TR-IST-049/%5Cprotect%20%5CT1%5Ctextdollar%20%5Cprotect%20%5CT1%5Ctextdollar%20TR-IST-049-ALL.pdf.
[3] https://www.fireeye.com/current-threats/annual-threat-report.html.

Algorithmic Contribution. We provide the respective semantics in terms of stochastic two-player games with reachability, cost-bounded reachability, and expected cost objectives. Subsequently, we present three algorithms to solve the game:

1. We translate the ADT into a game in the PRISM modelling language and evaluate it using PRISM-games [13].
2. We utilize the acyclic structure of the model and apply a simpler back-propagation algorithm for stochastic games.
3. For the case where events occur once only in the tree, we provide an algorithm for computing success probabilities directly on the tree in a bottom-up manner. Although this is in line with the tradition of ADT analysis, additional operators (such as TR and NAT) and the inherent sequential character of decision making require a more complex solution.

In all the cases, we also compute the respective attack and defense strategies optimizing the given criteria.

Structure of the Paper. In Sect. 2, we recall and illustrate ADT and in Sect. 3 we justify our approach, explaining the consequences on examples. Subsequently, in Sect. 4 we recall stochastic games and in Sect. 5 we provide the sequential formal semantics of ADT by translation to stochastic games. In Sect. 6, we discuss the analysis of the stochastic games resulting from ADT. Section 7 is devoted to a bottom-up computation of success probabilities directly on the tree. We discuss experimental results in Sect. 8. We review and compare to related work in Sect. 9 and conclude the paper in Sect. 10.

2   Attack-Defense Trees

In this section, we recall and illustrate the notion of attack-defense trees (ADT).[4]

Definition 1 (Syntax of ADT based on [18]). *An* attack-defense tree (ADT) *is a tuple* ADT = (V, E, t, TEdge) *where*

- (V, E) *is a directed acyclic graph with a designated* goal *sink vertex* att, *also called the* root *of the* ADT. *Source vertices* BE ⊆ V *are called* basic events. *Edges are directed from basic events towards the single root* att. *We assume* BE = BE$_A$ $\dot\cup$ BE$_D$, *where* BE$_A$ *and* BE$_D$ *are the events under the attacker's and defender's control, respectively. All other vertices* CE := V \ BE *are* gates *(or* composite events*). Direct predecessors of gates are called* inputs *and we denote the set of all inputs of a vertex* v *by* in(v).
- t: CE → O *is the* type function *assigning to each gate one of the operators in* O = {AND, OR, NOT, SAND, SOR, NAT, TR}.
- TEdge ⊆ {v ∈ CE | t(v) = TR} × BE *are* trigger *edges from* TR *gates to basic events. Note that* TEdge *is not a subset of* E.
- *We require* |in(v)| = 1 *for all vertices* v ∈ V *such that* t(v) ∈ {NAT, NOT, TR}.

---

[4] Since a step in the analysis transforms the trees into DAGs, we already introduce ADT more generally as a DAG, not necessarily a tree.

We let $\mathsf{BE_{TR}} := \{b \in \mathsf{BE} \mid \exists v \in \mathsf{V} : (v, b) \in \mathsf{TEdge}\}$ *denote the set of* triggerable *basic events. If* $(v, b) \in \mathsf{TEdge}$*, we say that* v *triggers* b*. We call an* ADT *attack tree* iff (V, E) *forms a tree rooted in* att *and* $\mathsf{BE_D} = \emptyset$.

The operators AND and OR have the usual logic meaning w.r.t. subgoals. Intuitively, SAND behaves like AND, but the subgoals need to be completed in order, while SOR behaves like OR, but subgoals need to fail in order. Formally, SAND requires not only both subgoals to be successfully completed, but the first one needs to be finished before the second one. In contrast, SOR requires only one subgoal to be successfully completed, but the second one may only be completed successfully if the first one is completed unsuccessfully. The operator TR enables basic events on being successfully completed (for instance, in Fig. 1 we can only get a correct password after we have installed the keylogger successfully). NOT allows us to express that certain events need to be completed unsuccessfully, while NAT turns out true if the subgoal is not attempted at all. We treat each subgoal this way till we reach the basic events.

Example 1. In Fig. 1, an attack-defense tree representing an attack on a company is depicted. The attacker tries to harm the company (7) by sending out phishing mails from a valuable employee's account (11). To do so, the attacker tries to either get physical access to the servers (4) or to install a keylogger to get the correct password (2 and 3). The defender can prevent an attack by blocking infected accounts (6). The attack turns out to be successful as long as the defender either failed in successfully completing the blocking or does not perform the blocking at all (13 and 8–10).

The set of events comprises the basic events 1 to 6 and the composed events labelled with operators. The goal sink at the top is labelled with AND and represents the ultimate goal of the attacker to harm the company by sending phishing mails from an employee's account. The AND reflects the idea that harm can only be done if the account has not been blocked by the defender. The basic events, which are controlled by the attacker, are red and horizontally striped, while the basic events under the defender's control are green and vertically striped. Trigger transitions are depicted as squiggled lines.

Basic events can be equipped, for instance, with probabilities, costs, average completion times etc. However, many scalars such as time can often be modelled as costs. Therefore, we only deal with *ADT*s, in which basic events are equipped with probabilities and costs. Hence, we assume that there exists a probability function $\mathsf{Pr} : \mathsf{BE} \to [0, 1]$ assigning to each basic event a success probability between 0 and 1 and a cost function $\mathsf{Cost} : \mathsf{BE} \to \mathbb{N}$ assigning each basic event a non-negative execution cost. In Fig. 1, we give the values of the cost and probability functions for basic events in the table next to the tree.

3   Examples and Modelling Approach

In this section, we explain our way of modelling in the presence of probabilities and costs, and argue that it is an adequate approach in this context.

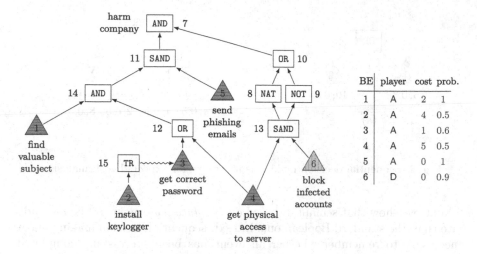

Fig. 1. An *ADT* representation of an attack on some email account

Firstly, in order to faithfully model and analyze expected costs, we stipulate that decisions are made *sequentially*, i.e. players do not fix upfront which events to attempt, but react to the outcome of events. This is in stark contrast with the traditional approach, which considers static valuations (subsets) of basic events that are executed.

Example 2. Consider the tree of Fig. 2a. In the classical static (non-sequential) analysis, there are three options to successfully attack. These correspond to the three valuations $\{1\}, \{2\}, \{1, 2\}$ that satisfy the OR gate, with the respective costs 5, 10, and $5 + 10 = 15$. In the probabilistic context, the maximum probability to succeed is $1 - (1 - 0.3)(1 - 0.8) = 0.86$ by attempting both events. Nevertheless, the cost may be reduced from 15 to $5 + (1 - 0.3) \cdot 10 = 12$ if we attempt event 1 first and only if it fails we attempt event 2. Similarly, the reverse order yields a cost of $10 + (1 - 0.8) \cdot 5 = 11$, which is the minimum. Similarly, in Example 1, it is better to first try events *(2), (3)* and before attempting event *(1)*, since event *(1)* cannot fail. In summary, the order of attempts affects the cost and should be reflected in the semantics of the model.

On the negative side, the sequential modelling implies increased difficulty in the analysis.

Example 3. Consider the tree in Fig. 2b. It is cheapest to first try 2 and only then 3, which is in a *different* subtree. Hence the optimal expected cost $1/4 \cdot 4 + 1 \cdot 1 + 1/2 \cdot 2 + 1/8 \cdot 8 = 4$ cannot be easily computed by a bottom-up analysis that summarizes each subtree (left and right one) into a single number, but a more global view is necessary.

Such *non-modularity* implies that the standard bottom-up analysis fails for expected cost analysis as indicated in [19, 23].

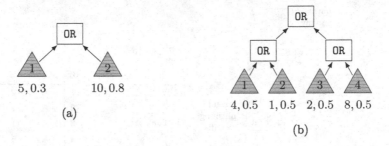

Fig. 2. Sequential decision making is necessary for expected cost analysis

Next, we show that semantics based on *3-valued logic* as in [18] is more adequate than the standard Boolean one. Indeed, sequential decision making makes it necessary to remember whether an event has been successfully completed, unsuccessfully completed or not attempted at all (yet).

Example 4. In Example 1, gate 10 checks whether the SAND gate 13 failed (by NOT gate 9) or has not even been attempted (by NAT gate 8) since the attack should succeed in both cases.

Now we show that *directed acyclic graphs* are more appropriate than trees in the analysis.

Example 5. Consider the event *(4) get physical access to server* in Example 1 and a model where this event is replicated for each successor so that it is a tree instead of a DAG. Since the two copies are not independent events, we only have one chance to attempt it and it either succeeds for both occurrences or none. Further, the incurred cost is only paid once. Hence, such a tree should be pre-processed into the DAG depicted in Fig. 1.

Finally, the presence of the *defense* results in an inherently more complex problem. While the success probability for reasonable strategies (not failing the attack by attempting SAND and SOR in the wrong order) in attack trees does *not* depend on the chosen strategy, the success probability in attack-defense trees depends on the defender's strategy.

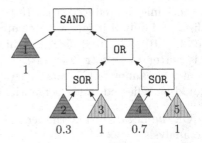

Fig. 3. The attacker's success probability depends on the defender's strategy.

Example 6. Consider the tree in Fig. 3. In any reasonable strategy, the attacker needs to attempt event *(1)* first. If the defender plays *(3)*, the success probability is 0.7 since the attacker wins if event *(4)* is successfully completed. In contrast, if the defender plays *(5)*, the attacker wins with success probability 0.3 since he wins if event *(2)* is successfully completed.

4   Stochastic Games

In this section, we recall (two-player turn-based) simple stochastic games [14] and equip them with costs. For a countable set X, $\mathcal{D}(X)$ is the set of all distributions over X, i.e. functions $\mu: X \to [0,1]$ with $\sum_{x \in X} \mu(x) = 1$. We also write $\mu = \bigoplus_{x \in \mathsf{Supp}(\mu)} \mu(x) \cdot x$ where $\mathsf{Supp}(\mu) = \{x \mid \mu(x) > 0\}$ is μ's *support*.

Definition 2 (Stochastic Game). *A* stochastic game (SG) *is a tuple* $\mathsf{G} = (\mathsf{S}, \mathsf{S_A}, \mathsf{S_D}, \mathsf{s_0}, \mathsf{M}, \mathsf{ava}, \mathsf{T}, \mathtt{1}, -\mathtt{1})$, *where* S *is a finite set of* states *partitioned into the sets* $\mathsf{S_A}$ *and* $\mathsf{S_D}$ *of states of the players* A *and* D, *respectively,* $\mathsf{s_0} \in \mathsf{S}$ *is the initial state,* M *is a finite set of* moves, *and* $\mathsf{ava}: \mathsf{S} \to 2^{\mathsf{M}}$ *assigns to every state a set of* available moves, *and* $\mathsf{T}: \mathsf{S} \times \mathsf{M} \to \mu(\mathsf{S})$ *is a* transition function *that given a state* s *and a move* $\mathsf{m} \in \mathsf{ava}(\mathsf{s})$ *yields a probability distribution over successor states,* $\mathtt{1}$ *and* $-\mathtt{1}$ *are the only sink states (with no available moves) winning for player* A *and* D, *respectively.*

The semantics is standard [14] and we only briefly recall some terminology. We denote transitions $(\mathsf{s}, \mathsf{m}, \mu) \in \mathsf{T}$ also by $\mathsf{s} \xrightarrow{\mathsf{m}} \mu$. By taking their turns, players generate an *execution*, i.e. an alternating sequence of states and moves $\rho = \mathsf{s_0} \dots \mathsf{m}_n \mathsf{s}_n$ with $\mathsf{s}_i \xrightarrow{\mathsf{m}_{i+1}} \mu$ and $\mathsf{s}_{i+1} \in \mathsf{Supp}(\mu)$ for every $0 \leq i < n$. We denote by $\mathsf{Exec}^{\mathsf{max}}(\mathsf{G})$ the set of all *maximum executions* of G, i.e. those ending[5] in $\mathtt{1}$ or $-\mathtt{1}$.

A *strategy* is a function $\sigma: \mathsf{S} \to \mathsf{M}$ assigning to each state one of its available moves[6]. A pair of strategies (σ, τ) induces a Markov chain $\mathsf{G}(\sigma, \tau)$ over $\mathsf{Exec}^{\mathsf{max}}(\mathsf{G})$ with the probability function $\mathbb{P}^{\sigma, \tau}$ using the standard cone construction [7, Ch. 10] where non-determinism in $\mathsf{S_A}$ and $\mathsf{S_D}$ is resolved using σ and τ, respectively.

Costs. To model costs of executing events, we add a cost function $\mathsf{Cost}: \mathsf{M} \to \mathbb{N}$ and lift it to executions by $\mathsf{Cost}(\mathsf{s_0}\mathsf{m}_1 \cdots \mathsf{m}_n\mathsf{s}_n) := \sum_{1 \leq i \leq n} \mathsf{Cost}(\mathsf{m}_i)$. Given a pair of strategies (σ, τ), restricting Cost to $\mathsf{Exec}^{\mathsf{max}}(\mathsf{G})$ results in a random variable C with the expected value $\mathbb{E}^{\sigma, \tau}[C]$.

5   Game Semantics of Attack-Defense Trees

In this section, we provide (i) formal semantics of attack-defense trees with shared subtrees in terms of stochastic games and (ii) game objectives corresponding to our probability-cost objectives on the attack-defense tree. Let $\mathsf{ADT} = (\mathsf{V}, \mathsf{E}, \mathsf{t}, \mathsf{TEdge})$ be an attack-defence tree with a probability function $\mathsf{Pr}: \mathsf{BE} \to [0,1]$ and a cost function $\mathsf{Cost}: \mathsf{BE} \to \mathbb{N}$.

---

[5] To simplify the presentation, we do not consider infinite executions since the games we deal with in this paper are finite and acyclic. Nevertheless, the theory would seamlessly extend to games with cycles and infinite executions if the need of such gates, e.g. [18], arises.

[6] In general, one can consider randomizing history-dependent strategies. However, in the context of our paper, positional strategies are sufficient even for cost-bounded objectives since the costs will be implicitly encoded in the states of the games.

Valuations. We capture the current status of events using valuations, which can be changed by moves of the players. Formally, we define *valuations* as functions $\upsilon \colon V \rightarrow \{1, 0, -1\}$ and denote the (finite) set of all valuations by Val. Initially, we assign to each basic event the value 0 (*not attempted yet*). While a basic event b has not been attempted, its value remains 0. After attempting, its value must be either 1 (*successfully completed*) with probability $\mathsf{Pr}(\mathsf{b})$ or -1 (*unsuccessfully completed*) with probability $1 - \mathsf{Pr}(\mathsf{b})$. When we restrict a valuation υ to an input set X, we denote it by $\upsilon|_X$.

In the following, we denote the single element of $\mathsf{in}(\mathsf{v})$ for a vertex $\mathsf{v} \in V$ with $\mathsf{t}(\mathsf{v}) \in \{\mathsf{NOT}, \mathsf{NAT}, \mathsf{TR}\}$ by in_v. To compute the new valuation after an attempt, we only need the previous valuation and the outcome of the attempt. Formally, let υ_p be the previous valuation and υ_{BE} be the new valuation of the basic events. The new valuation υ_n is inductively defined as

$$
\upsilon_\mathsf{n}(\mathsf{v}) = \begin{cases}
\upsilon_{\mathsf{BE}}(\mathsf{v}) & \text{if } \mathsf{v} \in \mathsf{BE} \\
1 & \text{if } \mathsf{t}(\mathsf{v}) = \mathsf{NAT} \text{ and } \upsilon_\mathsf{p}(\mathsf{in}_\mathsf{v}) = 0 \\
-1 & \text{if } \mathsf{t}(\mathsf{v}) = \mathsf{NAT} \text{ and } \upsilon_\mathsf{p}(\mathsf{in}_\mathsf{v}) = 1 \\
-1 & \text{if } \mathsf{t}(\mathsf{v}) = \mathsf{NAT} \text{ and } \upsilon_\mathsf{p}(\mathsf{in}_\mathsf{v}) = -1 \\
\upsilon_\mathsf{n}(\mathsf{in}_\mathsf{v}) & \text{if } \mathsf{t}(\mathsf{v}) = \mathsf{TR} \\
\mathsf{op}(\mathsf{in}(\mathsf{v})) & \text{if } \mathsf{op} = \mathsf{t}(\mathsf{v}) \in \{\mathsf{AND}, \mathsf{OR}, \mathsf{SAND}, \mathsf{SOR}, \mathsf{NOT}\}
\end{cases}
$$

Here op denotes the application of the respective three-valued operator: standard AND, OR and NOT or SAND and SOR according to Fig. 4. We denote the new valuation υ_n by $\mathsf{app}(\upsilon_\mathsf{p}, \upsilon_{\mathsf{BE}})$.

$\upsilon_\mathsf{n}(\mathsf{v_1})$	$\upsilon_\mathsf{n}(\mathsf{v_2})$	$\upsilon_\mathsf{p}(\mathsf{v})$	$\upsilon_\mathsf{n}(\mathsf{v})$	
–	–		1	1
–	–		-1	-1
–	-1		0	-1
-1	–		0	-1
0	1		0	-1
0	0		0	0
1	0		0	0
1	1		0	1 if $\upsilon_\mathsf{p}(\mathsf{v_1}) = 1$
				-1 otherwise

$\upsilon_\mathsf{n}(\mathsf{v_1})$	$\upsilon_\mathsf{n}(\mathsf{v_2})$	$\upsilon_\mathsf{p}(\mathsf{v})$	$\upsilon_\mathsf{n}(\mathsf{v})$	
–	–	1	1	
–	–	-1	-1	
-1	1	0	1	$\upsilon_\mathsf{p}(\mathsf{v_1}) = -1$
			-1	otherwise
-1	0	0	0	
0	0	0	0	
0	1	0	-1	
1	1	0	-1	
1	0	0	1	

Fig. 4. Definition of the semantics for vertex v with $\mathsf{t}(\mathsf{v}) = \mathsf{SAND}$ on the left and $\mathsf{t}(\mathsf{v}) = \mathsf{SOR}$ on the right. We denote the inputs with $\mathsf{v_1}$ and $\mathsf{v_2}$.

Available basic events. Initially, the set of available basic events is given by $\mathsf{BE} \setminus \mathsf{BE}_{\mathsf{TR}}$, i.e., all non-triggerable basic events are available. Whenever a composed event v labelled with TR is successfully completed, the basic events in $\{b \mid \exists(v,b) \in \mathsf{TEdge} \wedge \upsilon(b) = 0\}$ become available. While the first condition ensures that only basic events, which this event triggers, become available, the second one ensures that no basic event is available although it has been attempted already.

Finally, we can define the stochastic game induced by an ADT as follows. We limit the number of moves the attacker can attempt simultaneously by a and similarly for the defender by d.

Definition 3 (Induced stochastic game). *Let* $\mathsf{ADT} = (V, E, t, \mathsf{TEdge})$ *be an attack-defense tree. The* induced (a,d)-*stochastic game is defined as* $G_{\mathsf{a},\mathsf{d}}(\mathsf{ADT}) = (S, S_A, S_D, s_0, M, \mathsf{ava}, T, \mathbf{1}, -\mathbf{1})$

- $S = \mathsf{Val} \times \{A, D\}$, $S_A = \mathsf{Val} \times \{A\}$, $S_D = \mathsf{Val} \times \{D\}$
- $s_0 = (\mathsf{app}(\overline{0}, \overline{0}|_{\mathsf{BE}}), A)$, *where* $\overline{0}(v) = 0$ *for every* $v \in V$ *(note that the operator* NAT *might affect the initial valuation)*
- $M = 2^{\mathsf{BE}}$,
- $\mathsf{ava}((\upsilon, \mathsf{p})) = \mathsf{BE}_\mathsf{p} \setminus \{b \mid \upsilon(b) \neq 0 \vee (\forall(v,b) \in \mathsf{TEdge} : \upsilon(v) \neq 1)\}$ *for* $\mathsf{p} \in \{A, D\}$ *are the available basic events of the current player,*
- T *consists of exactly the transitions* $(\upsilon, \mathsf{p}) \xrightarrow{m} \mu$ *such that:*
  - $m \subseteq \mathsf{ava}((\upsilon, \mathsf{p}))$.
  - *If* $s \in S_A$, *then* $0 < |m| \leq \mathsf{a}$. *If* $s \in S_D$, *then* $|m| \leq \mathsf{d}$.
  - *Let*

$$V_{\mathsf{BE}}^{\mathsf{new}} := \{\upsilon_{\mathsf{BE}}^* \mid \upsilon_{\mathsf{BE}}^*|_{\mathsf{BE}\setminus m} = \upsilon|_{\mathsf{BE}\setminus m}, \forall b \in m : \upsilon_{\mathsf{BE}}^*(b) \neq 0\}$$

  be the set of all possible new valuations of basic events following valuation υ *and player* p *attempting the basic events in* m *and let*

$$\mathsf{Val}_\mathsf{p} := \{\upsilon^* \mid \upsilon^* = \mathsf{app}(\upsilon_\mathsf{p}, \upsilon_{\mathsf{BE}}), \upsilon_{\mathsf{BE}} \in V_{\mathsf{BE}}^{\mathsf{new}}\}$$

  be the set of all possible new valuations (of all events). The occurrence probability $\mathbb{P}(\upsilon^*)$ *of a valuation* $\upsilon^* \in \mathsf{Val}_\mathsf{p}$ *is given by*

$$\prod_{b \in m} \begin{cases} \mathsf{Pr}(b) & \text{if } \upsilon^*(b) = 1 \\ 1 - \mathsf{Pr}(b) & \text{if } \upsilon^*(b) = -1 \end{cases}.$$

  So finally the distribution μ *must satisfy* $\mu = \bigoplus_{\upsilon^* \in \mathsf{Val}_\mathsf{p}} \mathbb{P}(\upsilon^*) \cdot (\upsilon^*, \mathsf{p}')$, *where* $\mathsf{p}' = A$ *if* $\mathsf{p} = D$ *and vice versa.*
- $\mathbf{1}$ *results from grouping all winning states for the attacker, i.e.* (υ, p) *with* $\upsilon(\mathsf{att}) = 1$,
- $-\mathbf{1}$ *results from grouping all states where the attack definitely failed or no more basic attacks are possible, i.e. from states* (υ, p) *with* $\upsilon(\mathsf{att}) = -1$ *or* $\mathsf{p} = A, \mathsf{ava}((\upsilon, A)) = \emptyset$.

If we omit a, d, we assume the number of basic events each player can attempt simultaneously to be unbounded.

Example 7. A part of the game for our Example 1 is depicted in Fig. 5.

We lift the cost function from attack-defense trees to the induced stochastic games by defining $\mathsf{Cost}(m) := \sum_{b \in m} \mathsf{Cost}(b)$.

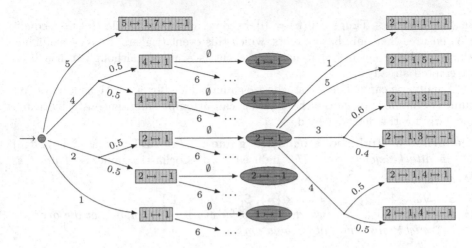

Fig. 5. A part of the induced stochastic game for the ADD in Fig. 1 with $a = 1 = d$. Note that initially $8 \mapsto 1$ and $10 \mapsto 1$. We omit outgoing transitions of all other attacker states in level 3 and the outgoing transitions of states in level 4. We also omit values of composed events since they can be computed using the inductive definition of υ_n.

Winning Objectives. Firstly, the optimal *probability of a succesful attack* is the value of the game $\sup_{\sigma \in \Sigma_A} \inf_{\tau \in \Sigma_D} \mathbb{P}(\Diamond 1)$ where $\Diamond 1$ is the set of all executions ending in 1. We denote the set of strategies realizing the supremum by $\Sigma_A^{\max}$. Secondly, we do *not* want to compute the (minimal) expected cost $\inf_{\sigma \in \Sigma_A} \sup_{\tau \in \Sigma_D} \mathbb{E}[C]$ as the respective success probability might even be 0, but rather the *minimal expected cost among all strategies guaranteeing the optimal success probability:* $\inf_{\sigma \in \Sigma_A^{\max}} \sup_{\tau \in \Sigma_D} \mathbb{E}[C]$. Thirdly, we are also interested in the *cost-bounded success probability*, counting only successful attacks with cost less than b, i.e. $\sup_{\sigma \in \Sigma_A} \inf_{\tau \in \Sigma_D} \mathbb{P}(\Diamond 1 \cap C < b)$. In addition, we want to compute optimal strategiesx

$$\arg\min_{\sigma \in \Sigma_A^{\max}} \sup_{\tau \in \Sigma_D} \mathbb{E}[C] \quad \text{and} \quad \arg\max_{\sigma \in \Sigma_A} \inf_{\tau \in \Sigma_D} \mathbb{P}(\Diamond 1 \cap C < b).$$

6   Probability–Cost Analysis of Acyclic Stochastic Games

Stochastic games induced by attack-defense tree are acyclic since once a basic event is attempted, it can never change its value back to 0. Hence, whenever the attacker or the defender executes a move, a new state is reached. However, the game is not tree-like since the order of some moves does not play a role. For instance, in Fig. 2b, for the resulting state it does not matter whether first event *(1)* and then *(2)* fail or the other way round.

Consequently, we can compute the maximal reachability probability and expected cost as well as the respective strategies efficiently with one iteration over the whole state space from sinks to the initial state, more precisely in increasing maximum distance to the goal vertices (so-called back-propagation, see e.g. [12]).

6.1   Success Probability

We use Bellman equations (e.g. [12]) to compute the reachability probability $V(s_0)$. The value of each state is only changed once if we update the vertices in the order of increasing maximum distance to the goal vertices

$$V(s) := \begin{cases} \max_{m \in \text{ava}(s)} V(s,m) & \text{if } s \in S_A \\ \min_{m \in \text{ava}(s)} V(s,m) & \text{if } s \in S_D \\ 1 & \text{if } s \in 1 \\ 0 & \text{if } s \in -1 \end{cases} \qquad V(s,m) := \sum_{s' \in S} T(s,m,s') \cdot V(s')$$

Restricting the moves of A in s to the set $\arg\max_{m \in \text{ava}(s)} V(s,m)$ yields a subgame where every strategy of player A is optimal, i.e. in $\Sigma_A^{\max}$, since the reachability game on an acyclic finite game is actually a safety game [7].

6.2   Expected Cost

In a similar manner, we compute expected costs of attempting an attack. However, we minimize in attacker's and maximize in defender's states. Moreover, we add costs of transitions taken only in the attacker's case.

$$C(s) := \begin{cases} \min_{m \in \text{ava}(s)} C(s,m) & \text{if } s \in S_A \\ \max_{m \in \text{ava}(s)} C(s,m) & \text{if } s \in S_D \\ 0 & \text{if } s \in -1 \text{ or } s \in 1 \end{cases}$$

$$C(s,m) := \sum_{s' \in S} T(s,m,s') \cdot C(s') \qquad \text{for defender's states } s \in S_D$$

$$C(s,m) := \text{Cost}(m) + \sum_{s' \in S} T(s,m,s') \cdot C(s') \qquad \text{for attacker's states } s \in S_A$$

Restricting moves of the attacker in s to $\arg\min_{m \in \text{ava}(s)} C(s,m)$ is optimal for the expected costs $\inf_{\sigma \in \Sigma_A} \sup_{\tau \in \Sigma_D} \mathbb{E}[C]$. Moreover, if we first restrict the game to the subgame of the previous subsection and then apply this algorithm, the obtained strategy is optimal for $\inf_{\sigma \in \Sigma_A^{\max}} \sup_{\tau \in \Sigma_D} \mathbb{E}[C]$.

6.3   Cost-Bounded Success Probability

The attacker's cost bound can efficiently be taken into account while generating the game by merging states that exceed the bound with -1. Whenever a transition only leads to states in the set -1, we do not add the transition to the game. Whenever a state does not have any successor after removing all these transitions, it is merged with -1. Inductively, we thus remove all paths, which fail the cost bound.

Additionally, we can include a separate cost bound for the defender: we need to allow only the moves m in defender's states such that $C + \mathsf{Cost}(m)$ does not exceed the cost bound, where C denotes the accumulated defender cost so far. Since we assign cost 0 to the move $\emptyset$, there is always at least one available move in any defender's state, ensuring the game is well defined.

7   Bottom-Up Analysis on Trees

7.1   Success Probability Computation

In this section we provide an algorithm for computing the exact success probability directly on the attack-defense trees under some restrictions. Let $\mathsf{ADT} = (\mathsf{V}, \mathsf{E}, \mathsf{t}, \mathsf{TEdge})$ be an ADT that is a tree, i.e. no shared subtrees and repeated labels. In addition, for each $\mathsf{v} \in \mathsf{V}$ with $\ell(\mathsf{v}) = \mathsf{TR}$, the set $\{\mathsf{v}' \mid (\mathsf{v}, \mathsf{v}') \in \mathsf{TEdge}\}$ is a singleton. For simplicity[7], we assume $\{\mathsf{v} \mid (\mathsf{v}, \mathsf{b}) \in \mathsf{TEdge}\}$ to be a singleton for every triggerable $\mathsf{b} \in \mathsf{BE}$. In addition, we assume that all basic events are (probabilistically) independent and no basic event controlled by the defender occurs in a subtree with root NAT.

For operators such as NOT and NAT, we are interested in the probability of achieving -1 or 0, respectively. Hence, we need to compute the probabilities of success, failure, and remaining 0; we denote them by $\mathbb{P}^1$, $\mathbb{P}^{-1}$, and $\mathbb{P}^0$, respectively. We can compute the success probability $\mathbb{P}^1(\mathsf{att})$ inductively on the tree as follows:

- $\mathsf{t}(\mathsf{v}) \in \{\mathsf{AND}, \mathsf{SAND}\}$ and v_1 and v_2 denote the inputs of v:
  - $\mathbb{P}^1(\mathsf{v}) := \mathbb{P}^1(\mathsf{v}_1) \cdot \mathbb{P}^1(\mathsf{v})$
  - $\mathbb{P}^0(\mathsf{v}) := \min\left(\mathbb{P}^0(\mathsf{v}_1) \cdot \mathbb{P}^0(\mathsf{v}_2) + \mathbb{P}^1(\mathsf{v}_1) \cdot \mathbb{P}^0(\mathsf{v}_2) + \mathbb{P}^1(\mathsf{v}_2) \cdot \mathbb{P}^0(\mathsf{v}_1), 1\right)$
  - $\mathbb{P}^{-1}(\mathsf{v}) := \mathbb{P}^{-1}(\mathsf{v}_1) + \mathbb{P}^{-1}(\mathsf{v}_2) - \mathbb{P}^{-1}(\mathsf{v}_1) \cdot \mathbb{P}^{-1}(\mathsf{v}_2)$
- $\mathsf{t}(\mathsf{v}) \in \{\mathsf{OR}, \mathsf{SOR}\}$ and v_1 and v_2 denote the inputs of v:
  - $\mathbb{P}^1(\mathsf{v}) := \mathbb{P}^1(\mathsf{v}_1) + \mathbb{P}^1(\mathsf{v}_2) - \mathbb{P}^1(\mathsf{v}_1) \cdot \mathbb{P}^1(\mathsf{v}_2)$
  - $\mathbb{P}^0(\mathsf{v}) := \min\left(\mathbb{P}^0(\mathsf{v}_1) \cdot \mathbb{P}^0(\mathsf{v}_2) + \mathbb{P}^{-1}(\mathsf{v}_1) \cdot \mathbb{P}^0(\mathsf{v}_2) + \mathbb{P}^{-1}(\mathsf{v}_2) \cdot \mathbb{P}^0(\mathsf{v}_1), 1\right)$
  - $\mathbb{P}^{-1}(\mathsf{v}) := \mathbb{P}^{-1}(\mathsf{v}_1) \cdot \mathbb{P}^{-1}(\mathsf{v}_2)$

---

[7] In principal, a basic event can be triggered by several events and is triggered as soon as one of these events is completed successfully, which is equivalent to a disjunction over all these triggers.

- $v \in BE_A$ and v is non-triggerable:
  - $\mathbb{P}^1(v) := Pr(v)$
  - $\mathbb{P}^{-1}(v) := 1 - Pr(v)$
  - $\mathbb{P}^0(v) := 1$.
- $v \in BE_D$ and v is non-triggerable:
  - $\mathbb{P}^1(v) := 0$
  - $\mathbb{P}^{-1}(v) := 0$
  - $\mathbb{P}^0(v) := 1$
- $v \in BE_A$ and v is triggered by v':
  - $\mathbb{P}^1(v) := Pr(v) \cdot \mathbb{P}^1(v')$
  - $\mathbb{P}^{-1}(v) := (1 - Pr(v)) \cdot \mathbb{P}^1(v')$
  - $\mathbb{P}^0(v) := 1$.
- $v \in BE_D$ and v is triggered by v':

- $\mathbb{P}^1(v) := 0$
- $\mathbb{P}^{-1}(v) := 0$
- $\mathbb{P}^0(v) := 1$
- $t(v) = $ NOT and let v_1 denote the input of v.
  - $\mathbb{P}^1(v) := \mathbb{P}^{-1}(v_1)$
  - $\mathbb{P}^0(v) := \mathbb{P}^0(v_1)$
  - $\mathbb{P}^{-1}(v) := \mathbb{P}^1(v_1)$
- $t(v) = $ NAT and let v_1 denote the input of v.
  - $\mathbb{P}^1(v) := \mathbb{P}^0(v_1)$
  - $\mathbb{P}^0(v) := \emptyset$
  - $\mathbb{P}^{-1}(v) := \mathbb{P}^{-1}(v_1)$

Proof Sketch. Under the given assumption, our bottom-up computation indeed yields the success probability, i.e. we show that $\mathbb{P}^1(att) = \sup_{\sigma \in \Sigma_A} \inf_{\tau \in \Sigma_D} \mathbb{P}^{\sigma,\tau}(\Diamond 1)$. In ADT, which satisfy the conditions above, each subtree is independent. For attack trees consisting of a basic event, the equation trivially holds. Inductively, we first compute optimal reachability strategies for induced games of the subtree(s). These strategies are also optimal in the game for the whole tree. Indeed, interleaving the strategies for the subtrees cannot increase the success probability and it cannot influence the availability of moves. The success probability of the tree is then given by the formulae above over the success probabilities for the subtrees.

7.2   Heuristics for Cost-Efficient Strategies

We now show how to compute a probability-optimal cost-efficient (not necessarily cost-optimal) winning strategy on attack trees, i.e. ADT without shared labels and without defender behavior. We first compute top-down a function tv: $V \rightarrow \{1, 0, -1\}$, which assigns each event the truth value it may need to turn to for a successful attack. tv is just an approximation since, for instance, for OR we add more events than actual necessary to turn OR to 1. We start the computation with tv(att) := 1 and proceed as following: Let $v \in V$ and let v_i denote a child of v.

We define:

- $tv(v_i) := tv(v)$ if $t(v) \in \{$AND, SAND, OR, SOR, TR$\}$.

- for $t(v) = $ NOT:
  - $tv(v_i) := 1$ if $tv(v) = -1$
  - $tv(v_i) := 0$ if $tv(v) = 0$
  - $tv(v_i) := -1$ if $tv(v) = 1$

- for $t(v) = $ NAT:
  - $tv(v_i) := 1$ if $tv(v) = 0$
  - $tv(v_i) := -1$ if $tv(v) = 1$ or if $tv(v) = -1$

Since we define the function tv top-down, we do not need a case for basic events.

To compute a cost-efficient strategy for the attacker, we rely on computing valuations as in Sect. 5, i.e. we start with the valuation $(\mathsf{app}(\overline{0}, \overline{0}|_{\mathsf{BE}}), \mathsf{A})$ and after attempting basic events, we denote the new valuation by $\mathsf{app}(\upsilon_\mathsf{p}, \upsilon_{\mathsf{BE}})$. A cost-efficient strategy for the attacker is then given by:

- do not attempt SAND and SOR in the wrong order ("suicide")
- only attempt available basic events
- never try basic events b such that $\mathsf{tv}(\mathsf{b}) = 0$
- always try the basic event b that has either the smallest expected cost for losing if $\mathsf{tv}(\mathsf{b}) = 1$ or the smallest expected cost for winning if $\mathsf{tv}(\mathsf{b}) = -1$ and the successor of b has still value 0 in the current valuation.

In attack trees, the success probability (except for suicidal moves) does not depend on the chosen strategy. Hence, following the strategy constructed above does not change the overall success probability of attacks, but just the expected cost following an attacker has to spend.

Example 8. Consider our running example in Fig. 1 without vertices 6, 8–10 and 13. We compute tv for each basic event b as $\mathsf{tv}(\mathsf{b}) = 1$. The cost of losing for the events is given by (event)-cost: (1)-0, (2)-2, (3)-0.4, (4)-2.5 and (5)-0. We cannot try (5) fist since its a suicidal move. Hence, we first try (1), then (2) (since (3) is not available yet). Depending on the outcome of (2), event (3) (if available) or (4) (if (2) or (3) failed), and depending on their outcomes finally event (5) are attempted. This strategy is not optimal since it is better to first try events (2), (3) and (4) since event (1) cannot fail.

For the attack trees in Fig. 2a, this heuristics lead to optimal strategies (attempting first event (2) then (1)) as well as for the attack tree in Fig. 2b ((2), (3), (1), (4)).

8 Implementation and Experiments

We implemented our analyses with arbitrary precision (using the library jscience)[8] using Java 8. Back-propagation on the game as well as the heuristics on the ADT and the valuation computation is implemented using the visitor pattern. All experiments are run using Arch Linux (Linux kernel 4.18.16-arch1-1-ARCH) with 7.7GiB available memory and an Intel Core i5-7200U. As an alternative approach to solve reachability on the simple stochastic games we generate an input file for the recent distribution of PRISM-games [13].

For performance testing, we used randomly generated attack-defense trees. We first generated a set of basic events with random costs and success probabilities. We then used a random number generator to determine which operator to use to connect randomly chosen subtrees into one larger attack tree.

Our experiments showed that number of states in the games is hard to predict solely from the number of vertices in the ADT; a detailed overview can be found

---

[8] http://jscience.org/.

in Fig. 6b. We have observed that operators AND and OR often cause a blow up since our prototypical implementation considers all possible interleavings, which could be alleviated with partial-order reduction techniques. For trees with only 16 to 18 vertices, but OR or AND close to the top, the generation of the game semantics may take hours (or even longer). This renders the generation of game the main bottleneck of the analysis.

The main difference between our back-propagation implementation and the game analysis using PRISM is then (1) the number of iterations: while back-propagation iterates once over the whole state space, PRISM mostly needs more iterations, e.g. nine in the running example and (2) that our implementation of back-propagation computes the results with arbitrary precision and PRISM uses floating-point arithmetics.

In contrast to the game approaches, success probability computation using bottom-up analysis on the tree scales very well (see Fig. 6a) and can even solve trees with more than 1600 states and thus, with far more states than used in typical real-world applications.

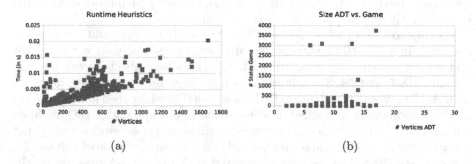

(a)   (b)

Fig. 6. On the left, we consider only the special case of tree-structured graphs with no further restriction (i.e. all available operators as well as both players are used). The diagram shows how the size of the tree determines the runtime of the heuristic. One the right, we depict the relation between the size of the tree and the number of states in the generated game.

9   Related Work

Attack trees were introduced two decades ago [29–31] and shortly after received a formal semantics in [25]. *Multi-parameter attack trees* [11,19] extend attack trees by attaching cost, probabilities and penalties to basic events. Recently, model checking exact cost for attack trees without shared subtrees, sequential aspects or defender's behavior has been explored in [3].

Sequential attack trees [2] add an operator SAND (sequential And) to specify the order of certain attack steps with an semantics based on acyclic phase-type distributions. In [1], sequential attack trees receive a semantics in terms of input/output Markov chains. In [24], a sequential attack tree analysis approach based on priced timed automata is presented taking temporal relations between attack steps, shared subtrees, and multi-dimensional resources into account.

Attack-defense trees (ADT) [21] allow to also represent the defender's behavior. In [6,21], *ADT* received a semantics in terms of a bottom-up traversal. In [20], it has been shown that they are equivalent to zero-sum extensive form games. Pareto-efficient solutions for *ADT*s with conflicting parameters have been discussed in [4].

For attack trees with dependent actions (but without sequential or defender behavior, without shared subtrees and no costs), success probabilities using Bayesian networks are treated in [22]. Quantitative analysis in the presence of repeated events is studied only recently in [10,18,23]. Strategy synthesis for attack trees with repeated labels has not been studied at all. [10] studies how to deal with attack-defense trees with repeated labels (but without sequential behavior) correctly.

The authors of [23] study *ADT*s with repeated labels. They devise conditions, under which the bottom-up traversal of the tree yields reliable results in the context of repeated labels. In comparison to our approach, they are *not* specifically interested in computing strategies, success probabilities or expected cost efficiently and do not treat sequential behavior.

In [26], cost and probabilities on *ADT* are analysed using statistical model checking. In [17], *ADT*s without repeated events are analysed w.r.t. to cost and success probabilities using UPPAAL [8]. *ADT*s have also been extended by sequential AND and OR. Stochastic games have been proposed as semantics for such *ADT* in [5] and translated to PRISM-games [13], where then multi-objective model-checking may be applied. In comparison to our approach, *ADT* in [5] do not consider such a rich set of operators, a three-valued semantics and do not allow to arbitrarily combine sequential and non-sequential version of AND and OR. However, the PRISM-games model generated in [5] from attack trees are acyclic directed graphs since intermediate results are shared in a similar manner as in our approach. While we do not apply multi-objective model checking to *ADT*s in this paper, any suitable analysis from the PRISM-games engine may be applied to our model.

In [18], an extension of attack-defense trees to attack-defense diagrams is proposed, which, for instance, allows to set back basic events to their initial value. However, optimal strategies or exact reachability probabilities and expected costs are not treated.

10   Conclusion and Future Work

We have provided a framework for analysis of attack-defense trees with costs and probabilities. The presence of probabilities significantly affects the nature of the problem and the techniques required to solve it. We have argued that decision making must be modelled sequentially even for the most basic attack trees. As a result, the need for a 3-valued logic and a non-tree-based evaluation arose. We have presented algorithms based on representing the problem as a stochastic game and the analysis of the game using either off-the-shelf solvers or a faster tailored computation. While some questions can be answered directly on the tree, demonstrating great scalability, other questions are inherently non-modular and the construction of the game slows down the analysis considerably. This could be alleviated using partial-order reduction techniques.

References

1. Arnold, F., Guck, D., Kumar, R., Stoelinga, M.: Sequential and parallel attack tree modelling. In: Koornneef, F., van Gulijk, C. (eds.) SAFECOMP 2015. LNCS, vol. 9338, pp. 291–299. Springer, Cham (2015). https://doi.org/10.1007/978-3-319-24249-1_25
2. Arnold, F., Hermanns, H., Pulungan, R., Stoelinga, M.: Time-dependent analysis of attacks. In: Abadi, M., Kremer, S. (eds.) POST 2014. LNCS, vol. 8414, pp. 285–305. Springer, Heidelberg (2014). https://doi.org/10.1007/978-3-642-54792-8_16
3. Aslanyan, Z., Nielson, F.: Model checking exact cost for attack scenarios. In: Maffei, M., Ryan, M. (eds.) POST 2017. LNCS, vol. 10204, pp. 210–231. Springer, Heidelberg (2017). https://doi.org/10.1007/978-3-662-54455-6_10
4. Aslanyan, Z., Nielson, F.: Pareto efficient solutions of attack-defence trees. In: Focardi, R., Myers, A. (eds.) POST 2015. LNCS, vol. 9036, pp. 95–114. Springer, Heidelberg (2015). https://doi.org/10.1007/978-3-662-46666-7_6
5. Aslanyan, Z., Nielson, F., Parker, D.: Quantitative verification and synthesis of attack-defence scenarios. In: Computer Security Foundations Symposium (CSF) (2016)
6. Bagnato, A., Kordy, B., Meland, P.H., Schweitzer, P.: Attribute decoration of attack-defense trees. In: IJSSE (2012)
7. Baier, C., Katoen, J.-P.: Principles of Model Checking (Representation and Mind Series). The MIT Press, Cambridge (2008). ISBN: 026202649X, 9780262026499
8. Behrmann, G., et al.: UPPAAL 4.0. In: Quantitative Evaluation of Systems (QEST) (2006)
9. de Bijl, M.: Using data analysis to enhance attack trees. In: Proceedings Twente Student Conference (2017)
10. Bossuat, A., Kordy, B.: Evil Twins: handling repetitions in attack–defense trees. In: Liu, P., Mauw, S., Stølen, K. (eds.) GraMSec 2017. LNCS, vol. 10744, pp. 17–37. Springer, Cham (2018). https://doi.org/10.1007/978-3-319-74860-3_2
11. Buldas, A., Laud, P., Priisalu, J., Saarepera, M., Willemson, J.: Rational choice of security measures via multi-parameter attack trees. In: Lopez, J. (ed.) CRITIS 2006. LNCS, vol. 4347, pp. 235–248. Springer, Heidelberg (2006). https://doi.org/10.1007/11962977_19
12. Chatterjee, K., Henzinger, T.A.: Value iteration. In: Grumberg, O., Veith, H. (eds.) 25 Years of Model Checking. LNCS, vol. 5000, pp. 107–138. Springer, Heidelberg (2008). https://doi.org/10.1007/978-3-540-69850-0_7
13. Chen, T., Forejt, V., Kwiatkowska, M., Parker, D., Simaitis, A.: PRISM-games: a model checker for stochastic multi-player games. In: Piterman, N., Smolka, S.A. (eds.) TACAS 2013. LNCS, vol. 7795, pp. 185–191. Springer, Heidelberg (2013). https://doi.org/10.1007/978-3-642-36742-7_13
14. Condon, A.: The complexity of stochastic games. Inf. Comput. 96(2), 203–224 (1992)

15. Edge, K.S., et al.: The use of attack and protection trees to analyze security for an online banking system. In: Systems Science (HICSS) (2007)

16. Fraile, M., Ford, M., Gadyatskaya, O., Kumar, R., Stoelinga, M., Trujillo-Rasua, R.: Using attack-defense trees to analyze threats and countermeasures in an ATM: a case study. In: Horkoff, J., Jeusfeld, M.A., Persson, A. (eds.) PoEM 2016. LNBIP, vol. 267, pp. 326–334. Springer, Cham (2016). https://doi.org/10.1007/978-3-319-48393-1_24

17. Gadyatskaya, O., Jhawar, R., Kordy, P., Lounis, K., Mauw, S., Trujillo-Rasua, R.: Attack trees for practical security assessment: ranking of attack scenarios with ADTool 2.0. In: Agha, G., Van Houdt, B. (eds.) QEST 2016. LNCS, vol. 9826, pp. 159–162. Springer, Cham (2016). https://doi.org/10.1007/978-3-319-43425-4_10. ISBN: 978-3-319-43425-4

18. Hermanns, H., Krämer, J., Krčál, J., Stoelinga, M.: The value of attack-defence diagrams. In: Piessens, F., Viganò, L. (eds.) POST 2016. LNCS, vol. 9635, pp. 163–185. Springer, Heidelberg (2016). https://doi.org/10.1007/978-3-662-49635-0_9

19. Jürgenson, A., Willemson, J.: Computing exact outcomes of multi-parameter attack trees. In: Meersman, R., Tari, Z. (eds.) OTM 2008. LNCS, vol. 5332, pp. 1036–1051. Springer, Heidelberg (2008). https://doi.org/10.1007/978-3-540-88873-4_8

20. Kordy, B., Mauw, S., Melissen, M., Schweitzer, P.: Attack–defense trees and two-player binary zero-sum extensive form games are equivalent. In: Alpcan, T., Buttyán, L., Baras, J.S. (eds.) GameSec 2010. LNCS, vol. 6442, pp. 245–256. Springer, Heidelberg (2010). https://doi.org/10.1007/978-3-642-17197-0_17

21. Kordy, B., Mauw, S., Radomirović, S., Schweitzer, P.: Foundations of attack-defense trees. In: Degano, P., Etalle, S., Guttman, J. (eds.) FAST 2010. LNCS, vol. 6561, pp. 80–95. Springer, Heidelberg (2011). https://doi.org/10.1007/978-3-642-19751-2_6

22. Kordy, B., Pietre-Cambacedes, L., Schweitzer, P.: DAG-based attack and defense modeling: don't miss the forest for the attack trees. CoRR (2013)

23. Kordy, B., Wideł, W.: On quantitative analysis of attack–defense trees with repeated labels. In: Bauer, L., Küsters, R. (eds.) POST 2018. LNCS, vol. 10804, pp. 325–346. Springer, Cham (2018). https://doi.org/10.1007/978-3-319-89722-6_14

24. Kumar, R., Ruijters, E., Stoelinga, M.: Quantitative attack tree analysis via priced timed automata. In: Sankaranarayanan, S., Vicario, E. (eds.) FORMATS 2015. LNCS, vol. 9268, pp. 156–171. Springer, Cham (2015). https://doi.org/10.1007/978-3-319-22975-1_11

25. Mauw, S., Oostdijk, M.: Foundations of attack trees. In: Won, D.H., Kim, S. (eds.) ICISC 2005. LNCS, vol. 3935, pp. 186–198. Springer, Heidelberg (2006). https://doi.org/10.1007/11734727_17

26. Mediouni, B.L., Nouri, A., Bozga, M., Legay, A., Bensalem, S.: Mitigating security risks through attack strategies exploration. In: Margaria, T., Steffen, B. (eds.) ISoLA 2018. LNCS, vol. 11245, pp. 392–413. Springer, Cham (2018). https://doi.org/10.1007/978-3-030-03421-4_25

27. Paul, S.: Towards automating the construction & maintenance of attack trees: a feasibility study. In: Graphical Models for Security (GramSec) (2014)

28. Ray, I., Poolsapassit, N.: Using attack trees to identify malicious attacks from authorized insiders. In: di Vimercati, S.C., Syverson, P., Gollmann, D. (eds.) ESORICS 2005. LNCS, vol. 3679, pp. 231–246. Springer, Heidelberg (2005). https://doi.org/10.1007/11555827_14

29. Salter, C., Saydjari, O.S., Schneier, B., Wallner, J.: Toward a secure system engineering methodolgy. In: New Security Paradigms (NSPW), New York, NY, USA (1998)
30. Schneier, B.: Attack trees. Dr. Dobb's J. (1999)
31. Schneier, B.: Secrets & Lies: Digital Security in a Networked World, 1st edn. Wiley, New York (2000). ISBN: 0471253111

A Process Algebra for (Delimited) Persistent Stochastic Non-Interference

Andrea Marin[1], Carla Piazza[2], and Sabina Rossi[1(✉)]

[1] Università Ca' Foscari Venezia, Venezia, Italy
{marin,sabina.rossi}@unive.it
[2] Università di Udine, Udine, Italy
carla.piazza@uniud.it

Abstract. In this paper, we consider the information flow security properties named *Persistent Stochastic Non-Interference* (*PSNI*) and *Delimited Persistent Stochastic Non-Interference* (*D_PSNI*) for stochastic cooperating processes described as terms of the Performance Evaluation Process Algebra (PEPA). A PEPA process P that satisfies *(D)_PSNI* admits only controlled information flows from the high, private, level of confidentiality to the low, public, one. In particular, the downgrading/declassification of information is permitted only when performed by a trusted component. Once a process has been defined one can only check whether it satisfies *(D)_PSNI* or not.

In this work, we contribute to the verification and construction of secure processes in two respects: (*i*) first we prove new compositionality properties for *(D)_PSNI* and then (*ii*) we exploit them in order to introduce a new process algebra which allows the definition of processes which are secure by construction, thus avoiding any further check.

1 Introduction

In this paper, we consider the information flow security properties named *Persistent Stochastic Non-Interference* (*PSNI*) [21,23,24] and *Delimited Persistent Stochastic Non-Interference* (*D_PSNI*) [22] which have been defined for stochastic cooperating processes expressed as terms of the Performance Evaluation Process Algebra (PEPA).

(Delimited) Non-Interference is an information flow security property which aims to protect confidential data by ensuring a complete (or controlled) absence of information flow from high level entities to low level ones. Hence, a PEPA process P that satisfies *(D)_PSNI* admits only controlled information flows from the high, private, level of confidentiality to the low, public, one. In particular, the downgrading/declassification of information is permitted only when it is under the control of a trusted component.

In [21–24], it has been proved that *PSNI* and *D_PSNI* can be verified by checking whether the system interacting with any high level component is *behaviourally equivalent* or not to the system in isolation. The notion of

D. Parker and V. Wolf (Eds.): QEST 2019, LNCS 11785, pp. 222–238, 2019.
https://doi.org/10.1007/978-3-030-30281-8_13

behavioural equivalence for *PSNI* and *D_PSNI* relies on the concept of lumpability ensuring that, for a secure process P, the steady-state probability of observing the system being in a specific state P' is independent of its possible high level interactions. In other words, we assume that the observer is able to observe any execution path with its delays and also to measure some timing properties like, e.g., the response time or the throughput. Hence, a system S is secure if any external observer is not able to distinguish the behaviour of S performing confidential, high level, activities from the behaviour of the same system but prevented from performing any high level action.

Although there exist efficient algorithms for verifying *PSNI* and *D_PSNI* (see, e.g., [21, 22]) which are polynomial with respect to the number of states and transitions of the labelled transition systems underlying PEPA components, we are not aware of any work dealing with the following two problems:

(i) The verification of behavioural equivalence often suffers of the so-called state space explosion problem, i.e., the number of states increases exponentially with respect to the degree of cooperations inside the considered system. The reason is that any interleaving among cooperating processes needs to be represented.

(ii) Once a process has been defined, one can only check whether the process satisfies *(D)_PSNI* or not. If the process is not secure then it is necessary to modify it, trying to preserve as much as possible its behaviour, while ensuring the security properties. A framework for defining processes which are *PSNI* or *D_PSNI* by construction would ease the development of secure systems.

In this work, we contribute to the verification and construction of secure processes in two respects:

(i) We prove new compositionality properties for the cooperation operator of PEPA which preserve both *PSNI* and *D_PSNI*.

(ii) We exploit such properties in order to introduce new process algebras which allow one to define processes which satisfy *PSNI* or *D_PSNI* by construction, thus avoiding any further check.

More precisely, the state space explosion problem intrinsic in the behavioural verification is avoided by exploiting the compositionality of *(D)_PSNI* with respect to the cooperation operator which is the source of the exponential growing of the number of states in a system. Indeed, if a property is preserved when secure systems are composed, then the analysis may be performed on subsystems and, in case of success, the system as a whole can be proved to be secure.

Moreover, the problem of defining systems which are secure by construction is tackled by exploiting the characterizations of *(D)_PSNI* given in [21–24] which are formulated through unwinding conditions [16]. Indeed, unwinding conditions allow us to express *PSNI* and *D_PSNI* in terms of a local property of high level activities requiring that whenever a high level activity is performed moving the system from state P to state P' then P and P' are behaviourally equivalent from a low level point of view. Intuitively, if this holds no high level activity h should be observable by a low level user, as there always exists a low level equivalent state that the system may reach without performing h. Here, we exploit these local properties in order to define two process algebras which allow us to statically derive secure *PSNI* or *D_PSNI* processes, i.e., by just following the syntax.

Related Work. Quantitative analysis of security is gaining new interest from the research community (see, e.g., [12,36]).

The concept of non-interference has been introduced by Goguen and Meseguer in [15]. Also the problem of modelling information flow policies admitting some forms of downgrading has first been addressed by Goguen and Meseguer in [16]. Since then, a large body of work has led to a variety of definitions for different application contexts [2,8,9,14,26,29,34,35,37,38,40,43]. A systematic overview is given in [17].

There are many approaches to the verification of information flow security properties. For instance, there are verification techniques for information flow security which are based on types (see, e.g., [10,18,41,42]) and control flow analysis (see, e.g., [3,11]). However, most of them are concerned with trace based models [27,28,31,32]. A proof system for, non persistent, bisimulation based security is proposed by Martinelli in [30]. Non persistent properties are usually harder to verify than persistent ones. Martinelli's approach deals with finite processes. Unwinding conditions for possibilistic security properties have been previously proposed in, e.g., [28,33,39]. All such conditions have been studied for trace based models and are, in most cases, only sufficient for the respective security properties. Necessary and sufficient unwinding conditions for bisimulation based security properties of CCS processes have been proposed in [4,5,7].

Our proof system extends that of [30] to the case of recursively defined processes and that of [6,13] to the case of stochastic processes.

Structure of the Paper. In Sect. 2 we introduce the process algebra PEPA, its semantics, and the observation equivalence. The notion of *Persistent Stochastic Non-Interference* (*PSNI*) and its characterization are briefly recalled in Sect. 3, where a process algebra for defining *PSNI* components is presented. In Sect. 4 the approach is extended to *Delimited Persistent Stochastic Non-Interference* (*D_PSNI*). Finally, Sect. 5 concludes the paper.

2   The Language

In this section, we briefly recall the *Performance Evaluation Process Algebra* (PEPA) [19] that we will use as a formal language to model and study quantitative properties of dynamic systems.

Syntax. The PEPA language [19] consists of two basic elements: *components* and *activities*. Activities are pairs (α, r) where α is called *action type* and belongs to

Table 1. Operational semantics for PEPA components

$$\frac{}{(\alpha,r).S \xrightarrow{(\alpha,r)} S} \qquad \frac{S \xrightarrow{(\alpha,r)} S'}{S+T \xrightarrow{(\alpha,r)} S'} \qquad \frac{T \xrightarrow{(\alpha,r)} T'}{S+T \xrightarrow{(\alpha,r)} T'}$$

$$\frac{P \xrightarrow{(\alpha,r)} P'}{P/A \xrightarrow{(\alpha,r)} P'/A}\ (\alpha \notin A) \qquad \frac{P \xrightarrow{(\alpha,r)} P'}{P/A \xrightarrow{(\tau,r)} P'/A}\ (\alpha \in A)$$

$$\frac{P \xrightarrow{(\alpha,r)} P'}{P \setminus A \xrightarrow{(\alpha,r)} P' \setminus A}\ (\alpha \notin A) \qquad \frac{S \xrightarrow{(\alpha,r)} S'}{X \xrightarrow{(\alpha,r)} S'}\ (X \stackrel{def}{=} S)$$

$$\frac{P \xrightarrow{(\alpha,r)} P'}{P \bowtie_A Q \xrightarrow{(\alpha,r)} P' \bowtie_A Q}\ (\alpha \notin A) \qquad \frac{Q \xrightarrow{(\alpha,r)} Q'}{P \bowtie_A Q \xrightarrow{(\alpha,r)} P \bowtie_A Q'}\ (\alpha \notin A)$$

$$\frac{P \xrightarrow{(\alpha,r_1)} P' \quad Q \xrightarrow{(\alpha,r_2)} Q'}{P \bowtie_A Q \xrightarrow{(\alpha,R)} P' \bowtie_A Q'} \qquad R = \frac{r_1}{r_\alpha(P)}\frac{r_2}{r_\alpha(Q)}\min(r_\alpha(P),r_\alpha(Q))\ (\alpha \in A)$$

a countable set $\mathcal{A}$, while r is called *activity rate* and belongs to the set $R^+ \cup \{\top\}$ where the symbol $\top$ is used to denote an *unspecified* rate. Hence, the duration of an activity is modelled as a negative exponential distribution with mean r^{-1}. The special action type $\tau \in \mathcal{A}$ is used to denote the *unknown* type. Activities of this type will be private to the component in which they occur.

In this paper, we slightly extend the syntax of PEPA by adding the empty component and the restriction operator which are not standard in PEPA. However, these are only shorthands for PEPA terms obtained by combing standard PEPA operators. The syntax of our language is given by the following grammar:

$$S ::= \mathbf{0} \mid (\alpha,r).S \mid S+S \mid X$$
$$P ::= S \mid P/A \mid P \setminus A \mid P \bowtie_A P$$

where $A \subseteq \mathcal{A} \setminus \{\tau\}$, S denotes a *sequential component*, X is a variable associated to a recursive definition of the form $X \stackrel{def}{=} S$, and P denotes a *model component*. We denote by $\mathcal{C}$ the set of all possible components.

Operational Semantics. Table 1 shows the operational semantics of PEPA. Component $\mathbf{0}$ does not carry out any activity. Component $(\alpha,r).S$ carries out the activity (α,r) of type α at rate r and subsequently behaves as component S. Term $S+T$ specifies a system which may behave either as S or as T. $S+T$ enables all the current activities of both S and T. The first activity to complete distinguishes one of the components, S or T. The other component of the choice is discarded. The continuous nature of the probability distributions ensures that the probability of S and T both completing an activity at the same time is zero. Component P/A behaves as P except that any activity of type within the set A are *hidden*, i.e., they are relabelled with the unknown type τ. Component $P \setminus A$ behaves as P except that any activity of type within the set A are forbidden, i.e., they are *restricted*. The meaning of a constant X is given by a defining equation such as $X \stackrel{def}{=} S$ which gives the constant S the behaviour of the sequential component S. Cooperation combinator $\underset{A}{\bowtie}$ is in fact an indexed family of combinators, one for each possible set of action types, $A \subseteq \mathcal{A} \setminus \{\tau\}$. The *cooperation set* A defines the action types on which the components must synchronise or *cooperate* (the unknown action type, τ, may not appear in any cooperation set). It is assumed that each component proceeds independently with the activities whose types do not occur in the cooperation set A (*individual activities*). However, activities with action types in A require the simultaneous involvement of both components (*shared activities*). Shared activities will only be enabled in $P \underset{A}{\bowtie} Q$ when they are enabled in both P and Q and have the same action type as the two contributing activities with a rate that reflects that of the slower component. If in a component an activity has rate $\top$, then we say that it is passive with respect to that action type. In this case, the rate of the shared activity will be that of the other component.

For a given component P and action type α, the *apparent rate* of α in P, $r_\alpha(P)$, is the sum of the rates of the α activities enabled in P. The semantics of each term in PEPA is given via a labelled *multi-transition system* where the multiplicities of arcs are significant. In the transition system, a state or *derivative* corresponds to each syntactic term of the language and an arc represents the activity which causes one derivative to evolve into another. The set of reachable states of a model P is termed the *derivative set* of P ($ds(P)$) and constitutes the set of nodes of the *derivation graph* of P ($\mathcal{D}(P)$) obtained by applying the semantic rules exhaustively.

We say that a component P is a *shorthand* for a component Q if their derivation graphs are *isomorphic* as defined in [19]. The empty component $\mathbf{0}$, which is not standard in PEPA, has a derivation graph with one node and no edges. Hence, it can be seen as the shorthand for $P \underset{\mathcal{A} \setminus \{\tau\}}{\bowtie} Q$ where P and Q are two components that do not share any activity and do not perform any τ action. Since P and Q cannot synchronize the derivation graph of $P \underset{\mathcal{A} \setminus \{\tau\}}{\bowtie} Q$ is formed by one node and no edges. Similarly, the restriction $P \setminus A$ can be seen as a shorthand for $P \underset{A}{\bowtie} \mathbf{0}$ for any $A \subseteq \mathcal{A} \setminus \{\tau\}$,

We denote the set of all the *current action types* of P by $\mathcal{A}(P)$. This is the set of action types which the component P may next engage in. We denote by $Act(P)$ the multiset of all the *current activities* of P.

Underlying Markov Chain. Let $P \stackrel{def}{=} P_0$ with $ds(P) = \{P_0, \ldots, P_n\}$ be a finite PEPA model. Then, the stochastic process $X(t)$ on the space $ds(P)$ is a continuous time Markov chain [19]. The *transition rate* between two states P_i and P_j is denoted by $q(P_i, P_j)$ and corresponds to the rate at which the system changes from behaving as component P_i to behaving as P_j, i.e.,

$$q(P_i, P_j) = \sum_{a \in \mathcal{A}ct(P_i | P_j)} r_a$$

with $P_i \neq P_j$ and the multiset $\mathcal{A}ct(P_i | P_j) = \{| a \in \mathcal{A}ct(P_i) | P_i \xrightarrow{a} P_j |\}$. When P_j is not a one-step derivative of P_i we set $q(P_i, P_j) = 0$.

Another notion that will be used in the paper is that of *conditional transition rate* from P_i to P_j via an action type α, denoted by $q(P_i, P_j, \alpha)$. This is the sum of the activity rates labelling arcs connecting the corresponding nodes in the derivation graph which are also labelled by the action type α. It is the rate at which a system behaving as component P_i evolves to behaving as component P_j as the result of completing a type α activity. The *total conditional transition rate* from P to $C \subseteq \mathcal{C}$, $q[P, C, \alpha]$, is defined as

$$q[P, C, \alpha] = \sum_{P' \in C} q(P, P', \alpha).$$

Observation Equivalence. We consider a bisimulation-like equivalence notion for PEPA components, named *lumpable bisimilarity*, that we previously introduced in [20]. Two PEPA components are *lumpably bisimilar* if there exists an equivalence relation between them such that, for any action type α, with the exception of τ actions internal to the classes, the total conditional transition rates from those components to any equivalence class, via activities of this type, are the same.

Definition 1. (Lumpable bisimulation [20])   *An equivalence relation over PEPA components, $\mathcal{R} \subseteq \mathcal{C} \times \mathcal{C}$, is a lumpable bisimulation if whenever $(P, Q) \in \mathcal{R}$ then for all $\alpha \in \mathcal{A}$ and for all $C \in \mathcal{C}/\mathcal{R}$ such that*

- *either $\alpha \neq \tau$,*
- *or $\alpha = \tau$ and $P, Q \notin C$,*

it holds $q[P, C, \alpha] = q[Q, C, \alpha]$.

Notice that, in contrast with the notion of strong equivalence [19], lumpable bisimulation allows arbitrary activities with type τ among components belonging to the same equivalence class, and therefore it is less strict.

We are interested in the relation which is the largest lumpable bisimulation, that is the union of all lumpable bisimulations.

Definition 2. (Lumpable bisimilarity [20])   *Two PEPA components P and Q are lumpably bisimilar, written $P \approx_l Q$, if $(P, Q) \in \mathcal{R}$ for some lumpable bisimulation $\mathcal{R}$, i.e., $\approx_l = \bigcup \{\mathcal{R} \mid \mathcal{R}$ is a lumpable bisimulation$\}$.*
$\approx_l$ is called lumpable bisimilarity and it is the largest symmetric lumpable bisimulation over PEPA components.

In [20], we proved that for any PEPA component P, lumpable bisimilarity induces a partition of the derivative set $ds(P)$ of P into equivalence classes that is a *strong lumpability* [25] for the underlying Markov chain. In [1] algorithms and applications for lumpable bisimilarity have been studied. Moreover, the aggregated process satisfies the property that the steady-state probability of each aggregated macro-state is equal to the sum of the steady-state probabilities of the corresponding equivalent states in the initial CTMC. Finally, we proved that lumpable bisimilarity is a congruence, i.e., if $P_1 \approx_l P_2$ then

- $(\alpha, r).P_1 \approx_l (\alpha, r).P_2$ for all $\alpha \in \mathcal{A}$
- $P_1/A \approx_l P_2/A$ for all $A \subseteq \mathcal{A} \setminus \{\tau\}$
- $P_1 \bowtie_A Q \approx_l P_2 \bowtie_A Q$ for all $A \subseteq \mathcal{A} \setminus \{\tau\}$.

Moreover, it is immediate to prove that the compositionality property can be extended also to the restriction operator as stated by the following proposition.

Proposition 1. *Let P_1 and P_2 be two PEPA components and $A \subseteq \mathcal{A} \setminus \{\tau\}$. If $P_1 \approx_l P_2$, then*

$$P_1 \setminus A \approx_l P_2 \setminus A.$$

Even more interestingly, the restriction operator commutes with the synchronization one as stated by the following theorem which will be at the basis of the results in the next sections.

Theorem 1. *Let P, Q be two PEPA components and $A, B \subseteq \mathcal{A} \setminus \{\tau\}$ be two set of actions. It holds that*

$$(P \bowtie_A Q) \setminus B \approx_l (P \setminus B) \bowtie_A (Q \setminus B).$$

Notice that this result holds also if we replace $\approx_l$ with strong equivalence [19]. The property holds for PEPA, but not for other process algebras such as CCS. This is due to the fact that PEPA synchronization does not replace the synchronizing actions with τ.

Finally, we will exploit also the following trivial result.

Lemma 1. *Let P and Q be PEPA components. If for each $R \in ds(P) \cup ds(Q)$ it holds that $\mathcal{A}(R) = \{\tau\}$, then $P \approx_l Q$.*

3   A Process Algebra for *PSNI*

In this section, we recall the security property named *Persistent Stochastic Non-Interference (PSNI)* for PEPA components which aims at characterizing classes of processes having no information flows from high to low [24]. Then, we introduce a process algebra which allows us to define only PEPA components that satisfy *PSNI* by construction.

Property *PSNI* tries to capture every possible information flow from a *classified (high)* level of confidentiality to an *untrusted (low)* one. The definition of *PSNI* is based on the basic idea of Non-Interference [15]: "No information flow is possible from high to low if what is done at the high level *cannot interfere* in any way with the low level". Hence, the notion of *PSNI* consists of checking all the states reachable by the system against all high level potential interactions.

In order to formally define this security property, we partition the set $\mathcal{A} \setminus \{\tau\}$ of visible action types, into two sets, $\mathcal{H}$ and $\mathcal{L}$, of high and low level action types, respectively. A high level PEPA component H is a PEPA term such that for all $H' \in ds(H)$, $\mathcal{A}(H') \subseteq \mathcal{H}$, i.e., every derivative of H may perform in only high level actions. We denote by $\mathcal{C}_H$ the set of all high level PEPA components.

A system P satisfies property *PSNI* if for every state P' reachable from P and for every high level process H a low level user cannot distinguish P' running in isolation, denoted by $P' \bowtie_{\mathcal{H}} 0$, from $P' \bowtie_{\mathcal{H}} H$ where H is a high level component cooperating with P'. In other words, a system P satisfies *PSNI* if what a low level user sees of the system is not modified when it cooperates with any high level process H.

The observation equivalence at the base of our definition relies on the notion of lumpable bisimilarity and this ensures that, for a secure process P, the steady-state probability of observing the system being in a specific state P' is independent of its possible high level interactions.

The low level view of the system is the one in which high activities are hidden, i.e., they are performed, but the low level observer is not able to distinguish them.

Definition 3 (*PSNI* [24]). *Let P be a PEPA component. $P \in PSNI$ iff $\forall P' \in ds(P)$, $\forall H \in \mathcal{C}_H$,*

$$(P' \bowtie_{\mathcal{H}} 0)/\mathcal{H} \approx_l (P' \bowtie_{\mathcal{H}} H)/\mathcal{H} .$$

Notice that $(P' \bowtie_{\mathcal{H}} 0)$ does not engage in any high level activity and hence the derivation graphs of $(P' \bowtie_{\mathcal{H}} 0)$ and $(P' \bowtie_{\mathcal{H}} 0)/\mathcal{H}$ are isomorphic. Moreover, the derivation graph of $(P' \bowtie_{\mathcal{H}} 0)$ is isomorphic to that of $P' \setminus \mathcal{H}$. So, we immediately get the following characterization of *PSNI*.

Proposition 2. *Let P be a PEPA component. $P \in PSNI$ iff $\forall P' \in ds(P)$, $\forall H \in \mathcal{C}_H$,*

$$P' \setminus \mathcal{H} \approx_l (P' \bowtie_{\mathcal{H}} H)/\mathcal{H} .$$

The definition of *PSNI* involves a first universal quantification over the derivatives of P and a second one over all high level components in $\mathcal{C}_H$. Hence, even if P has a finite set of derivatives, in order to verify *PSNI* we would have to perform an infinite set of tests. Luckily, there are different characterizations of *PSNI* that show its polynomial time decidability with respect to the size of $\mathcal{D}(P)$. In particular, in this paper it is useful to recall the following characterization expressed in terms of an unwinding condition.

Theorem 2 ([24]). *Let P be a PEPA component. $P \in PSNI$ iff $\forall P' \in ds(P)$,*

$$P' \xrightarrow{(h,r)} P'' \text{ implies } P' \setminus \mathcal{H} \approx_l P'' \setminus \mathcal{H} .$$

The process algebra we define relies on the compositionality properties of *PSNI*. Some of these have been proved in [24]. However, here we observe that, exploiting Theorem 2 we can prove stronger results on the compositionality of the synchronization operator. In particular, the following lemma has been proved only for low level cooperation sets in [24].

Lemma 2. *Let P, Q be two PEPA components and $A \subseteq \mathcal{A} \setminus \{\tau\}$. If $P, Q \in$ PSNI, then also $P \bowtie_A Q \in PSNI$.*

Therefore, we can prove the following compositionality properties for *PSNI*.

Lemma 3. *Let P_i with $i \in \mathbb{N}$ be PEPA components. If for all $i \in \mathbb{N}$ it holds $P_i \in PSNI$, then:*

a. $\mathbf{0} \in PSNI$
b. $Q \setminus \mathcal{L} \in PSNI$ for all PEPA component Q
c. $Q \setminus \mathcal{H} \in PSNI$, for all PEPA component Q
d. $(\ell, r).P_i \in PSNI$ for all $\ell \in \mathcal{L} \cup \{\tau\}$
e. $P_i / A \in PSNI$ for all $A \subseteq \mathcal{A} \setminus \{\tau\}$
f. $P_i \setminus A \in PSNI$ for all $A \subseteq \mathcal{A} \setminus \{\tau\}$
g. $P_i \bowtie_A P_j \in PSNI$ for all $A \subseteq \mathcal{A} \setminus \{\tau\}$
h. $Q_1, Q_2, \ldots, Q_p \in PSNI$, where for all $c \in [1, p]$:

$$Q_c \stackrel{\text{def}}{=} \sum_{i \in I} (\ell_i, r_i).P_i + \sum_{k \in K} (\ell_k, r_k).Q_k + \sum_{j \in J} (h_j, r_j).Q_c \setminus H_j + \sum_{m \in M} (h_m, r_m).Q'_c$$

$$Q'_c \stackrel{\text{def}}{=} \sum_{i \in I} (\ell_i, r_i).P_i + \sum_{k \in K} (\ell_k, r_k).Q_k$$

with I, J, K, M set of indices, $\ell_i, \ell_k \in \mathcal{L} \cup \{\tau\}$, $h_j, h_m \in \mathcal{H}$, and $H_j \subseteq \mathcal{H}$.

From the above compositionality properties we can define a process algebra in which only PEPA components that satisfy *PSNI* can be constructed.

Definition 4 (*PSNI* Process Algebra ($\mathcal{C}_{PSNI}$). *Let Q be PEPA component, $A \subseteq \mathcal{A} \setminus \{\tau\}$, I, J, K, M be set of indexes, $\ell, \ell_i \in \mathcal{L}$, $h_j, h_m \in \mathcal{H}$, and $H_j \subseteq \mathcal{H}$. $\mathcal{C}_{PSNI}$ is the set of PEPA components defined by the following grammar:*

$$S ::= \mathbf{0} \mid Q \setminus \mathcal{H} \mid Q \setminus \mathcal{L} \mid (\ell, r).S \mid X$$
$$P ::= S \mid P/A \mid P \setminus A \mid P \bowtie_A P$$

where X has a recursive definition of the form

$$X \stackrel{\text{def}}{=} \sum_{i \in I} (\ell_i, r_i).S_i + \sum_{j \in J} (h_j, r_j).X \setminus H_j + \sum_{m \in M} (h_m, r_m).X'$$

$$X' \stackrel{\text{def}}{=} \sum_{i \in I} (\ell_i, r_i).S_i .$$

Theorem 3. *Let $P \in \mathcal{C}_{PSNI}$. It holds that $P \in PSNI$.*

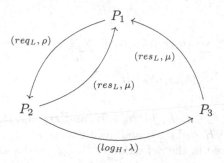

Fig. 1. A simple client/server model.

It is easy to prove that the converse of Theorem 3 is not true. Indeed, for the processes belonging to $\mathcal{C}_{PSNI}$ a stronger version of the property in Theorem 2 holds, where $\approx_l$ is replaced by isomorphism.

Example 1. Consider a web server which can serve the requests coming from different customers. A customer first makes a request through the low level activity req_L with rate ρ. This may either originate a direct response, represented by the low level activity res_L with rate μ, or the customer is required to access a confidential method (e.g., a payment service) through a high level authentication with type log_H and rate λ and then a reply is sent to the customer with the low level activity res_L.

We desire the low level observer not to be able to understand if a payment took place. In order to achieve this goal, the definition of *PSNI* requires that the response sent by the service to customer has a rate which is independent of whether the customer has accessed to the payment service or not. We can formalize it as a PEPA term throught the following system of recursive equations:

$$P_1 \stackrel{def}{=} (req_L, \rho).P_2$$
$$P_2 \stackrel{def}{=} (res_L, \mu).P_1 + (log_H, \lambda).P_3$$
$$P_3 \stackrel{def}{=} (res_L, \mu).P_1$$

The corresponding derivation graph is represented as in Fig. 1.

It is immediate to see that this system belongs to $\mathcal{C}_{PSNI}$, hence it is *PSNI*. In particular, the probability for a low level user to observe, in steady-state, the system being in state P_1 is independent of the behaviour of P_2, e.g., of whether the high level activity (log, λ_H) is performed or not.

In the next example we present a *PSNI* component that does not belong to $\mathcal{C}_{PSNI}$. However, it is possible to consider more complex rules for recursive definitions and enlarge the subclass of *PSNI* components that can be defined. Let us consider systems of recursive components defined for $k \in K$ expressed as follows:

$$X_k \stackrel{def}{=} \sum_{i \in I_k} (\ell_i, r_i).S_i + \sum_{j \in JX_k} (h_j, r_j).X_k \backslash H_j + \sum_{z \in Z_k} (\ell_z, r_z).X_z + \sum_{x \in WX_k} (h_x, r_x).Y_k \backslash H_x$$

$$Y_k \stackrel{def}{=} \sum_{i \in I_k} (\ell_i, r_i).S_i + \sum_{j \in JY_k} (h_j, r_j).Y_k \backslash H_j + \sum_{z \in Z_k} (\ell_z, r_z).Y_z + \sum_{y \in WY_k} (h_y, r_y).X_k \backslash H_y$$

where $S_i \in \mathcal{C}_{PSNI}$, $\ell_i, \ell_z \in \mathcal{L}$, $h_j, h_x, h_y \in \mathcal{H}$, and $H_j, H_x, H_y \subseteq \mathcal{H}$. It is immediate to prove that $X_k \backslash \mathcal{H} \approx_l Y_k \backslash \mathcal{H}$ and, as a consequence, such kind of systems generalize the ones used in the definition of $\mathcal{C}_{PSNI}$ and allow one to build only *PSNI* components. Here there is an interplay in the recursion between the X_k's and the Y_k's which again can be generalized by introducing other variables.

Example 2. We consider the case study proposed in [21] and we show that we can prove that it satisfies *PSNI* by construction. The example considers a distributed system with $K \geq 2$ servers. Ordinary jobs arrive at the system according to a Poisson process with rate λ_L. All the events associated with arrivals and departures of ordinary jobs can be observed by a malicious user. Moreover, the system executes a control task that alternates a sleeping and a working phase. The durations of these two phases are independent and exponentially distributed with rates λ_H, and μ_H, respectively. During the working phase, the control job uses one of the K servers. Ordinary customers require independent and exponentially distributed service times with rate μ_L. If the internal job becomes active and none of the servers is free, then one random ordinary job is preempted and the internal job is executed immediately. Notice that since the distributions of the service times have the memoryless property, it is not necessary to discuss the resume policy for the preempted jobs. The queue has infinite capacity. In [21], we showed that, in order to hide the state of the system when the internal process is being executed to the external, possibly malicious, observers, we had to devote one server to the execution of the control job. In this way, it is impossible to observe the difference in the service times of ordinary jobs when the control job is being executed. We can model the system as follows:

$$P_0 \stackrel{def}{=} (L, \lambda_L).P_1 + (H, \lambda_H).P_{0H}$$
$$P_{0H} \stackrel{def}{=} (L, \lambda_L).P_{1H} + (H, \mu_H).P_0$$
$$P_n \stackrel{def}{=} (L, \mu_L).P_{n-1} + (L, \lambda_L).P_{n+1} + (H, \lambda_H).P_{nH}$$
$$P_{nH} \stackrel{def}{=} (L, \mu_L).P_{(n-1)H} + (L, \lambda_L).P_{(n+1)H} + (H, \mu_H).P_n$$

Figure 2 shows the derivation graph of the model for $K = 2$. In contrast with the approach proposed in [21] where we propose an explicit state algorithm to verify *PSNI*, here we do not need to check that our model satisfies *PSNI*, since its syntactical specification follows the grammar for *PSNI* processes defined above. Notice that since this example has an infinite number of states we would need a symbolic algorithm to verify it, instead thanks to the syntactical approach proposed in this paper we can deal with infinite state systems without any change in the method.

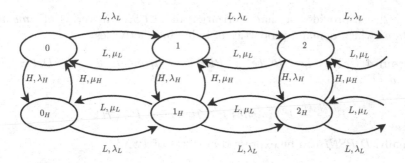

Fig. 2. LTS of the model that satisfies $PSNI$ for $K = 2$.

4 A Process Algebra for D_PSNI

In this section, we consider the security property named *Delimited Persistent Stochastic Non-Interference (D_PSNI)* for PEPA components which relaxes the conditions imposed by $PSNI$ allowing for limited flows of information that are deliberated by a trusted component. Then, we extend the process algebra introduced in the previous section in order to obtain a process algebra which allows the definition of PEPA components that satisfy D_PSNI by construction.

The notion of $PSNI$ is too demanding when dealing with practical applications: indeed no real policy ever guarantees a total absence of information flow. In many concrete applications, confidential data can flow from high to low provided that the flow is not direct and it is controlled by the system, i.e., a trusted part of the system can control the downgrading of sensitive information. In [22], we showed how our security property can be generalized in order to obtain a notion of stochastic non-interference for PEPA components which allows systems to intentionally release some information.

To model downgrading we now partition the set $\mathcal{A} \setminus \{\tau\}$ of visible action types, into three sets, $\mathcal{H}$, $\mathcal{L}$ and $\mathcal{D}$ of high, low and downgraded action types. Downgraded action types are used to specify the behaviour of trusted components interacting with the system. We assume that the low level users cannot observe the actions performed by the trusted part.

The notion of *Delimited Persistent Stochastic Non-Interference* (D_PSNI) can be formalized as follows.

Definition 5 (D_PSNI [22]). *Let P be a PEPA component. $P \in D\_PSNI$ iff $\forall P' \in ds(P)$, $\forall H \in \mathcal{C}_H$,*

$$((P' \bowtie_{\mathcal{H}} \mathbf{0})/\mathcal{H}) \setminus \mathcal{D} \approx_l ((P' \bowtie_{\mathcal{H}} H)/\mathcal{H}) \setminus \mathcal{D}.$$

Notice that this definition states that a system P satisfies D_PSNI if *whenever it does not cooperate with a trusted part*, what a low level user sees of the system is not modified when it cooperates with any high level process H. Hence, flows from the high level to the trusted part and flows from the trusted part to the low level are admissible, while direct flows from high to low are not allowed.

In [22], we provided a characterization of D_PSNI in terms of *unwinding conditions* which demand properties of individual activities.

Theorem 4 ([22]). *Let P be a PEPA component. $P \in D\_PSNI$ iff $\forall P' \in ds(P)$,*

$$P' \xrightarrow{(h,r)} P'' \text{ implies } P' \setminus \mathcal{H} \cup \mathcal{D} \approx_l P'' \setminus \mathcal{H} \cup \mathcal{D}.$$

Finally, D_PSNI can be expressed in terms of $PSNI$.

Proposition 3 ([22]). *Let P be a PEPA component. $P \in D\_PSNI$ iff $\forall P' \in ds(P)$, $P' \setminus \mathcal{D} \in PSNI$.*

We can prove the following compositionality properties for D_PSNI.

Lemma 4. *Let P, Q be two PEPA components and $A \subseteq \mathcal{A} \setminus \{\tau\}$. If P and Q are D_PSNI, then also $P \bowtie_A Q$ is D_PSNI.*

Lemma 5. *Let P_i with $i \in \mathbb{N}$ be PEPA components. If for all $i \in \mathbb{N}$ it holds $P_i \in D\_PSNI$, then:*

a. $\mathbf{0} \in D\_PSNI$,
b. $Q \setminus \mathcal{L} \in D\_PSNI$ *for all Q PEPA component;*
c. $Q \setminus \mathcal{H} \in D\_PSNI$, *for each Q PEPA component;*
d. $(\ell, r).P_i \in D\_PSNI$ *for all $\ell \in \mathcal{L} \cup \{\tau\}$;*
e. $P_i / A \in D\_PSNI$ *for all $A \subseteq \mathcal{A} \setminus \{\tau\}$;*
f. $P_i \setminus A \in D\_PSNI$ *for all $A \subseteq \mathcal{A} \setminus \{\tau\}$;*
g. $P_i \bowtie_A P_j \in D\_PSNI$ *for all $A \subseteq \mathcal{A} \setminus \{\tau\}$;*
h. $Q_1, Q_2, \ldots, Q_p \in D\_PSNI$, *where for all $c \in [1, p]$:*

$$Q_c \stackrel{\text{def}}{=} \sum_{i \in I} (\ell_i, r_i).P_i + \sum_{k \in K} (\ell_k, r_k).Q_k + \sum_{j \in J} (h_j, r_j).Q_c \setminus B_j + \sum_{m \in M} (h_m, r_m).Q'_c$$

$$Q'_c \stackrel{\text{def}}{=} \sum_{i \in I} (\ell_i, r_i).P_i + \sum_{k \in K} (\ell_k, r_k).Q_k$$

with I, J, K, M set of indexes $\ell_i, \ell_k \in \mathcal{L} \cup \{\tau\}$, $h_j, h_m \in \mathcal{H}$, and $B_j \subseteq \mathcal{H} \cup \mathcal{D}$;
i. $(h, r).(d, \delta).P_i \in D\_PSNI$ *for all $h \in \mathcal{H}$ and $d \in \mathcal{D}$.*

From the above compositionality properties we can define a process algebra in which only PEPA components that are D_PSNI are definable.

Definition 6 (D_PSNI Process Algebra ($\mathcal{C}_{D\_PSNI}$)). *Let Q be PEPA component, $A \subseteq \mathcal{A} \setminus \{\tau\}$, I, J, K, M be set of indexes, $\ell, \ell_i \in \mathcal{L}$, $h, h_j, h_m \in \mathcal{H}$, $d \in \mathcal{D}$ and $B_j \subseteq \mathcal{H} \cup \mathcal{D}$. $\mathcal{C}_{D\_PSNI}$ is the set of PEPA components defined by the following grammar:*

$$S ::= \mathbf{0} \mid Q \setminus \mathcal{H} \mid Q \setminus \mathcal{L} \mid (\ell, r).S \mid (h, r).(d, \delta).S \mid X$$
$$P ::= S \mid P/A \mid P \setminus A \mid P \bowtie_A P$$

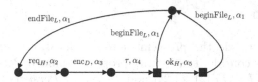

Fig. 3. Secure database query.

where X has a recursive definition of the form

$$X \stackrel{def}{=} \sum_{i \in I}(\ell_i, r_i).S_i + \sum_{j \in J}(h_j, r_j).X \setminus B_j + \sum_{m \in M}(h_m, r_m).X'$$

$$X' \stackrel{def}{=} \sum_{i \in I}(\ell_i, r_i).S_i.$$

Theorem 5. *Let $P \in \mathcal{C}_{D\_PSNI}$. It holds that $P \in D\_PSNI$.*

Example 3. In this example, we review the case study proposed in [22] that models a confidential query to a database as shown in Fig. 3. The left-most state models the system waiting for a request req_H. This uses a private channel based on a asymmetric cryptography protocol. Once the request is received, the system negotiates a symmetric key that will be used for the transmission of the query answer (enc_D). This phase is observable by a malicious user, but by using the downgrading we are stating that we tolerate the information flow that happens up to this point. The transition labelled τ models the computation of the query answer which can be either successful or unsuccessful. In the former case, the system transmits on the private channel the acknowledge ok_H, then begins ($beginFile_L$) and ends ($endFile_L$) the transmission of the reply encrypted with the shared key negotiated before. For this reason, these activities are modelled by means of low-level action types. In the case of failure, the system transmits an error message ($beginErr_L$, $endErr_L$). In [22], we studied some conditions to make this system secure. First, we avoid the use of a different action type for signalling the error. Second, we use the same distribution for the sizes of the reply and error message. In our case, the transmission of both an error message or a query answer will take place with an exponentially distributed time with rate α_1. The model can be expresses as a $\mathcal{C}_{D\_PSNI}$ term as follows:

$$P_0 \stackrel{def}{=} (req_H, \alpha_2).(enc_D, \alpha_3).P_1$$
$$P_1 \stackrel{def}{=} (\tau, \alpha_4).P_2$$
$$P_2 \stackrel{def}{=} (beginFile_L, \alpha_1).P_3 + (ok_H, \alpha_5).P_2'$$
$$P_3 \stackrel{def}{=} (endFile, \alpha_1).P_0$$
$$P_2' \stackrel{def}{=} (beginFile_L, \alpha_1).P_3$$

5 Conclusion

In this paper, we tackle the problem of the verification and construction of secure systems. We consider the security properties named *Persistent Stochastic Non-Interference* (*PSNI*) [21,23,24] and *Delimited Persistent Stochastic Non-Interference* (*D_PSNI*) [22], which have been defined for stochastic cooperating processes expressed as terms of PEPA. These properties aim at protecting the flow of high level information also against possible timing attacks. Indeed, the notion of behavioural equivalence for *PSNI* and *D_PSNI* relies on the concept of lumpability ensuring that, for a secure process P, the steady state probability of observing the system being in a specific state P' is independent of its possible high level interactions.

Moreover, we exploit compositionality to simplify the definition of secure systems. Indeed, the development of large and complex systems strongly depends on the ability of dividing the task of the system into subtasks that are solved by system subcomponents. Thus, it is useful to define properties which are compositional in the sense that if the properties are satisfied by the system subcomponents then the system as a whole will satisfy the desired property by construction. We also face the state space explosion problem by exploiting new compositionality results for *(D)_PSNI* which ensure that the cooperation of secure PEPA components is again a secure component. Finally, we exploit the characterizations of *PSNI* and *D_PSNI* in terms of unwinding conditions in order to define two new process algebras, one for *PSNI* and a generalized one for *D_PSNI*, which allow us to define processes which are secure by construction in an incremental way.

References

1. Alzetta, G., Marin, A., Piazza, C., Rossi, S.: Lumping-based equivalences in markovian automata: algorithms and applications to product-form analyses. Inf. Comput. **260**, 99–125 (2018)
2. Backes, M., Pfitzmann, B.: Intransitive non-interference for cryptographic purposes. In Proceedings of the IEEE Symposium on Security and Privacy (SSP 2003), IEEE, pp. 140–152 (2003)
3. Bodei, C., Degano, P., Nielson, F., Nielson, H.R.: Static analysis for secrecy and non-interference in networks of processes. In: Malyshkin, V. (ed.) PaCT 2001. LNCS, vol. 2127, pp. 27–41. Springer, Heidelberg (2001). https://doi.org/10.1007/3-540-44743-1_3
4. Bossi, A., Focardi, R., Macedonio, D., Piazza, C., Rossi, S.: Unwinding in information flow security. Electron. Notes Theor. Comput. Sci. **99**, 127–154 (2004)
5. Bossi, A., Focardi, R., Piazza, C., Rossi, S.: Bisimulation and unwinding for verifying possibilistic security properties. In: Zuck, L.D., Attie, P.C., Cortesi, A., Mukhopadhyay, S. (eds.) VMCAI 2003. LNCS, vol. 2575, pp. 223–237. Springer, Heidelberg (2003). https://doi.org/10.1007/3-540-36384-X_19
6. Bossi, A., Focardi, R., Piazza, C., Rossi, S.: A proof system for information flow security. In: Leuschel, M. (ed.) LOPSTR 2002. LNCS, vol. 2664, pp. 199–218. Springer, Heidelberg (2003). https://doi.org/10.1007/3-540-45013-0_16

7. Bossi, A., Focardi, R., Piazza, C., Rossi, S.: Verifying persistent security properties. Comput. Lang. Syst. Struct. **30**(3–4), 231–258 (2004)
8. Bossi, A., Macedonio, D., Piazza, C., Rossi, S.: Information flow in secure contexts. J. Comput. Secur. **13**(3), 391–422 (2005)
9. Bossi, A., Piazza, C., Rossi, S.: Compositional information flow security for concurrent programs. J. Comput. Secur. **15**(3), 373–416 (2007)
10. Boudol, G., Castellani, I.: Noninterference for concurrent programs. In: Orejas, F., Spirakis, P.G., van Leeuwen, J. (eds.) ICALP 2001. LNCS, vol. 2076, pp. 382–395. Springer, Heidelberg (2001). https://doi.org/10.1007/3-540-48224-5_32
11. Braghin, C., Cortesi, A., Focardi, R.: Control flow analysis of mobile ambients with security boundaries. In: Proceedings of IFIPM International Conference on Formal Methods for Open Object-Based Distributed Systems, Kluwer, pp. 197–212 (2002)
12. Cheh, C., Chen, B., Temple, W.G., Sanders, W.H.: Data-driven model-based detection of malicious insiders via physical access logs. In: Bertrand, N., Bortolussi, L. (eds.) QEST 2017. LNCS, vol. 10503, pp. 275–291. Springer, Cham (2017). https://doi.org/10.1007/978-3-319-66335-7_17
13. Focardi, R., Piazza, C., Rossi, S.: Proofs methods for bisimulation based information flow security. In: Cortesi, A. (ed.) VMCAI 2002. LNCS, vol. 2294, pp. 16–31. Springer, Heidelberg (2002). https://doi.org/10.1007/3-540-47813-2_2
14. Focardi, R., Rossi, S.: Information flow security in dynamic contexts. J. Comput. Secur. **14**(1), 65–110 (2006)
15. Goguen, J.A., Meseguer, J.: Security policies and security models. In: Proceedings of the IEEE Symposium on Security and Privacy (SSP 1982), pp. 11–20 (1982)
16. Goguen, J.A., Meseguer, J.: Unwinding and inference control. In: Proceedings of the IEEE Symposium on Security and Privacy (SSP 1984), pp. 75–86 (1984)
17. Hedin, D., Sabelfeld, A.: A perspective on information-flow control. In: Software Safety and Security - Tools for Analysis and Verification, IOS Press, pp. 319–347 (2012)
18. Hennessy, M., Riely, J.: Information flow vs. resource access in the asynchronous pi-calculus. ACM Trans. Program. Lang. Syst. (TOPLAS) **24**(5), 566–591 (2002)
19. Hillston, J.: A Compositional Approach to Performance Modelling. Cambridge Press, Cambridge (1996)
20. Hillston, J., Marin, A., Piazza, C., Rossi, S.: Contextual lumpability. In: Proceedings of Valuetools 2013 Conference, ACM Press, pp. 194–203 (2013)
21. Hillston, J., Marin, A., Piazza, C., Rossi, S.: Information flow security for stochastic processes. In: Computer Performance Engineering - 15th European Workshop, EPEW, pp. 142–156 (2018)
22. Hillston, J., Marin, A., Piazza, C., Rossi, S.: Delimited persistent stochastic non-interference. In: Proceedings of the 12th EAI International Conference on Performance Evaluation Methodologies and Tools, VALUETOOLS 2019, pp. 135–142 (2019)
23. Hillston, J., Marin, A., Piazza, C., Rossi, S.: Persistent stochastic non-interference. Information and Computation (2019, Submitted)
24. Hillston, J., Piazza, C., Rossi, S.: Persistent stochastic non-interference. In: Proceedings of Combined 25th International Workshop on Expressiveness in Concurrency and 15th Workshop on Structural Operational Semantics and 15th Workshop on Structural Operational Semantics, EXPRESS/SOS 2018, pp. 53–68 (2018)
25. Kemeny, J.G., Snell, J.L.: Finite Markov Chains. D. Van Nostrand Company Inc. (1960)

26. Lafrance, S., Mullins, J.: Bisimulation-based non-deterministic admissible interference and its application to the analysis of cryptographic protocols. Electron. Notes Theor. Comput. Sci. **61**, 1–24 (2002)
27. Mantel, H.: Possibilistic definitions of security - an assembly kit -. In: Proceedings of the IEEE Computer Security Foundations Workshop (CSFW 2000), IEEE Computer Society Press, pp. 185–199 (2000)
28. Mantel, H.: Unwinding possibilistic security properties. In: Cuppens, F., Deswarte, Y., Gollmann, D., Waidner, M. (eds.) ESORICS 2000. LNCS, vol. 1895, pp. 238–254. Springer, Heidelberg (2000). https://doi.org/10.1007/10722599_15
29. Mantel, H.: Information flow control and applications — bridging a gap —. In: Oliveira, J.N., Zave, P. (eds.) FME 2001. LNCS, vol. 2021, pp. 153–172. Springer, Heidelberg (2001). https://doi.org/10.1007/3-540-45251-6_9
30. Martinelli, F.: Partial model checking and theorem proving for ensuring security properties. In: Proceedings of the IEEE Computer Security Foundations Workshop (CSFW 1998), IEEE Computer Society Press, pp. 44–52 (1998)
31. McCullough, D.: A hookup theorem for multilevel security. IEEE Trans. Software Eng. **16**, 563–568 (1990)
32. McLean, J.: A general theory of composition for trace sets closed under selective interleaving functions. In: Proceedings of the IEEE Symposium on Security and Privacy (SSP 1994), IEEE Computer Society Press, pp. 79–93 (1994)
33. Millen, J.K.: Unwinding forward correctability. In: Proceedings of the IEEE Computer Security Foundations Workshop (CSFW 1994), IEEE Computer Society Press, pp. 2–10 (1994)
34. Mullins, J.: Nondeterministic admissible interference. J. Univers. Comput. Sci. **11**, 1054–1070 (2000)
35. Piazza, C., Pivato, E., Rossi, S.: Cops - checker of persistent security. In: Tools and Algorithms for the Construction and Analysis of Systems (TACAS 2004), pp. 144–152 (2004)
36. Rausch, M., Fawaz, A., Keefe, K., Sanders, W.H.: Modeling humans: a general agent model for the evaluation of security. In: McIver, A., Horvath, A. (eds.) QEST 2018. LNCS, vol. 11024, pp. 373–388. Springer, Cham (2018). https://doi.org/10.1007/978-3-319-99154-2_23
37. Roscoe, A.W., Goldsmith, M.H.: What is intransitive noninterference? In: Proceedings of the IEEE Computer Security Foundations Workshop (CSFW 1999), pp. 228–238 (1999)
38. Rushby, J.: Noninterference, transitivity, and channel-control security policies. Technical Report CSL-92-02, SRI International, December 1992
39. Ryan, P.Y.A.: A CSP Formulation of Non-Interference and Unwinding. Cipher, pp. 19–27 (1991)
40. Ryan, P.Y.A., Schneider, S.: Process algebra and non-interference. J. Comput. Secur. **9**(1/2), 75–103 (2001)
41. Sabelfeld, A., Sands, D.: Probabilistic noninterference for multi-threaded programs. In: Proceedings of the IEEE Computer Security Foundations Workshop (CSFW 2000), IEEE Computer Society Press, pp. 200–215 (2000)
42. Smith, G., Volpano, D.M.: Secure information flow in a multi-threaded imperative language. In: Proceedings of ACM SIGPLAN-SIGACT Symposium on Principles of Programming Languages (POPL 1998), ACM Press, pp. 355–364 (1998)
43. von Oheimb, D.: Information flow control revisited: noninfluence = noninterference + nonleakage. In: Computer Security (ESORICS 2004), pp. 225–243 (2004)

Automated Game-Theoretic Verification
of Security Systems

Chunyan Mu[(⊠)]

Department of Computer Science, Teesside University, Middlesbrough, UK
c.mu@tees.ac.uk

Abstract. Security-sensitive computerised communication systems are
of increasing importance, however checking that they function correctly
can be non-trivial. We propose automated verification techniques for the
formal analysis of quantitative properties of such systems. Since communication networks typically require the collaboration of their participants
to work effectively, we adopt a game-theoretic approach. Utility functions
for each player, such as the degree of security offered and the communication costs incurred, are formally specified using quantitative temporal
logics. Then, building upon probabilistic verification techniques for parametric Markov chains, we develop methods to identify Nash equilibria
representing stable strategies for the participants. We implement our
methods as an extension of the PRISM model checker, and illustrate
their applicability by studying anonymity-cost trade-offs in the Crowds
anonymity protocol.

Keywords: Quantitative verification · Game theory · Security

1 Introduction

Security properties have become essential requirements in today's computerised communication systems. Absolute guarantees on such properties are often
impractical in real life, so we may instead tolerate a loss of anonymity or privacy in a system with low probability. Furthermore, system designs often need
to trade off the degree of security offered against other practical concerns such
as response time or power consumption. So, effective methods for the analysis
of security also need to take quantitative aspects into account.

In this work, we present novel automatic verification techniques for the formal
modelling and analysis of security properties in communication networks. Since
such systems generally rely on the collaboration of their participants to work
effectively, we adopt a game-theoretic approach to verification. We propose a
new framework in which systems are modelled as *n-player parametric Markov*

The author thanks David Parker for many suggestions, help and insightful discussions.
This work is supported in part by EPSRC (EP/K038575/1), and was partially performed when the author was at University of Birmingham.

chain games, where each decision-maker chooses the value of a parameter, which is used to define the transition probabilities of a Markov chain. These parameters allow participants in the system to make strategic probabilistic decisions about their behaviour, such as the probability which they will act in a co-operative or unfriendly or malicious fashion during the execution of a security protocol. Since we model system behaviour in a probabilistic fashion, we can also capture a variety of other important stochastic aspects of a system, such as message loss, failures, or other sources of randomisation.

We apply game-theoretic notions and methods to study the behaviour of interacting decision-makers. We define utility functions for players capturing their preferences regarding the different system outcomes, and then define the expected value of these utility functions using our Markov chain models. Individual players can decide how to behave in order to maximise their utility, although their choices will typically influence the outcomes for other players too. In order to investigate the effectiveness of a protocol (e.g., for anonymity) that requires some cooperation between multiple individuals with conflicting objectives, we use the concept of Nash equilibria [14]. These represent the existence of situations where no system player can benefit by changing their own strategy, assuming that the other players keep their strategies unchanged.

We propose techniques to formally specify games and utility functions and to automatically compute Nash equilibria for them. We build upon existing techniques and tools for probabilistic model checking, which is a widely used technique for modelling and automatically verifying quantitative properties of systems with stochastic behaviour. In particular, we build upon *parametric probabilistic model checking* methods [6], in which transition probabilities of models can be given as functions over parameters, and an analysis of these models can yield results expressed as symbolic functions over parameters. We use probabilistic temporal logics as a means of specifying utility functions for individual players and then use parametric model checking to determine functions representing the expected utility. We then generate and solve a set of polynomial equations, the solutions to which yield Nash equilibria for the system.

We developed an implementation of this approach using the parametric model checking functionality of the PRISM model checker [10] and solving polynomial systems using the polyhedral homotopy continuation based PHC-pack [21]. We describe how our approach can be used to model and analyse properties of security-sensitive communicating networks. In particular, we illustrate this on the Crowds [16] anonymity protocol, considering the trade-offs between the degree of anonymity provided and the corresponding communication cost.

Related Work. Multiple efforts have been made to develop methods for game-theoretic analysis of communicating networks for security concern. Yang et al. [22] proposed a game theoretic framework to analyse users' behaviours in anonymity networks. Performance utilities were modelled as a combination of weighted cost and anonymity utilities. Simulations were performed to show the impact of users' cooperation level and the weights of the anonymity and cost factors to the optimisation of their utilities. However, this work did not compute the Nash equilibria in an automatic way. In our work, we use exact solution methods, rather than simulation, and focus on automated methods to find equilibria.

Venkitasubramaniam [20] investigated the problem of maximising security properties from a game-theoretic perspective. The problem was formalised as a two-player zero-sum game between the network designer and the adversary. Given the adversary's observation, the anonymity degree was measured using conditional entropy of the routes. The adversary tried to choose a subset of the nodes to monitor in order to minimise the anonymity of the routes, while the designer aimed to maximise the anonymity. They did not deal with the problem of computing equilibria of games with multiple players with regard to performance analysis, however, which has been considered in our work.

Formal methods have played a significant important role in modelling and analysing security protocols. They include two main categories: proof-based theorem proving and state-exploration based model checking. Specifically, we focus on quantitative analysis of security properties. Mhamdi et al. [12] introduced two measures of information leakage: the information leakage degree and the conditional information leakage degree, to evaluate the anonymity and privacy properties of protocols. A theorem prover was applied to conduct a probabilistic and information-theoretic analysis for the evaluation of the anonymity and privacy properties. However, they did not tackle the problem of how the users should behave in order to optimise the security and performance properties from a game-theoretic point of view.

On the other hand, Shmatikov [17,18] applied the PRISM model checker to model and analyse the Crowds protocol for anonymity properties. By modelling the system behaviour as a discrete-time Markov chain, and formalising the anonymity properties in PCTL, the PRISM model checker was employed to perform automated probabilistic analysis and verify anonymity properties quantitatively. However, they did not study the problem of strategy decision-making or attempt a game-theoretic analysis of the system. Approaches of computing a Nash equilibrium in Stochastic games have been studied in [3,19]. It was not straightforward to adapt them in modelling and analysing security systems. Our framework can be naturally applied in such systems to synthesize optimal decision strategies.

2 Preliminaries

2.1 Game Theory

We first recall some required definitions from game theory [2,15], beginning with the definitions of convexity and some related notions.

Definition 1 (Convex set). *A set S of vectors over real numbers is convex if $(1 - \lambda)x + \lambda x' \in S$ whenever $x, x' \in S$, and $\lambda \in [0,1]$.*

Definition 2 (Upper level set). *Let f be a multivariate function defined on a set S. For $a \in \mathbb{R}$, the upper level set of f for a is $P_a = \{x \in S : f(x) \geq a\}$.*

Definition 3 (Quasi-concave function). *A multivariate function f defined on a convex set S is* quasi-concave *if every upper level set of f is convex.*

Definition 4 (Preference relation). *A preference relation $\succeq$ over a set S is a total, transitive and reflexive binary relation over S.*

Definition 5 (Strategic game). *A strategic game $\langle N, (A_i), (\succeq_i) \rangle$ consists of: a finite set of players N; for each player $i \in N$: a non-empty set of actions A_i available to i, and a preference relation $\succeq_i$ on $A = \times_{j \in N} A_j$ of i. The game is finite if the sets A_i of actions for player i are all finite.*

A *strategy* for player i is a choice of an action $a_i \in A_i$ and a *strategy profile* $\sigma = (a_1, \ldots, a_n) \in A$ is a choice of actions for all players. We write σ_i for the choice of player i in σ, σ_{-i} for the choices of all players except i, and (σ_{-i}, a_i) for the strategy profile that combines the choices from σ_{-i} and some $a_i \in A_i$.

Definition 6 (Nash equilibrium). *A* Nash equilibrium *of a strategic game $\langle N, (A_i), (\succeq_i) \rangle$ is a strategy profile $\sigma^* \in A$ with the property that, for every player $i \in N$ and strategy $a_i \in A_i$, we have $(\sigma^*_{-i}, \sigma^*_i) \succeq_i (\sigma^*_{-i}, a_i)$.*

Definition 7 (Best response function). *For any σ_{-i}, the* best response function *$B_i(\sigma_{-i})$ is defined as the set of player i's best actions given σ_{-i}:*

$$B_i(\sigma_{-i}) = \{a_i \in A_i : \forall a'_i \in A_i . (\sigma_{-i}, a_i) \succeq_i (\sigma_{-i}, a'_i)\}.$$

Note that, in the above, players' preferences with respect to strategies are defined by a preference relation $\succeq_i$. In the remainder of the paper, for modelling convenience and in order to allow players' choices to be probabilistic, we will instead use a *utility function* $u_i : A \to \mathbb{R}$ for each player i, which it aims to maximise. The basic strategic game can then be rewritten as a tuple $\langle N, (A_i), (u_i) \rangle$.

Proposition 1 ([15]). *The strategic game $\langle N, (A_i), (\succeq_i) \rangle$ has a Nash equilibrium if, for all $i \in N$: (i) the set A_i of actions of player i is a non-empty compact convex subset of Euclidian space; and (ii) the corresponding utility function u_i of the preference relation $\succeq_i$ is continuous and quasi-concave on A_i.*

In this paper, we model the communication systems in Markov chain games and discuss the existence of Nash equilibria for this model. Proposition 1 presents the requirements for the game to have a Nash equilibrium.

2.2   Parametric Markov Chains

In this paper, we build on parametric model checking techniques for the model of parametric Markov chains [6]. We briefly review some relevant background.

Definition 8 (Rational function). *Let $V = \{x_1, \ldots, x_n\}$ be a finite set of variables with domain $\mathbb{R}$. A polynomial g over V is constructed via the following grammar:*

$$g ::= c \mid x \mid g + g \mid (g) \cdot (g)$$

where $c \in \mathbb{R}$, $x \in V$, $+$ and $\cdot$ are the standard addition and multiplication respectively. A rational function f over V is a fraction of two polynomials g_1 and g_2 over V such as: $f = \frac{g_1}{g_2}$ where g_2 is not reducible to 0.

Definition 9 (Evaluation). Let $\mathbb{R}[x_1, \ldots, x_n]$ be the set of polynomials over the set of variables $V = \{x_1, \ldots, x_n\}$, let $\mathcal{F}_V : \mathbb{R}[x_1, \ldots, x_n] \to \mathbb{R}$ denote the set of rational functions, and $dom(f)$ denote the domain of function f. An evaluation $\mathcal{V} : X \to \mathbb{R}$ is a function for a subset $X \subseteq V$. For a rational function $f \in \mathcal{F}_V$, $f[X/\mathcal{V}]$ denotes the function obtained by substituting each $x \in (X \cap dom(\mathcal{V}))$ with its evaluation $\mathcal{V}(x)$.

A *parametric Markov chain* is an extension of a discrete-time Markov chain, using rational functions instead of real numbers to label transition probabilities.

Definition 10 (Parametric Markov chains). A *parametric Markov chain is a tuple* $\mathcal{M} = (S, I, \Delta, V, AP, L)$, where S is a countable set of states, $I : S \to \mathcal{F}_V$ is the initial distribution such that $\sum_{s \in S} I(s) = 1$, $\Delta : S \times S \to \mathcal{F}_V$ is the parametric transition probability matrix such that $\forall s \in S. \sum_{s' \in S} \Delta(s, s') = 1$, $V = \{x_1, \ldots, x_n\}$ is a finite set of parameters with domain $\mathbb{R}$, AP is a finite set of atomic propositions, and $L : S \to 2^{AP}$ is a labelling function mapping each state to a set of atomic propositions taken from a set AP.

For an evaluation $\mathcal{V}$ and a parametric Markov chain $\mathcal{M}$, an *induced* parametric Markov chain $\mathcal{M}_\mathcal{V}$ is defined by substituting each variable in $dom(\mathcal{V})$ with its evaluation. By applying a total evaluation $\mathcal{V}$ with $dom(\mathcal{V}) = V$, we obtain real values for each probability instead of rational functions. Let $\text{Prob}^\mathcal{M} \subseteq \mathcal{F}_V$ denote the set of probabilities of $\mathcal{M}$, such as

$$\text{Prob}^\mathcal{M} := \{I(s) | s \in S\} \cup \{\Delta(s, s') | s, s' \in S\},$$

and similarly $\text{Prob}_\mathcal{V}^\mathcal{M} \subseteq \mathcal{F}_{V/dom(\mathcal{V})}$ for an evaluation $\mathcal{V}$. A total evaluation $\mathcal{V}$ of $\mathcal{M}$ is called *well-defined* if:

- $\forall f \in \text{Prob}_\mathcal{V}^\mathcal{M} : f[dom(\mathcal{V})/\mathcal{V}] \in [0, 1]$, which ensures every possible evaluation is a probability;
- $\forall f \in \text{Prob}_\mathcal{V}^\mathcal{M} : f \neq 0 \Leftrightarrow f[dom(\mathcal{V})/\mathcal{V}] \neq 0$, which ensures every non-zero rational function does not evaluate to 0.

A parametric Markov chain $\mathcal{M}$ can be viewed as a state transition system in which transitions are associated with parametric probabilities indicating their likelihood. We say there is a *transition* from state $s \in S$ to $s' \in S$ iff $\Delta_\mathcal{V}(s, s') > 0$ for all well-defined evaluations $\mathcal{V}$. A (finite or infinite) *path* describes one possible execution of $\mathcal{M}$, and is defined as a sequence of states $\rho = s_0, s_1, \ldots$ such that $\forall i \geq 0. \Delta(s_i, s_{i+1}) = f(X) > 0$, where $f(X)$ is a polynomial over $X \subseteq V$.

Let $\Omega_{\mathcal{M}, s}$ denote the set of paths of $\mathcal{M}$ starting from state s (we omit the s when referring to all such paths). In order to reason about the behaviour of $\mathcal{M}$, it is required to formalise the probability of sets of paths taken. The construction is based on calculating the probability of individual finite paths induced by the parametric transition probability matrix Δ. The probability of the path $\rho = s_0, \ldots, s_k$ is given by: $Pr(\rho) \triangleq \prod_{i=0}^{k-1} \Delta(s_i, s_{i+1})$, where $\rho \in \Omega_{\mathcal{M}, s_0}$.

2.3   Probabilistic Temporal Logics

For the purposes of probabilistic verification, properties to be checked against a model are typically specified in probabilistic extensions of temporal logic. In this paper we use the property specification language of the PRISM tool [9], the basis for which is the logic PCTL [7], plus various notions of reward.

Definition 11 (PCTL with rewards). *The syntax of PCTL with rewards is given by the grammar:*

$$\phi ::= \mathtt{true} \mid a \mid \neg\phi \mid \phi \wedge \phi \mid P_{\bowtie q}[\psi] \mid R^r_{\bowtie x}[F\phi]$$
$$\psi ::= X\phi \mid \phi\, \mathcal{U}^{\leq k}\, \phi \mid \phi\, \mathcal{U}\, \phi$$

where $a \in AP$, r is a reward structure, $\bowtie\, \in \{<, \leq, >, \geq\}$, $q \in \mathbb{R} \cap [0,1]$, $x \in \mathbb{R}_{\geq 0}$, and $k \in \mathbb{N}$.

For example, $P_{\geq 0.5}[F\phi]$ means the probability of eventually reaching states satisfying ϕ is at least 0.5; $R^1_{\leq 10}[F\phi]$ means the expected cumulated reward (or cost) of reward structure 1 until reaching states satisfying ϕ is at most 10. We also often use notation such as $P_{=?}[\cdot]$ and $R^r_{=?}[\cdot]$ to represent numerically-valued probability or reward properties. We omit a full definition of the semantics of PCTL with rewards. Further details can be found in [9].

3   Game-Theoretic Verification with Parametric Probabilistic Model Checking

We now present a framework for game-theoretic verification of quantitative properties, based on parametric model checking techniques for discrete-time Markov chains. In this section, we introduce a model called *parametric Markov chain games* (PMCGs), in which parameters represent (probabilistic) decisions taken by players, and then we discuss the existence of Nash equilibria for this model. Subsequently, we will show to automatically synthesise these Nash equilibria, based on formal specifications of a system model and utility functions, and by building upon existing parametric model checking techniques.

3.1   Parametric Markov Chain Games

We model systems as n-player games. We will assume a fixed set of players $N = \{1, \ldots, n\}$, each of which has m possible actions $A = \{a_1, \ldots, a_m\}$ (i.e., all players share the same action space (A). Each $a_j \in A$ corresponds to a different possible course of action which can typically be decided upon multiple times during the execution of the system. We are generally interested in *mixed* strategies which are defined as a probability distribution π_i over A. Every time that player i needs to decide which action from A to take, it will do so by selecting action $a_j \in A$ with probability $\pi_i(a_j)$.

Given such strategies π_i for each player i, the subsequent behaviour of the system is necessarily probabilistic. Furthermore, we usually want to model other stochastic aspects of the system, for example, message transmission failure. So, we will assume that the behaviour of the system, under strategies π_i, can be modelled as a discrete-time Markov chain. In general, the transition probabilities of this Markov chain will be defined as expressions in terms of the probabilities $\pi_i(a_j)$ for each player i selecting each action a_j. This can be modelled as a *parametric* Markov chain whose parameters correspond to the probabilities $\pi_i(a_j)$. We refer to the resulting model as a *parametric Markov chain game* (PMCG), which is formally defined as follows.

Definition 12 (Parametric Markov chain game). *A parametric Markov chain game (PMCG) is a tuple of the form:*
$$\mathcal{G} = ((S, I, \Delta, V, AP, L), N, A, \{V_i\}_{i \in N}, \{u_i\}_{i \in N}), \text{ where:}$$

- (S, I, Δ, V, AP, L) *is a parametric Markov chain, which we will denote* $\mathcal{M}_{\mathcal{G}}$;
- N *is a finite set of players;*
- $A = \{a_1, \ldots, a_m\}$ *is a finite set of m actions;*
- (V_i) *is a partition of the parameter set* V *of* $\mathcal{M}_{\mathcal{G}}$, *assigning a subset* $V_i = \{x_{i,1}, \ldots, x_{i,m}\} \subseteq V$ *to each player* $i \in N$;
- $u_i : \Omega_{\mathcal{M}_{\mathcal{G}}} \to \mathbb{R}$ *is a utility function for each player* i.

A PMCG $\mathcal{G}$ incorporates a parametric Markov chain $\mathcal{M}_{\mathcal{G}}$, whose parameter set V is partitioned into subsets $V_i = \{x_{i,1}, \ldots, x_{i,m}\}$ for each player i. Each individual parameter $x_{i,j}$ represents the probability with which player i will choose action $a_j \in A$. The PMCG also defines a utility function u_i for each player i, represented as a function from (infinite) paths in $\mathcal{M}_{\mathcal{G}}$ to a real value.

3.2   Mixed Strategies and Nash Equilibria for PMCGs

Given a PMCG $\mathcal{G}$, the set of *pure strategies* for player i is the set of available actions A. A *mixed strategy* for player i, denoted by π_i, is given as a probability distribution over the set of pure strategies and written as a vector $\pi_i = (\pi_{i,1}, \pi_{i,2}, \ldots, \pi_{i,m})$ where $\pi_{i,j}$ denotes the probability of player i choosing action a_j. A (mixed) *strategy profile* $\pi = (\pi_1, \ldots, \pi_n) = (\pi_{i,1}, \pi_{i,2}, \ldots, \pi_{n,m})$ comprises a strategy for all players in the game.

The parametric Markov game $\mathcal{M}_{\mathcal{G}}$ of $\mathcal{G}$ represents the behaviour of the system under any possible strategy profile, with parameter $x_{i,j}$ representing the probability of player i choosing a_j. Thus, for a fixed strategy profile π, the resulting behaviour of the system is modelled by the induced Markov chain $\mathcal{M}_{\mathcal{G}, V/\pi}$, in which each $v_{i,j}$ is assigned value $\pi_{i,j}$. This Markov chain gives us a probability measure, denoted $\text{Prob}_{\mathcal{G}}^{\pi}$ over the set of all paths in $\Omega_{\mathcal{G}}$ through $\mathcal{G}$.

Now, to reason about Nash equilibria of $\mathcal{G}$, we first need to specify a preference ordering over strategies. As discussed earlier, we do so implicitly by defining a utility function u_i whose expected value each player i aims to maximise. In a PMCG, a utility function u_i assigns a real value to each *path* through the model. The expected value of u_i under a mixed strategy profile π is then defined by the Markov chain induced from $\mathcal{G}$ by π. More precisely, the expected utility for player i is $\mathbb{E}_{\mathcal{G}}^{\pi}(u_i)$, i.e., the expected value of function u_i with respect to the probability measure $\text{Prob}_{\mathcal{G}}^{\pi}$ over paths through $\mathcal{G}$. Abusing notation, we will often simply write $u_i(\pi)$ instead of $\mathbb{E}_{\mathcal{G}}^{\pi}(u_i)$. This allows us to give the following formal definition of a mixed strategy Nash equilibrium for a PMCG.

Definition 13 (Mixed strategy Nash equilibrium of $\mathcal{G}$). *Given a PMCG $\mathcal{G} = (\mathcal{M}_{\mathcal{G}}, N, A, (V_i), (u_i))$, a mixed strategy profile π for $\mathcal{G}$ is a Nash equilibrium if, for any player i and any mixed strategy π'_i of player i, we have $u_i(\pi_{-i}, \pi_i) \geq u_i(\pi_{-i}, \pi'_i)$, where u_i gives the expected utility for player i under a strategy profile, as explained above.*

Theorem 1. *Let $\mathcal{G} = (\mathcal{M}_{\mathcal{G}}, N, A, (V_i), (u_i))$ be a PMCG. If, for all $i \in N$, the utility function u_i is continuous and quasi-concave over the set of mixed strategies for player i, then $\mathcal{G}$ has a mixed strategy Nash equilibrium.*

Proof. Consider the set of mixed strategies for player i, which is a set of distributions over the set A of m actions. This is a non-empty, convex and compact subset of $\mathbb{R}^m$. By Proposition 1, as long as each utility function u_i (and thus the preference relation $\succeq_i$) is continuous and quasi-concave over the set of mixed strategies, then $\mathcal{G}$ satisfies all the requirements to have a Nash equilibrium. $\square$

The following result gives an important property of mixed strategy Nash equilibria for PMCGs $\mathcal{G}$ when calculating such equilibria:

Lemma 1. *Given a PMCG $\mathcal{G} = (\mathcal{M}_{\mathcal{G}}, N, A, \{V_i\}_{i \in N}, \{u_i\}_{i \in N})$, if u_i is monotonic on player i's mixed strategies, then: an n-tuple of mixed strategies profile $\pi = (\pi_1, \ldots, \pi_n)$ is a mixed strategy Nash equilibrium of $\mathcal{G}$ iff for every player $i \in N$, every pure strategy in the support of π is a best response to π_{-i}.*

Proof. ("$\Rightarrow$"): Assume that there is an action a in the support of π_i which is not a best response to π_{-i}. Then by the monotonicity of the utility function u_i, player i can increase his utility by switching probability from a to an action that is a best response, so π_i is not a best response to π_{-i}, which leads to a contradiction.

("$\Leftarrow$"): Suppose that there is a mixed strategy π'_i that gives a higher expected utility than π_i does in response to π_{-i}. Then at least one action in the support of π'_i must give a higher utility than some action in the support of π_i, so that not all actions in the support of π_i are best responses to π_{-i}, which leads to a contradiction. $\square$

One can imagine that, if a mixed strategy π_i is a best response, then each of the pure strategies involved in the mix must itself be a best response. Hence, all the pure strategies in the mix must yield the same expected utility. That is to say, every choice in the support of any player's equilibrium mixed strategy must yield the same utility value: $u_i(\pi_{-i}, a) = u_i(\pi_{-i}, a')$ for any two pure strategies $a, a' \in A$ such that the probabilities of player i choosing pure strategies a and a' are positive: $\pi_{i,a} > 0$ and $\pi_{i,a'} > 0$. We will use this fact to find mixed strategy

Nash equilibria of $\mathcal{G} = (\mathcal{M}_{\mathcal{G}}, N, A, \{V_i\}_{i \in N}, \{u_i\}_{i \in N})$ since we can therefore write the Nash equilibria conditions as follows:

$$\begin{cases} u_i(\pi_{-i}, a) = u_i(\pi_{-i}, a') \ \ \forall a, a' \in A, \ s.t. \ \pi_{i,a}, \pi_{i,a'} > 0 \\ \sum_{j=1}^{m} \pi_{i,j} = 1 \ \ \forall i \\ 0 \leq \pi_{i,j} \leq 1 \ \ \forall i, j. \end{cases} \tag{1}$$

By solving the equations above, we can find the equilibria.

4 Finding All Nash Equilibria

In this section we consider practical approaches to the synthesis of Nash equilibria for systems modelling using the parametric Markov chain game formalism introduced above. First, we need a formal specification of both the model of the system and the utility functions that are being used to define equilibria. Our work builds upon functionality in the PRISM model checker [10] for modelling and constructing parametric Markov chains, so systems are specified using the PRISM modelling language and utility functions for the models specified using PRISM's temporal logic notation, summarised in Sect. 2.3.

Using PRISM, we apply probabilistic verification to a parametric Markov chain, yielding rational functions that represent the expected values of utility functions. These rational functions are over the variables $x_{i,j}$ corresponding to the probabilities in the mixed strategies of each player. At this point we apply a simple optimisation to reduce the number of variables required. Given that we know $x_{i,1} + \cdots + x_{i,m} = 1$ for any i, one of the variables is redundant and we can rewrite, for example, $x_{i,m}$ as $1 - (x_{i,1} + \cdots + x_{i,m-1})$.

Next, we check the monotonicity of each player's utility function on the variables for its own mixed strategy.

Finally, we construct and solve, from the computed utility functions, a set of equalities in order to determine the set of Nash equilibria for the model. This is done based on the Nash equilibria conditions identified previously in (1), and is described in more detail below.

We assume the cases that each player's utility function is monotonic (and thus quasi-concave) for his own strategy from then on. The details of each step are discussed in the following subsections.

4.1 Nash Equilibria Conditions as Polynomial Equations

Given a set of players' utility properties from a PMCG, each of the form

$$u_i(x_{1,1}, \ldots, x_{1,m}, \ldots, x_{i,1}, \ldots, x_{i,m}, \ldots x_{n,1}, \ldots, x_{n,m}).$$

the Nash equilibria conditions can be used to construct a polynomial system of equations. Since player i's optimal choices of $x_{i,j}$ should equal the utility of the other players from playing pure strategies, by Lemma 1, we can build a set of equations for each i:

$$
\begin{aligned}
&u_i(x_{1,1}, \ldots, x_{i-1,1}, \ldots, x_{i-1,m}, 1, 0, \ldots, 0, x_{i+1,1}, \ldots, x_{n,m}) \\
&= u_i(x_{11}, \ldots, x_{i-1,1}, \ldots, x_{i-1,m}, 0, 1, \ldots, 0, x_{i+1,1}, \ldots, x_{n,m}) \\
&= \ldots \quad \ldots \\
&= u_i(x_{11}, \ldots, x_{i-1,1}, \ldots, x_{i-1,m}, 0, 0, \ldots, 1, x_{i+1,1}, \ldots, x_{n,m}).
\end{aligned}
$$

In addition, for all $i \in N$, $\{x_{i,j}|1 \leq j \leq m\}$ is a distribution so: $0 \leq x_{i,j} \leq 1$ and $\sum_{j=1}^{m} x_{i,j} = 1$. By solving the set of equations obtained above, we can find a set of complex solutions to $\{x_{ij}|i = 1, \ldots, n, \ j = 1, \ldots, m\}$, which, if they exist, yield the Nash equilibria, as required.

Solving the above system is non-trivial since the problem is typically non-linear. Herings and Peeters [8] show the feasibility of computing all Nash equilibria of general finite games in theory, and Datta provides an implementation in [5]. Homotopy continuation methods have been proven to be a reliable and powerful mathematical method to compute all isolated complex solutions [11,13] of polynomial systems. The method includes a number of main steps: use the algebraic structure to count the roots and to construct a start system, the root count determines the number of solution paths to be traced; the target system is embedded to solve in *homotopy* system, i.e., a family of systems connecting the start and the target system; following the solution paths of the homotopy system, the *continuation* methods are applied to extend the solutions of the start system to the desired solutions of the target system.

There are a number of software packages devoted to solving polynomial systems by using *homotopy continuation methods*. In this paper, we exploit the PHCpack [21] platform since it performs better than other software packages in terms of computational stability and capacity [4].

We conclude the Nash equilibrium generation with a few final checks. First, all of the mixing probabilities we have constructed must indeed be probabilities: $\forall i \in N. \sum_{j=1}^{m} x_{i,j} = 1$. Second, if there are no probability solutions, we need to check whether the player has a strictly profitable deviation. In the following sections, we illustrate the process on some examples.

5   Game-Theoretic Modelling of Security Systems

Now, we move on to describe how game-theoretic verification approach described in the previous two sections can be applied specifically in the context of security-sensitive communicating systems. In particular, we show to model such systems as parametric Markov chain games (see Definition 12) equipped with appropriate utility functions. In the next section, we will demonstrate the approach on a case study: the Crowds anonymity protocol.

We consider computerised communication systems consisting of a set of *players* and a set of destinations. We focus on the core procedure of transmitting messages through the system. This typically involves randomisation, and potentially other stochastic aspects such as message loss. Individual players may make certain strategic decisions about how to participate in the system. These are modelled as probabilistic actions, whose probabilities are parameters controlled by the player in question. The *adversary* is the set of malicious players who partially observe or participate in the transmission of messages and try to learn the sensitive information stored in the system states.

Players' *utility functions* are used to indicate preferences between strategies. Using our PMCG model, utility functions are defined as real-valued functions over paths, i.e., a mapping from each possible system execution to a real value. Strategies are then compared based on the expected value of this function. In the context of communication systems, we consider utility functions comprising two parts: *security* measurement and *cost* measurement. It is reasonable to expect that conflicts or trade-offs might exist between these two: for example, additional relaying of messages might improve security but at additional cost.

Security Measurement. In a communicating network, the adversaries act as a player and make a series of partial observations over the communication network. There is a set of execution paths, and some of them release information to the adversaries. We say an execution path is *bad* to i if there is a transition from a sensitive node controlled by i to a node controlled by a malicious player along the path, which cause information leakage. In our quantitative setting, we consider (the information of) player i to be *secure* if the probability of *bad* paths for i is small enough. Specifically, letting $\psi \subseteq \Omega_G$ denote the set of paths reaching the destination, and $\psi_i^* \subseteq \Omega_G$ denote the set of *bad* paths of the player i, we define the measurement of security for mixed strategy π to be:

$$u_i^s(\pi) = \text{Prob}^\pi(\psi \setminus \psi_i^*) = \text{Prob}^\pi(\psi) - \text{Prob}^\pi(\psi_i^*), \tag{2}$$

where ψ and ψ^* can be specified as PCTL formulas. Note that the bigger the security metric the more secure the system is with regard to the security property of interest.

Example 1. Consider players $N = \{0, 1, 2\}$ and assume $0, 1$ are honest players and 2 is a malicious player. Consider a PMCG in which states s_i ($i \in \{0, 1, 2\}$) are controlled by player i (say i is sending or forwarding a message for instance), s_d denotes the message reaching its destination, and parameters x_i represent probabilities controllable by player i. Assume that player 0 starts a message at s_0 (sensitive), and any transitions from s_0 to s_2 will violate the security policy. The *bad* paths ψ^* contain all the paths including the transition from s_0 to s_2. The probabilistic transition graph and security metric computation by (2) is presented in Fig. 1.

Cost Measurement. Markov chains with reward (or cost) structures allow us to specify two distinct types of rewards: *state* rewards, which are assigned to states by means of a function of the form $S \to \mathbb{R}_{\geq 0}$, and *transition* rewards, which

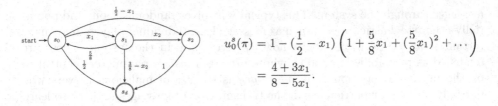

Fig. 1. Example: security measurement

are assigned to transitions by means of the function of the form $S \times S \to \mathbb{R}_{\geq 0}$. The state reward is the reward acquired in each state per time-step, and transition rewards are acquired each time a transition between states occurs. For the cost measurement, we consider the "expected cost consumed until a message reaches its destination". Let $r_i = (r_{i,\text{state}}, r_{i,\text{action}})$ be a reward structure for player i, and ϕ represent a set of target states (e.g., where a message has reached the destination). The required expected cost can be specified using the reward operator from PRISM's temporal logic (see Sect. 2.3): $R^{r_i}_{=?}[F\phi]$.

Example 2. Consider again the example presented in Fig. 1. Let $r_{0,\text{state}}(s_0) = 0$, assume the cost of player 0 forwarding a message is 1. Letting $u^c_{i,s}$ represent the expected cost for player i from state s, we can calculate the cost metric by solving:

$$\begin{cases} u^c_{0,s_2} = u^c_{0,s_d} = 0 \\ u^c_{0,s_0} = (1 + x_1 u^c_{0,s_1}) + (\frac{1}{2} - x_1)u^c_{0,s_2} + \frac{1}{2}u^c_{0,s_d} \\ u^c_{0,s_1} = (0 + \frac{5}{8}u^c_{0,s_0}) + x_2 u^c_{0,s_2} + (\frac{3}{8} - x_2)u^c_{0,s_d} \end{cases}$$

We have $u^c_{0,s_0} = \frac{8}{8-5x_1}$, i.e., the parametric expected cost of player 0 is $\frac{8}{8-5x_1}$.

Expected Utilities. Note that the performance of the system is in direct proportion to the security metric, and is in inverse proportion to the cost metric. So we define the *utility* of player i as a ratio of the security metric function over the cost metric function:

$$u_i(\pi) = \frac{u^s_i(\pi) * w_s}{u^c_{i,s_0}(\pi) * w_c} \tag{3}$$

where w_s denotes the weight of the security property, and w_c denotes the weight of the cost property.

Example 3. Let $w_s = 3$, $w_c = 1$, we calculate the expected utility of the previous example as: $u_0(\pi) = \frac{(4+3x_1)/(8-5x_1)*3}{24/(8-5x_1)} = \frac{3(4+3x_1)}{8}$.

6   Experimental Results: The Crowds Protocol

We have implemented our parametric model checking based approach to game-theoretic verification as an extension of the PRISM model checker [1]. Our tool can find the mixed strategy Nash equilibria for the players in a game when each player's utility function is monotonic w.r.t. his own mixed strategy. Building on the ideas set out in the previous section, we have used this approach to analyse the anonymity and cost of the Crowds protocol [16].

Crowds is a protocol allowing users to forward messages anonymously. The idea is that each user randomly chooses a user to forward a message rather than send their message to the destination directly. A forwarding route is therefore established within a collection of network members. In our game-theoretic model, each honest forwarder makes a decision whether to be cooperative or to be selfish. If he decides to be cooperative, then he flips a coin to decide to either send the message to the destination directly (with probability $1 - \mathtt{PF}$), or relay it to another crowd member randomly (with probability $\mathtt{PF}$); otherwise he discards the message directly. The malicious player behaves like a normal player but he will send the message he received to the destination directly. A malicious user can never be certain whether the observed user is the actual sender, or is simply forwarding another user's message.

Suppose in the protocol, there arc two honest members (player-1, 2) and two malicious members (player-3, 4). Assume that sending and relaying a message costs $c_{s_1} = 1$ and $c_{r_1} = 2$ respectively for player-1, and $c_{s_2} = 2$ and $c_{r_2} = 3$ for player-2. Without loss of generality, we choose different cost values for different operations and players here for demonstration. Let x_i denote the probability of player i being cooperative, and $1 - x_i$ the probability of being selfish. Figure 3 in Appendix B presents the transition graph of the model. We define the property specification for initiator (honest) player k as:

$$\frac{(\mathtt{P_{=?}}[F\ destination] - \mathtt{P_{=?}}[F(\mathtt{to{=}3} \vee \mathtt{to{=}4}) \wedge (\mathtt{from{=}}k) \wedge (\mathtt{sender{=}}k)]) * w_s}{(\mathtt{R}_{k{=?}}[F\ deadlock]) * w_c},$$

where $k \in \{1,2\}$ is an honest player, $\mathtt{to} \in \{1,2,3,4\}$ is the player whom the message is sent to, $\mathtt{from} \in \{1,2,3,4\}$ is the player who is sending/forwarding the message, $\mathtt{P_{=?}}[F\ destination]$ calculates the probability of the set of runs ϕ reaching the destination; $\mathtt{P_{=?}}[F\ (\mathtt{to} = 3 \vee \mathtt{to} = 4) \wedge (\mathtt{from} = k) \wedge (\mathtt{sender} = k)]$ calculates the probability of the set of runs ϕ^* violating anonymity property, i.e., sender k is sending or forwarding the message to a malicious player (3 or 4) directly; $\mathtt{R}_{k{=?}}[F\ deadlock]$ calculates the accumulated costs u_k^c reaching a terminating state; and w_s and w_c denote the weights for anonymity and costs, respectively. Assume $w_s = 3$, $w_c = 1$, let $u_1(x_1, x_2)$ and $u_2(x_1, x_2)$ denote the polynomial utility functions generated for player-1 and player-2 respectively, we get the parametric model checking results given as a set of polynomials as follows:

$$u_1(x_1, x_2) = \frac{15 * x_2 * x_1 - 36 * x_1 - 30 * x_2 - 144}{6 * x_2 * x_1 - 28 * x_1 + 12 * x_2 - 120}$$

$$u_2(x_1, x_2) = \frac{15 * x_2 * x_1 - 30 * x_1 - 36 * x_2 - 144}{8 * x_2 * x_1 + 24 * x_1 - 36 * x_2 - 216}$$

First let us check the monotonicity of $u_1(x_1, x_2)$ for player 1's mixed strategy:

$$\frac{\partial u_1}{\partial x_1} = \frac{6(15x_2^2 - 92x_2 + 12)}{(3x_1x_2 - 14x_1 + 6x_2 - 60)^2}$$

Let $\frac{\partial u_1}{\partial x_1} = 0$, we have $x_2 = \frac{2}{15}, 6$. Considering $x_2 \in [0,1]$, we have: when $0 \leq x_2 \leq \frac{2}{15}$, $\forall x_1. \frac{\partial u_1}{\partial x_1} \geq 0$ i.e., u_1 is nondecreasing with x_1; similarly, when $\frac{2}{15} \leq x_1 \leq 1$, $\forall x_1. \frac{\partial u_1}{\partial x_1} \leq 0$ i.e., u_1 is nonincreasing with x_1. Therefore, for $x_1, x_2 \in [0,1]$, u_1 is always monotonic w.r.t. player 1's mixed strategy; similarly, u_2 is monotonic w.r.t. player 2's mixed strategy.

To find the equilibria, we want to choose x_1 (x_2) so as to equalise the utility of player-2 (player-1) receives from playing either pure strategies. We therefore write the equations as: $u_1(0, x_2) = u_1(1, x_2)$, $u_2(x_1, 0) = u_2(x_1, 1)$, i.e.,

$$(-30 * x_2 - 144) * (18 * x_2 - 148) = (-15 * x_2 - 180) * (12 * x_2 - 120)$$
$$(-30 * x_1 - 144) * (32 * x_1 - 252) = (-15 * x_1 - 180) * (24 * x_1 - 216).$$

By solving the above equations, we get the Nash equilibria and the relevant utility values: $x_1 = 0.72$; $x_2 = 0.13$; $u_1 = 1.25$; $u_2 = 0.83$.

In order to check the solutions obtained, let us look at the experiment results of the utilities of layer $i = 1, 2$ with constant variables $x_1 = 0 : 1$ and $x_2 = 0 : 1$ produced by PRISM. Figure 2 presents the utilities to player 1's (cf. player 2's) pure strategies as functions of player 2's (cf. player 1's) mixed strategy. One can see that the intersection of the two lines are mixed strategy equilibria: i.e., player-1 chooses $x_1 = 0.72$, player-2 chooses $x_2 = 0.13$, which also meets our equilibria results produced by our PMCG analyser. In addition, if we set up the reward structure symmetrically for player 1 and 2, say both transitions s_1 and s_2 cost 2, both transitions r_1 and r_2 cost 3, we obtain the symmetric Nash equilibria for player 1 and 2: $x_1 = x_2 = 0.72$; $u_1 = u_2 = 0.83$. Table 1 lists a group of experimental results produced by our PMCG analyser regarding to different number of players, honest players, reward structures (in which c_{s_i} and c_{r_i} denote the cost of player i for sending and relaying a message respectively) with the size of the state space of the parametric Markov chain and the total computation time spent to achieve the final equilibria results. This demonstrates our proposed approach can be used to automatically find the mixed Nash equilibria for the Crowds protocol with multiple players. The computation time increases with the number of honest players, and the result of the Nash equilibria is mainly affected by the specified reward structures.

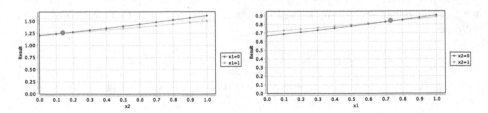

Fig. 2. Utilities to player 1's (left) and 2's (right) pure strategies as functions of player 2's and 1's mixed strategy

Table 1. Experimental results with size of state space and time cost

N	N_h	Reward structure $(c_{s_1}, c_{r_1}; \ldots; c_{s_{N_h}}, c_{r_{N_h}})$	Size of PMC	Total time (sec.)	Nash equilibria $(x_1, x_2, \ldots)$
3	2	$(1, 2; 2, 3)$	19	1.18	$(0.62, 0.46)$
4	2	$(1, 2; 2, 3)$	23	1.181	$(0.72, 0.13)$
4	3	$(1, 1; 1, 1; 1, 1)$	48	1.857	$(0.32, 0.32, 0.32)$
5	2	$(2, 3; 2, 3)$	35	1.878	$(0.29, 0.29)$
5	3	$(1, 0.8; 1, 0.8; 1, 0.9)$	57	2.003	$(0.038, 0.038, 0.407)$
7	2	$(1, 1.2; 1, 1.3)$	35	1.131	$(0.186, 0.608)$
7	3	$(1, 0.6; 1, 0.6; 1, 0.6)$	75	1.964	$(0.469, 0.469, 0.469)$
7	4	$(1, 0.5; 1, 0.5; 1, 0.5; 1, 0.5)$	131	116.145	$(0.15, 0.15, 0.15, 0.15)$

We present more details of how the reward structures affect the Nash equilibria results in Table 2 in Appendix A. It can be seen that the players tend to be more cooperative (larger equilibria) when the ratio of the cost of sending a message s_i and relaying a message r_i is bigger (see Table 2(a)). In the cases investigated, a range of such ratios results in *mixed* Nash equilibria, while the ratios outside of that range lead to *pure* Nash equilibria. The strategy of each player is also affected by other players' reward structures (see Table 2(b and c)).

7 Conclusions

We have presented a new automated game-theoretic approach for quantitative verification of security properties of software systems. Security-sensitive communication networks typically require the collaboration of their participants to work effectively. We study the problem of how the participants *should* react regarding to collaborating strategies in order to improve the overall performance with a balance between security and cost. We apply a game-theoretic approach to capture such a balance represented as Nash equilibria. We propose methods to automatically find the equilibria under which no participants can benefit by changing their strategies.

To achieve our goal, we propose the model of parametric Markov chain games and apply parametric model checking techniques to compute utility functions, using models described in the PRISM modelling language and utilities specified in probabilistic temporal logic. We generate and solve a polynomial equation system, from which we identify the Nash equilibria. To illustrate the applicability of our approach, we have implemented our approach as an extension to the tool of PRISM model checker, and analysed the Crowds protocol, studying the trade-offs between anonymity/cost.

Both theoretical and experimental evidence are presented for the utility of the approach for quantitative security analysis. We believe this is a significant contribution to automatically analysing security systems from a *quantitative* and *game-theoretic* view. For future work, we plan to study the precise computational complexity issues of the presented approach, and adapt our method in wider cases in addition to security systems.

Appendix A: Sensitivity Study of the Reward Structures

Table 2. Sensitivity study of the reward structures to Nash equilibria (N.E.s) & utilities ($N = 4, N_h = 3$)

(a)		
Reward structure $(c_{s_1}, c_{r_1}; c_{s_2}, c_{r_2}; c_{s_3}, c_{r_3})$	N.E. (x_1, x_2, x_3)	Utility (u_1, u_2, u_3)
$(1.0, 0.2;\ 1.0, 0.2;\ 1.0, 0.2)$	$(1.0, 1.0, 1.0)$	$(6.06, 6.06, 6.06)$
$(1.0, 0.4;\ 1.0, 0.4;\ 1.0, 0.4)$	$(1.0, 1.0, 1.0)$	$(4.92, 4.92, 4.92)$
$(1.0, 0.6;\ 1.0, 0.6;\ 1.0, 0.6)$	$(0.908, 0.908, 0.908)$	$(3.93, 3.93, 3.93)$
$(1.0, 0.8;\ 1.0, 0.8;\ 1.0, 0.8)$	$(0.55, 0.55, 0.55)$	$(2.753, 2.753, 2.753)$
$(1.0, 1;\ 1.0, 1;\ 1.0, 1.0)$	$(0.32, 0.32, 0.32)$	$(2.113, 2.113, 2.113)$
$(1.0, 1.2;\ 1.0, 1.2;\ 1.0, 1.2)$	$(0.16, 0.16, 0.16)$	$(1.713, 1.713, 1.713)$
$(1.0, 1.4;\ 1.0, 1.4;\ 1.0, 1.4)$	$(0.048, 0.048, 0.048)$	$(1.44, 1.44, 1.44)$
$(1.0, 1.6;\ 1.0, 1.6;\ 1.0, 1.6)$	$(0.0, 0.0, 0.0)$	$(1.304, 1.304, 1.304)$
$(1.0, 1.8;\ 1.0, 1.8;\ 1.0, 1.8)$	$(0.0, 0.0, 0.0)$	$(1.25, 1.25, 1.25)$
$(1.0, 2.0;\ 1.0, 2.0;\ 1.0, 2.0)$	$(0.0, 0.0, 0.0)$	$(1.2, 1.2, 1.2)$

(b)		
Reward structure $(c_{s_1}, c_{r_1}; c_{s_2}, c_{r_2}; c_{s_3}, c_{r_3})$	N.E. (x_1, x_2, x_3)	Utility (u_1, u_2, u_3)
$(1.0, 0.8;\ 1.0, 0.8;\ 1.0, 0.5)$	$(1.0, 1.0, 1.0)$	$(3.58, 3.58, 4.5)$
$(1.0, 0.8;\ 1.0, 0.8;\ 1.0, 0.6)$	$(0.908, 0.908, 0.191)$	$(2.78, 2.78, 3.93)$
$(1.0, 0.8;\ 1.0, 0.8;\ 1.0, 0.7)$	$(0.708, 0.708, 0.393)$	$(2.76, 2.76, 3.24)$
$(1.0, 0.8;\ 1.0, 0.8;\ 1.0, 0.8)$	$(0.55, 0.55, 0.55)$	$(2.75, 2.75, 2.75)$
$(1.0, 0.8;\ 1.0, 0.8;\ 1.0, 0.9)$	$(0.42, 0.42, 0.68)$	$(2.756, 2.756, 2.39)$
$(1.0, 0.8;\ 1.0, 0.8;\ 1.0, 1.0)$	$(0.321, 0.321, 0.779)$	$(2.763, 2.763, 2.113)$
$(1.0, 0.8;\ 1.0, 0.8;\ 1.0, 1.1)$	$(0.235, 0.235, 0.864)$	$(2.77, 2.77, 1.89)$
$(1.0, 0.8;\ 1.0, 0.8;\ 1.0, 1.2)$	$(0.16, 0.16, 0.94)$	$(2.78, 2.78, 1.713)$
$(1.0, 0.8;\ 1.0, 0.8;\ 1.0, 1.3)$	$(0.1, 0.1, 0.9997)$	$(2.79, 2.79, 1.56)$
$(1.0, 0.8;\ 1.0, 0.8;\ 1.0, 1.4)$	$(1.0, 1.0, 1.0)$	$(3.58, 3.58, 2.54)$

(c)		
Reward structure $(c_{s_1}, c_{r_1}; c_{s_2}, c_{r_2}; c_{s_3}, c_{r_3})$	N.E. (x_1, x_2, x_3)	Utility (u_1, u_2, u_3)
$(1.0, 1.0;\ 1.0, 1.0;\ 1.0, 0.5)$	$(1.0, 1.0, 1.0)$	$(3.15, 3.15, 4.5)$
$(1.0, 1.0;\ 1.0, 1.0;\ 1.0, 0.6)$	$(1.0, 1.0, 1.0)$	$(3.15, 3.15, 4.14)$
$(1.0, 1.0;\ 1.0, 1.0;\ 1.0, 0.7)$	$(1.0, 1.0, 1.0)$	$(3.15, 3.15, 3.84)$
$(1.0, 1.0;\ 1.0, 1.0;\ 1.0, 0.8)$	$(0.55, 0.55, 0.09)$	$(2.12, 2.12, 2.75)$
$(1.0, 1.0;\ 1.0, 1.0;\ 1.0, 0.9)$	$(0.42, 0.42, 0.22)$	$(2.11, 2.11, 2.39)$
$(1.0, 1.0;\ 1.0, 1.0;\ 1.0, 1.0)$	$(0.32, 0.32, 0.32)$	$(2.113, 2.113, 2.113)$
$(1.0, 1.0;\ 1.0, 1.0;\ 1.0, 1.1)$	$(0.235, 0.235, 0.407)$	$(2.114, 2.114, 1.892)$
$(1.0, 1.0;\ 1.0, 1.0;\ 1.0, 1.2)$	$(0.16, 0.16, 0.48)$	$(2.116, 2.116, 1.713)$
$(1.0, 1.0;\ 1.0, 1.0;\ 1.0, 1.3)$	$(0.1, 0.1, 0.54)$	$(2.12, 2.12, 1.56)$
$(1.0, 1.0;\ 1.0, 1.0;\ 1.0, 1.4)$	$(1.0, 1.0, 1.0)$	$(2.123, 2.123, 1.44)$

Appendix B: The PRISM Model of Crowds Protocol

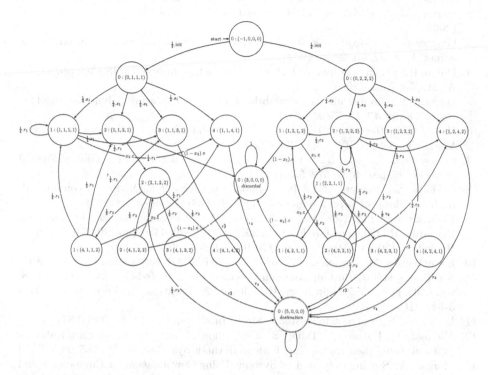

Fig. 3. Model of Crowds with 2 honest players and 2 malicious player with $PF = 0.5$. For $i \in \{1, 2, 3, 4\}$, transition label r_i denotes relaying a message by player i; for $i \in \{1, 2\}$, s_i denotes sending a message by (honest) player i, c_i denotes the player i decide to be cooperative, n_i denotes the i choose to be selfish. Label **init** denotes randomly pick up an honest player as a initiator to send out a message. State labelled as $i : (\text{status}, \text{from}, \text{to}, \text{sender})$ implies state $(\text{status}, \text{from}, \text{to}, \text{sender}) \in S_i$ for player $i \in \{0, 1, 2, 3, 4\}$, player $i = 0$ is used to model a coordinator, where the **status** $= 0, 1, 2, 3, 4, 5$ denotes that the sender is randomly picked up, the message is sent, the player decides to be cooperative, the player decides to be selfish and the message is discarded, and the message reaches the destination respectively.

(1) Cost structures for honest players $i = 1, 2$: assigns a cost of 1 and 2 to all transitions labelled with 's_1' and 'r_1' to player 1 respectively; and assigns a cost of 2 and 3 to all transitions labelled with 's_2' and 'r_2' to player 2 respectively.

(2) Property specification for honest players: the utility function of player i is defined as the probability of *good* behaviours/costs. We say a run is *good* if it reaches the destination without violating the anonymity properties.

References

1. Prototype tool and case studies. http://www.prismmodelchecker.org/files/qest19/
2. Carter, M.: Foundations of Mathematical Economics. MIT Press, Cambridge (2001)
3. Chatterjee, K., Majumdar, R., Jurdzinski, M.: On nash equilibria in stochastic games. In: CSL, pp. 26–40 (2004)
4. Datta, R.S.: Using computer algebra to find nash equilibria. In: ISSAC, pp. 74–79. ACM, New York (2003)
5. Datta, R.S.: Finding all nash equilibria of a finite game using polynomial algebra. Econ. Theory **42**(1), 55–96 (2009)
6. Hahn, E.M., Hermanns, H., Zhang, L.: Probabilistic reachability for parametric Markov models. STTT **13**(1), 3–19 (2011)
7. Hansson, H., Jonsson, B.: A logic for reasoning about time and reliability. Formal Aspects Comput. **6**(5), 512–535 (1994)
8. Herings, P.J., Peeters, R.J.A.P.: A globally convergent algorithm to compute all nash equilibria for n-person games. Ann. OR **137**(1), 349–368 (2005)
9. Kwiatkowska, M., Norman, G., Parker, D.: Stochastic model checking. In: Bernardo, M., Hillston, J. (eds.) SFM 2007. LNCS, vol. 4486, pp. 220–270. Springer, Heidelberg (2007). https://doi.org/10.1007/978-3-540-72522-0_6
10. Kwiatkowska, M., Norman, G., Parker, D.: PRISM 4.0: verification of probabilistic real-time systems. In: Gopalakrishnan, G., Qadeer, S. (eds.) CAV 2011. LNCS, vol. 6806, pp. 585–591. Springer, Heidelberg (2011). https://doi.org/10.1007/978-3-642-22110-1_47
11. Li, T.: Solving polynomial systems. Math. Intelligencer **8**(3), 33–39 (1987)
12. Mhamdi, T., Hasan, O., Tahar, S.: Evaluation of anonymity and confidentiality protocols using theorem proving. Formal Methods Syst. Des. **47**(3), 265–286 (2015)
13. Morgan, A.: Solving Polynominal Systems Using Continuation for Engineering and Scientific Problems. Society for Industrial and Applied Mathematics, Philadelphia (2009)
14. Nash, J.: Equilibrium points in n-person games. Proc. Nat. Acad. Sci. **36**(1), 48–49 (1950)
15. Osborne, M.J.: An Introduction to Game Theory. Oxford University Press, Oxford (2009)
16. Reiter, M.K., Rubin, A.D.: Crowds: anonymity for web transactions. ACM Trans. Inf. Sys. Secur. **1**, 66–92 (1998)
17. Shmatikov, V.: Probabilistic analysis of anonymity. In: CSFW, pp. 119–128. IEEE Computer Society Press (2002)
18. Shmatikov, V.: Probabilistic model checking of an anonymity system. J. Comput. Secur. **12**(3/4), 355–377 (2004)
19. Ummels, M., Wojtczak, D.: The complexity of nash equilibria in stochastic multiplayer games. Logic. Methods Comput. Sci. **7**(3) (2011)
20. Venkitasubramaniam, P., Tong, L.: A game-theoretic approach to anonymous networking. IEEE/ACM Trans. Netw. **20**(3), 892–905 (2012)
21. Verschelde, J.: Algorithm 795: PHCpack: a general-purpose solver for polynomial systems by homotopy continuation. ACM Trans. Math. Softw. **25**(2), 251–276 (1999)
22. Yang, M., Sassone, V., Hamadou, S.: A game-theoretic analysis of cooperation in anonymity networks. In: POST, pp. 269–289 (2012)

Probabilistic Modelling and Abstraction

Bayesian Abstraction of Markov Population Models

Luca Bortolussi[1,2] and Francesca Cairoli[1(✉)]

[1] DMG, University of Trieste, Trieste, Italy
`lbortolussi@units.it, francesca.cairoli@phd.units.it`
[2] MOSI, Saarland University, Saarbrücken, Germany

Abstract. Markov Population Models are a widespread formalism, with applications in Systems Biology, Performance Evaluation, Ecology, and many other fields. The associated Markov stochastic process in continuous time is often analyzed by simulation, which can be costly for large or stiff systems, particularly when simulations have to be performed in a multi-scale model (e.g. simulating individual cells in a tissue). A strategy to reduce computational load is to abstract the population model, replacing it with a simpler stochastic model, faster to simulate. Here we pursue this idea, building on previous work [3] and constructing an approximate kernel for a Markov process in continuous space and discrete time, capturing the evolution at fixed Δt time steps. This kernel is learned automatically from simulations of the original model. Differently from [3], which relies on deep neural networks, we explore here a Bayesian density regression approach based on Dirichlet processes, which provides a principled way to estimate uncertainty.

Keywords: Model abstraction · Markov Population Models · Bayesian density regression · Dirichlet processes

1 Introduction

Stochastic models are undoubtedly one of the most powerful frameworks to describe and reason about complex systems. Due to the severe state space explosion of these models, simulation is often the only viable tool to analyse them. Even simulation, however, can face severe computational limits, in particular when the systems of interest have a multi-scale nature. Consider for example a biological scenario, in which we want to model the effect of a drug targeting individual cells in a tissue, for instance a tumour. In order to build an accurate model of such a system, both the dynamics at the individual cells and the one at the tissue level have to be described and simulated. Unfortunately, tissues typically contain millions of cells, each requiring the simulation of complex interaction pathways [5]. This complexity defies also modern High Performance Computing resources, and can be tackled only by simplifying the model of each individual cell, i.e. resorting to model abstraction.

© Springer Nature Switzerland AG 2019
D. Parker and V. Wolf (Eds.): QEST 2019, LNCS 11785, pp. 259–276, 2019.
https://doi.org/10.1007/978-3-030-30281-8_15

Typical approaches in this direction require a large dose of experience and ingenuity to hand-craft a suitable abstraction. Recent alternatives rely on modern artificial intelligence to learn the best abstraction from a given class of models. The more general the class, the more flexible the method, the higher the learning cost.

Related Work. Some approaches in literature try to introduce as much knowledge in the abstraction as possible, thus reducing the complexity of the learning problem. A notable example is [13], in which authors exploit knowledge about the key drivers of bacterial chemotaxis, combining an abstract model of the dynamic of such drivers with a simple model with few states describing the decision of the bacteria, i.e. whether to rotate or proceed straight. The final model is a Continuous Time Markov Chain (CTMC) where transition rates are learned using Gaussian Process Regression from some simulations of the full model. On the other hand of the spectrum, we find the work of Palaniappan et al. [16], in which the authors start from a bunch of simulations of the original model, using information theoretic ideas extract a subset of relevant variables, discretize them and then learn a dynamic Bayesian network in discrete time. The abstract model was used for fast approximate simulation of the original model. The work on this paper follows these lines, starting from the approach of [3], in which we abstracted a CTMC model by discretizing time, choosing a time step Δt relevant for the dynamics of the higher organisational scale (e.g. the time-scale of the diffusion dynamics at the tissue level). The so obtained Discrete Time Markov Process is defined in continuous space, approximating the exact transition matrix by a transition kernel modelled as a mixture of Gaussian distributions with means and covariances taken as functions of the current state of the model. These functions are learned using Deep Neural Networks. Despite the method was effective in the examples studied, it is not without drawbacks. First of all, there is no quantification of the uncertainty in the so-learned kernel, hence no measure of confidence on the accuracy of the abstraction. Secondly, the number of mixture components is a hyperparameter strongly affecting the performance of the method.

Contributions. In this work, we continue the investigation started in [3], exploring the use of non-parametric Bayesian machine leaning [1] to provide a consistent estimate of uncertainty and to self-tune kernel complexity from data.

Our idea is to work with probability distributions in the space of transition kernels, defining a prior distribution and computing a posterior by conditioning on observed simulation data. If we fix the current state $\mathbf{x}$ in the model, distributions over kernels reduce to distributions over probability distributions of the next state $\mathbf{x}'$ after Δt units of time. To simulate the abstract model, we first need to sample a distribution, and then sample the next state from this distribution. Provided these sampling operations can be implemented efficiently, this could considerably speed-up simulation. Importantly, having a distribution of distributions allows us to incorporate uncertainty in the reconstruction of the kernel and identify initial states from which the reconstruction is more problematic. This information can be used to tune the accuracy of the abstraction, e.g. improve it in areas of the state space visited more often by the process.

The technical details of this approach are non-trivial. Fortunately, the problem we want to solve has been studied in statistics and machine learning and goes under the name of density regression [6,7]. Borrowing from this literature, we extend and tailor existing methods that use Dirichlet processes (essentially distributions over probability densities) to finally obtain a posterior distribution which is a mixture of Gaussian with a variable number of components, learned from data. The posterior distribution cannot be computed analytically, hence we rely on Gibbs sampling, a Monte Carlo method which has to be run only at training time. In fact, during training we will sample from the posterior and approximate it by its empirical distribution which permits a very fast sampling from the transition kernel. The method is validated on few experimental case studies, which we use to discuss potentials and limitations of this approach.

Paper Structure. In Sect. 2 we introduce some background material on stochastic models and case studies. Section 3 is devoted to present the model abstraction framework and formulate the learning problems, while density regression and the algorithm we use is described in detail in Sect. 4. Experiments are reported in Sect. 5, while the final discussion is in Sect. 6.

2 Background

2.1 Chemical Reaction Networks

Chemical Reaction Networks (CRNs) use the formalism of chemical equations to capture the dynamics of population models, including biological systems and epidemic spreading scenarios. Let $X_1, \ldots, X_m$ be a collection of m species and $\eta_{t,i}$, $i = 1, \ldots, m$, denotes the population size of species X_i present in the system at time t. The dynamics of a CRN is described by a set of reactions $\mathcal{R} = \{R_1, \ldots, R_p\}$. The firing of reaction R_i results in a transition of the system from state $\eta_t = (\eta_{t,1}, \ldots, \eta_{t,m}) \in \mathcal{S} = \mathbb{N}^m$ to state $\eta_t + \nu_i$, with ν_i being the update vector. A general reaction R_i is identified by the tuple (f_{R_i}, ν_i), where f_{R_i}, known as propensity function of reaction R_i, depends on the state of the system.

The time evolution of a CRN can be modelled as a Continuous Time Markov Chain (CTMC [14]) on the discrete space $\mathcal{S}$. Motivated by the well-known memoryless property of CTMC, let $\mathbb{P}_{s_0}(\eta_t = s)$ denote the probability of finding the system in state s at time t given that it was in state s_0 at time t_0. This probability satisfies a system of ODEs known as Chemical Master Equation (CME):

$$\partial_t \mathbb{P}_{s_0}(\eta_t = s) = \sum_{j=1}^{p} \left[f_{R_j}(s - \nu_j) \mathbb{P}_{s_0}(\eta_t = s - \nu_j) - f_{R_j}(s) \mathbb{P}_{s_0}(\eta_t = s) \right]. \quad (1)$$

Since the CME is a system in general with countably many differential equations, its analytic or numeric solution is almost always unfeasible. An alternative computational approach is to generate trajectories using stochastic algorithms for simulation, like the well-known the Gillespie's SSA [10].

The SIR Model. The SIR epidemiological model describes a population of N individuals divided in three mutually exclusive groups: susceptible (S), infected (I) and recovered (R). The system state at time t is $\eta_t = (S_t, I_t, R_t)$. The possible reactions, given by the interaction of individuals (representing the molecules of a CRN), are the following:

- $R_1 : S + I \xrightarrow{k_1 \cdot I_t S_t / N} 2I$ (infection),
- $R_2 : I \xrightarrow{k_2 \cdot I_t} R$ (recovery).

The model describes the spread, in a population with fixed size N, of an infectious disease that grants immunity to those who recover from it. As the SIR model is well-known and stable, we use it as a testing ground for our Bayesian abstraction procedure.

The Gene Regulatory Network Model. Consider a simple self-regulated gene network [2] in which a single gene G is transcribed to produce copies of a mRNA molecule M; each mRNA molecule can then be translated into a protein P. In addition P acts as a repressor with respect to the gene G. In other words, the gene activity is regulated through a negative-feedback loop, a common pattern in biological systems. The reactions are the following:

- $R_1 : G^{ON} \xrightarrow{k_{prodM} \cdot G^{ON}} G^{ON} + M$ (transcription)
- $R_2 : M \xrightarrow{k_{prodP} \cdot M} M + P$ (translation)
- $R_3 : G^{ON} \xrightarrow{k_{deact} \cdot G^{ON}} G^{OFF}$ (protein binding)
- $R_4 : G^{OFF} \xrightarrow{k_{act} \cdot G^{OFF}} G^{ON} + P$ (protein unbinding)
- $R_5 : M \xrightarrow{k_{degM} \cdot M} \varnothing$ (mRNA degradation)
- $R_6 : P \xrightarrow{k_{degP} \cdot P} \varnothing$ (protein degradation).

The system dynamic varies significantly accordingly with the choice of reaction rates. We refer to [2] for a detailed exploration of different behavioural patterns. According to our choice, see Sect. 5, the system exhibits several well-separated stable configurations. If we look at trajectories in Fig. 1 on a smaller scale, we notice that each stable point is actually noisy, i.e. there is a high number of small amplitude oscillations. Starting from an initial state η_0, we can sequentially generate a large number of instances of the state of the system at a fixed time t_1, η_{t_1}. The empirical probability distribution of such η_{t_1} approximates, for a number of samples sufficiently large, the density function $\mathbb{P}_{\eta_0}(\eta_t)$, shown in Fig. 3 (blue lines). The system exhibits up to 5 distinguishable modes, with some Gaussian noise affecting each one of them.

Motivated by the multimodality of this system we seek a Bayesian non-parametric density regression method, able to capture such hierarchical structure.

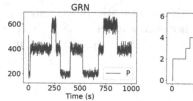
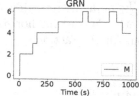
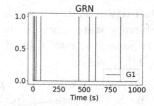

Fig. 1. Example of a trajectory, for the gene regulation network model. The overall time is 10,000 s. Left: trajectory of proteins (P), which present a multistable configuration. Middle: trajectory for the mRNA molecules (M). Right: trajectory for the active gene G^{ON}.

3   Model Abstraction

Let's now introduce the concept of model abstraction as presented in [3]. The underlying idea is the following: given a stochastic process $\{\eta_t\}_{t\geq 0}$ with transition probabilities $\mathbb{P}_{s_0}(\eta_t = s)$, we aim at finding another stochastic process whose trajectories are similar to the first one, but faster to simulate. Instead of working with transition probabilities themselves we rather use transition kernels. This requires a discretization of time. In other words, the process is considered only at time points with a fixed temporal distance. If we fix a time step Δt and an initial time instant $t_0 \in \mathbb{R}$, the states can be expressed as $\tilde{\eta}_i := \eta_{t_0+i\Delta t}$. The CTMC, $\{\eta_t\}_{t\geq 0}$, is now expressed as a time-homogeneous Discrete Time Markov Chain $\{\tilde{\eta}_i\}_i$ with transition kernel

$$K_d(s \mid s_0) = \mathbb{P}_{s_0}(\eta_{\Delta t} = s), \tag{2}$$

for all $s, s_0 \in \mathcal{S}$. Two additional approximations are required:

1. The abstract model takes values in $\mathcal{S}' = \mathbb{R}^m_{\geq 0}$, a continuous space in which the state space $\mathcal{S} = \mathbb{N}^m$ is embedded.
2. The kernel K_d, Eq. 2, is approximated by a new kernel $K(s' \mid s'_0)$ taking values in the continuous space $\mathcal{S}'$.

In constructing the approximate kernel $K(s' \mid s'_0)$, rather than trying to preserve the full behavior of the process, we restrict our attention to a time-bounded *reward function* r from $\mathcal{S}^M$ to an arbitrary space $\mathcal{T}$ (i.e. $\mathbb{R}$, $\mathbb{N}$, $\mathbb{B}$, or $\mathbb{R}^k$). Here M is an upper bound on the duration of discrete time trajectories we consider to evaluate the reward; we indicate time-bounded trajectories by $\tilde{\eta}_{[0,M]}$. Such a function r can be a projection, monitoring the number of molecules belonging to a certain subset of chemical species at a certain time step, or it can take Boolean values in $\mathbb{B} = \{0, 1\}$, representing the truth of a linear temporal property, for example checking if the system has entered into a dangerous region. Note that $r(\tilde{\eta}_{[0,M]})$ is a probability distribution on $\mathcal{T}$. The formal definition of an abstract model is the following.

Definition 1. Let $\eta = \{\eta_i\}_{i=0}^M$ be a discrete time stochastic process over an arbitrary state space $\mathcal{S}$, with $M \in \mathbb{N}_+$ a time horizon, and let $r : \mathcal{S}^M \to \mathcal{T}$ be the associated reward function. An abstraction of (η, r) is a triple $(\mathcal{S}', p, r', \eta' = \{\eta_i'\}_{i=0}^M)$ where:

- $\mathcal{S}'$ is the *abstract state space*;
- $p : \mathcal{S} \to \mathcal{S}'$ is the *abstraction function*;
- $r' : \mathcal{S}'^M \to \mathcal{T}$ is the *abstract reward*;
- $\eta' = \{\eta_i'\}_{i=0}^M$ is the *abstract discrete time stochastic process* over $\mathcal{S}'$.

Definition 2. Let $\varepsilon > 0$, η' is said to be ε-close to η with respect to d if, for almost any $s_0 \in \mathcal{S}$,

$$d\big(r(\eta_{[0,M]}), r'(\eta'_{[0,M]})\big) < \varepsilon \qquad \text{conditioned on } \eta_0 = s_0, \eta_0' = p(s_0). \tag{3}$$

Dataset Generation. The model abstraction procedure can be translated into a supervised learning problem. Choose n random initial states $\{s_0^{(j)}\}_{j=1}^n$ from $\mathcal{S}$. Starting from each of these states we run a simulation from t_0 to $t_1 := t_0 + \Delta t$. $\eta_{t_1}^{(j)}$ denotes the system state at time t_1 for each one of these simulations. By defining $\mathbf{x}_j := s_0^{(j)}$ and $\mathbf{y}_j := \eta_{t_1}^{(j)}$ for all $j \in \{1, \ldots, n\}$, we have thus built a dataset $\mathcal{D} := \{(\mathbf{x}_j, \mathbf{y}_j)\}_{j=1}^n$, where each $\mathbf{y}_j$ is a sample from the probability distribution $\mathbb{P}_{\mathbf{x}_j}(\eta_{\Delta t})$.

In order to validate the abstraction procedure, we choose a high number of different initial settings, different from the initial states of the training set, and from them a very large number of SSA trajectories is simulated. The empirical distribution obtained can be compared with the distribution estimated with the abstract kernel at these points.

Abstract Model Simulation. We now have an abstract model that can be used to simulate a trajectory. We just need to sample up to time horizon $M > 0$ from the approximate kernel K, starting from the initial state s_0 and initial time t_0. The simulated trajectory lies on the continuous state space $\mathcal{S}'$. Each time step of our simulations has thus a fixed computational cost that does not depend on the Δt chosen. This saves a lot of computational resources when simulating long trajectories. This algorithm can be easily employed in a multi-scale setting: we just need to train the kernel of the abstract model once, but after that a high number of simulations can be performed at a very high speed.

Measuring the Error. The error introduced by the abstract model, i.e. how much the abstract distribution differs from $r(\eta'_{[0,M]})$, is a fundamental ingredient to quantify. In general, the distance among two random distributions, X and Y, can be computed using the L_1 norm. In practice, this metric will be evaluated statistically, resulting in the so called histogram distance [4]

$$D(X, Y) = \sum_{i=1}^K \frac{|h_X(I_i) - h_Y(I_i)|L}{K}, \tag{4}$$

where $I_1, \ldots, I_K$ are K bins of size L and $h_*(I_i)$ indicates the number of samples in bin I_i. We call *self distance* the histogram distance between two set of samples drawn from the same distribution.

Confidence Intervals: The big advantage of keeping a Bayesian non-parametric perspective is the possibility to estimate the uncertainty in the approximation procedure. The transition kernel is estimated using a trace of predictive densities, from which we can reconstruct a confidence interval. We expect this interval to be tight when the abstract kernel well approximate the original kernel, and wide when the reconstruction is poor.

4 Bayesian Density Regression

Density estimation [9] is a well known process to model the density from which a given set of observations is drawn. If data are assumed to be distributed hierarchically, meaning each point belongs to a randomly chosen cluster and members of a cluster are further distributed randomly within that cluster, we are dealing with a data clustering problem. In order to place a prior probability on the structure of data, we may assume, for instance, that there are K normally distributed clusters, each cluster with its own parameters. The Bayesian non-parametric intuition is to work without pre-specifying the number of clusters K and select instead a random prior over an infinite set of clusters with infinitely many parameters. Dirichlet processes [8], which are the infinite-dimensional generalization of Dirichlet distributions, are used as prior on such unknown distribution. They are denoted as $DP(\alpha G_0)$, where α is the precision parameter and G_0 is the base measure. A parametric form, Gaussian in our case, with unknown parameters is usually chosen for G_0. Realizations of such process, $G \sim DP(\alpha G_0)$, are random distributions. In order to fit the model based on data, we should compute the posterior distribution over cluster probabilities and their associated parameters. Since we cannot write the posterior explicitly, we are going to draw samples from the posterior using a Gibbs sampling algorithm.

However, our problem, which is embedded in the supervised learning scenario, as presented in Sect. 3, is to estimate the conditional distribution of a variable $y \in \mathcal{Y}$, a one dimensional projection of the state space $\mathcal{S}$, that depends on a vector of covariates $\mathbf{x} \in \mathcal{X}$, which represents the entire state space $\mathcal{S}$. This task is called conditional density estimation or *density regression*. In other words, given n observations $\mathcal{D} = (\mathbf{x}_1, y_1), \ldots, (\mathbf{x}_n, y_n) \in \mathcal{X} \times \mathcal{Y}$, we would like to estimate, for a generic $\mathbf{x} \in \mathcal{X}$, the density on $\mathcal{Y}$ of the response variable y, i.e. the conditional density $f(y|\mathbf{x})$.

A simple solution under the assumption that $G \sim DP(\alpha G_0)$, would be to use the dependent DP approach of MacEachern [11], which relies on the stick-breaking representation [17] of DP:

$$G = \sum_{h=1}^{\infty} \pi_h \cdot \delta_{\theta_h}, \tag{5}$$

with $\pi_h / \prod_{l=1}^{h-1}(1 - \pi_l) \sim Beta(1, \alpha)$. In the formula above, $\pi = (\pi_h, h = 1, \ldots, \infty)$ is an infinite sequence of stick breaking weights, δ_θ is a degenerate distribution with all its mass at θ and $\theta = (\theta_h, h = 1, \ldots, \infty)$ are atoms sampled from G_0. In the form of a DP mixture of normal linear regression models, the stick-breaking process of (5) becomes

$$f(y_i|\mathbf{x}_i) = \sum_{h=1}^{\infty} \pi_h \cdot \mathcal{N}(y_i|\mathbf{x}_i\beta_h, \tau_h^{-1}), \tag{6}$$

where $\theta_h = (\beta_h, \tau_h)$ is still sampled from G_0.

The challenge is that unexpected changes in the shape of the density depending on the predictor values $\mathbf{x}$ may occur, therefore we cannot assume the distribution G to be constant over $\mathcal{X}$. In this more complex scenario, priors for a collection of dependent random distributions, $G_\mathbf{x}$ with $\mathbf{x} \in \mathcal{X}$, must be considered. Dunson, Pillai and Park [7] proposed a kernel-weighted mixture of DPs (WMDP), using a non-parametric mixture of linear regression models for the conditional density of y given $\mathbf{x}$. The conditional density function is expressed as a mixture of parametric densities:

$$f(y|\mathbf{x}) = \int_\Phi f(y|\mathbf{x}, \phi) \cdot dG_\mathbf{x}(\phi), \tag{7}$$

where $f(y|\mathbf{x}, \phi)$ is a known density on $\mathcal{Y}$ that depends on a parameter $\phi \in \Phi$ and $G_\mathbf{x}$ is a random mixing distribution on Φ indexed by the predictor $\mathbf{x} \in \mathcal{X}$. The unknown collection of mixture distributions is allowed to vary with predictors by defining a WMDP prior. See [7] for a detailed treatment.

Since mixtures of a sufficiently large number of Gaussian distributions have been proved to be able to approximate any distribution accurately, we focus on the following mixture of regression models:

$$f(y_i|\mathbf{x}_i) = \int \mathcal{N}(y_i|\mathbf{x}_i'\beta_i, \tau_i^{-1}) \cdot dG_\mathbf{x}(\phi_i), \tag{8}$$

with $\phi_i = (\beta_i, \tau_i)$. In order to limit the number of clusters, the WMDP prior proposed in [7] set restrictions on the uncountable collection of mixture distributions $G_\mathcal{X} = \{G_\mathbf{x} : \mathbf{x} \in \mathcal{X}\}$. For every $\mathbf{x} \in \mathcal{X}$, they express $G_\mathbf{x}$ as

$$G_\mathbf{x} = \sum_{i=1}^{n} \pi_i(\mathbf{x})G_{\mathbf{x}_i}^*, \qquad G_{\mathbf{x}_i}^* \sim DP(\alpha G_0), \tag{9}$$

where $\pi(\mathbf{x}) = [\pi_1(\mathbf{x}), \ldots, \pi_n(\mathbf{x})]$ is a vector of probability weights with $\sum_i \pi_i(\mathbf{x}) = 1$. This formulation introduces independent DP random basis distributions at each of the predictor values in the sample, and then mixes across these basis distributions to obtain a prior for the unknown mixture distribution, $G_\mathbf{x}$, at each possible predictor value, $\mathbf{x} \in \mathcal{X}$. Suppose that $(\phi_i|\mathbf{x}_i) \sim G_{\mathbf{x}_i}$, for $i = 1, \ldots, n$, then, by marginalizing out the infinite-dimensional WMDP prior, it is possible to obtain a generalization of the so called DP Polya urn scheme:

$$(\phi_i|\phi^{(i)}, X, \alpha) = \left(\frac{\alpha}{\alpha + w_i}\right) G_0 + \sum_{j \neq i} \left(\frac{w_{ij}}{\alpha + w_i}\right) \delta_{\phi_j}, \tag{10}$$

where w_{ij} are weights that depend on the function π, the DP parameter α and the set of observed predictors X, and $w_i = \sum_{j \neq i} w_{ij}$. In [6] the explicit specification of π was avoided, using a simpler and more interpretable form: $w_{ij} = w_\psi(x_i, x_j)$, where $w_\psi : \mathcal{X} \times \mathcal{X} \to [0,1]$ is a bounded kernel measuring how close two predictors are in terms of a distance measure d, with ψ a smoothing parameter controlling how rapidly $w_\psi(x, x') \to 0$ as $d(x, x')$ decreases. In the limit as $\psi \to \infty$, $w_\psi(x, x') = 0$ for any $x, x' \in \mathcal{X}$ having $d(x, x') > 0$. In addition, for all finite ψ, $\lim_{x \to x'} w_\psi(x, x') = 1$. Under this simplification, Eq. (10) can be written as

$$(\phi_i|\phi^{(i)}, X, \alpha, \psi) = \left(\frac{\alpha}{\alpha + w_i(\psi)}\right) G_0 + \sum_{h=1}^{k^{(i)}} \left(\frac{w_{ij}^*(\psi)}{\alpha + w_i(\psi)}\right) \delta_{\theta_h}^{(i)}, \tag{11}$$

where $\theta^{(i)} = (\theta_1^{(i)}, \ldots, \theta_{k^{(i)}}^{(i)})$ denotes the unique values of $\phi^{(i)}$ and $w_{ij}^*(\psi) = \sum_{j \neq i} 1_{(\phi_j = \theta_h^{(i)})} w_\psi(x_i, x_j)$ and $w_i(\psi) = \sum_{j \neq i} w_\psi(x_i, x_j)$. The prior in (11) automatically allocates the n subjects into $k \leq n$ clusters according to their ϕ_i values. Because subjects located close together are more likely to be clustered together, the prior tends to penalize changes across $\mathcal{X}$ in the values of parameters. The hyperparameters α and ψ control the speed at which the prior introduce new clusters as n increase: new clusters are added more rapidly as α increases and ψ decreases.

A natural choice for w_ψ, at least for continuous x, is the Gaussian kernel $w_\psi(x, x') = \exp(-\psi||x - x'||_2^2)$. Note that, with this kernel choice, it is important to standardize data, avoiding sensitivity to scales.

4.1 Posterior Computation

Following [7], the posterior distribution needs to be computed in order to integrate out the latent cluster parameters. Since the posterior distribution is not known explicitly, but the conditional distribution of each cluster variables is known, the Gibbs sampling algorithm results being an efficient technique to estimate the posterior. Let θ be a vector of length k containing the parameter values of each of the k Gaussians distributions in the mixture, i.e. $\theta_h = (\beta_h, \tau_h)$ for $h = 1, \ldots, k$. The vector $S = (S_1, \ldots, S_n)$ maps each subject to the cluster it is allocated to. In other words, $S_i = h$ if $\phi_i = \theta_h$. Excluding the i-th subject, $\theta^{(i)}$ denotes the $k^{(i)}$ unique values of $\phi^{(i)}$ and $S^{(i)}$ denotes the configuration of subjects $\{1, \ldots, n\} \setminus i$ to these values.

The full conditional posterior distribution of ϕ_i is

$$(\phi_i|\phi^{(i)}, \mathcal{D}, \alpha, \psi, \gamma) \propto q_{i,0} G_{i,0} + \sum_{h=1}^{k^{(i)}} q_{i,h} \delta_{\theta_h^{(i)}}. \tag{12}$$

In the formula above, $G_{i,0}$ is the posterior obtained by updating the prior $G_0(\phi|\gamma)$, where γ indicates the hyperparameters of the base measure G_0, with the likelihood $f(y_i|\mathbf{x}_i, \phi)$:

$$G_{i,0}(\phi) = \frac{G_0(\phi|\gamma)f(y_i|\mathbf{x}_i, \phi)}{\int f(y_i|\mathbf{x}_i, \phi)dG_0(\phi|\gamma)} := \frac{G_0(\phi|\gamma)f(y_i|\mathbf{x}_i, \phi)}{h_i(y_i|\mathbf{x}_i, \gamma)}. \tag{13}$$

In addition, $q_{i,0} = c\alpha h_i(y_i|\mathbf{x}_i, \gamma)$ and $q_{i,h} = cw_{ih}^*(\psi)f(y_i|\mathbf{x}_i, \theta_h)$, where c is the normalization constant. Note that α and ψ only appear in the expressions for the configuration probabilities $q_{i,h}$.

Conditional on α and ψ, posterior computation can proceed via a Gibbs sampling algorithm, which alternates between the following three steps.

(1) Updating the configuration of subjects to clusters, $\mathbf{S}$, and the number of clusters, k, by sequentially sampling from the full conditional posterior distribution of each S_i: $\mathbb{P}(S_i = h|\phi^{(i)}, \mathcal{D}) = q_{i,h}$, for $h = 0, 1, \ldots, k$. When $S_i = 0$ a new cluster is generated sampling from from $G_{i,0}$.

(2) Updating the cluster-specific parameters θ by sampling from the full conditional posterior given the configuration, i.e. $\mathbf{S}$ and k:

$$(\theta_h|\theta^{(h)}, \mathbf{S}, k, \mathcal{D}) \propto \left(\prod_{i:S_i=h} f(y_i|x_i, \theta_h) \right) G_0(\theta_h). \tag{14}$$

(3) Updating the hyperparameters γ by sampling from their full conditional posteriors.

4.2 Implementation

As already expressed in Eq. (8), we are considering the case in which $f(y_i|\mathbf{x}_i, \phi_i) = \mathcal{N}(y_i|\mathbf{x}_i'\beta_i, \tau_i^{-1})$, with $\phi_i = (\beta_i', \tau_i)'$ and $\beta_i = (\beta_{i1}, \ldots, \beta_{id})'$, so that both the regression coefficients and variance can vary across clusters. This generalizes [7] where only the mean parameter was allowed to vary, while τ was kept constant. For this particular choice, the posteriors distributions, needed at step (1), (2) and (3) of the Gibbs sampler, have simple closed forms. The detailed derivation of the following equations is described in the Appendix.

A natural choice for G_0 is the multivariate normal-gamma density

$$G_0(\beta_h, \tau_h) = \mathcal{NG}(\beta_h, \tau_h|\beta, \Sigma_\beta, a_\tau, b_\tau) = \mathcal{N}(\beta_h|\beta, \tau_h^{-1}\Sigma_\beta) \cdot \mathcal{G}(\tau_h|a_\tau, b_\tau).$$

Let $\gamma = \{\beta, \Sigma_\beta, a_\tau, b_\tau\}$ denote the set of hyperparameters. In order to provide more flexibility, we allow uncertainty in γ by choosing hyper-prior densities for β, Σ_β and b_τ, while fixing a_τ. More precisely, we choose multivariate normal prior for β, $p(\beta) = \mathcal{N}(\beta|\beta_0, \Sigma_0)$, a multivariate inverse-gamma prior for Σ_β, $p(\Sigma_\beta) = \mathcal{IW}(\Sigma_\beta|\nu_0, V_0)$, also known as inverse Wishart distribution with ν_0 degrees of freedom and mean V_0, and, finally, a gamma prior for b_τ, $p(b_\tau) = \mathcal{G}(b_\tau|a_0, b_0)$.

In step (1) of the Gibbs sampler we need to compute $h_i(y_i|\mathbf{x}_i, \gamma)$, which takes the following simple form:

$$h_i(y_i|\mathbf{x}_i, \beta, \Sigma_\beta, a_\tau, b_\tau) = \frac{C(a_\tau, b_\tau) \cdot \det \Sigma_\beta}{\sqrt{2\pi} C(\bar{a}_i, \bar{b}_i) \cdot \det \bar{\Sigma}_i},$$

where $\bar{\Sigma}_i = \mathbf{x}_i\mathbf{x}_i' + \Sigma_\beta^{-1}$, $\bar{a}_i = a_\tau + \frac{1}{2}$, $\bar{b}_i = b_\tau + \frac{1}{2}[(y_i - \mathbf{x}_i'\bar{\beta}_i)^2 + (\bar{\beta}_i - \beta)'\Sigma_\beta^{-1}(\bar{\beta}_i - \beta)]$ with $\bar{\beta}_i = \bar{\Sigma}_i^{-1}(\mathbf{x}_i y_i + \Sigma_\beta^{-1}\beta)$ and $C(a, b) = \frac{b^a}{\Gamma(a)}$, where $\Gamma(a)$ is the gamma function.

In addition, the full conditional posterior distribution of $\theta_h = (\beta_h', \tau_h)$, required at step (2), is

$$(\beta_h, \tau_h|\theta^{(h)}, \mathbf{S}, k, \gamma, \mathcal{D}) \sim \mathcal{N}_d(\beta_h|\tilde{\beta}, \tau_h^{-1}\tilde{\Sigma}_h^{-1}) \cdot \mathcal{G}(\tau_h|a_\tau + \frac{n_h}{2}, b_\tau + \frac{\xi_h^2}{2}).$$

If we denote with $\mathbf{X}_h$ and $\mathbf{Y}_h$ the vectors containing the values of predictors and responses for the n_h subjects in cluster h, the terms $\tilde{\Sigma}_h$, $\tilde{\beta}_h$ and ξ_h^2, in the formula above, are defined as follow: $\tilde{\Sigma}_h = (\mathbf{X}_h'\mathbf{X}_h + \Sigma_\beta^{-1})$, $\tilde{\beta}_h = \tilde{\Sigma}_h^{-1}(\mathbf{X}_h' \cdot \mathbf{Y}_h + \Sigma_\beta^{-1}\beta)$ and $\xi_h^2 = (\mathbf{Y}_h - \mathbf{X}_h\tilde{\beta}_h)' \cdot (\mathbf{Y}_h - \mathbf{X}_h\tilde{\beta}_h) + (\tilde{\beta}_h - \beta)' \cdot \Sigma_\beta^{-1} \cdot (\tilde{\beta}_h - \beta)$.

Finally, the conditional posterior distributions of hyper-parameters (β, Σ_β, b_τ), needed at step (3), are defined as follows.

The posterior for β is $(\beta|\theta, \beta_0, \Sigma_0) \sim \mathcal{N}(\beta|\hat{\beta}, \hat{\Sigma}^{-1})$ where $\hat{\Sigma} = (\sum_h \tau_h) \cdot \Sigma_\beta^{-1} + \Sigma_0^{-1}$ and $\hat{\beta} = \hat{\Sigma}^{-1}(\Sigma_0^{-1}\beta_0 + \sum_h \tau_h\Sigma_\beta^{-1}\beta_h)$.

The posterior for Σ_β is $(\Sigma_\beta|\beta_1, \ldots, \beta_k) \sim \mathcal{IW}(\Sigma_\beta|\eta_0 + k, V_0 + S)$, where $S = \sum_h \tau_h \cdot (\beta_h - \beta)(\beta_h - \beta)'$.

The b_τ posterior is $(b_\tau|\tau_1, \ldots, \tau_k) \sim \mathcal{G}(b_\tau|\tilde{a}, \tilde{b})$, where $\tilde{a} = a_0 + k \cdot a_\tau$ and $\tilde{b} = b_0 + \sum_h \tau_h$.

4.3   Conditional Predictive Density

Once the posterior distribution has been computed, we can finally estimate the response density for new subjects $\mathbf{x}_* \in \mathcal{X}$, i.e. $f(y_*|\mathbf{x}_*)$. This can be done using the simple form of the conditional predictive density:

$$f(y_*|\mathbf{x}_*, Y, X, \mathbf{S}, k, \theta, \gamma, \alpha, \psi) = \left(\frac{\alpha}{\alpha + w_*(\psi)}\right) h_*(y_*|\mathbf{x}_*, \gamma) +$$

$$+ \left(\frac{1}{\alpha + w_*(\psi)}\right) \sum_{h=1}^{k} \bar{w}_{*,h}(\psi)\mathcal{N}(y_*|\mathbf{x}_*'\beta_h, \tau_h^{-1}),$$

$$(15)$$

where $w_*(\psi) = \sum_{i=1}^{n} w_{*,i}(\psi)$ and $\bar{w}_{*,h} = \sum_{i=1}^{n} 1_{(S_i=h)} w_{*,i}(\psi)$. In words, this means that each normal component has a weight that depends on the number of subjects in the dataset allocated to that component and on the cumulative distance from these subjects and $\mathbf{x}_*$. Instead, if α is large or only few subjects in the dataset are close to $\mathbf{x}_*$, we will observe a shrinkage towards the first component.

Measuring Uncertainty in the Approximation. After convergence, each iteration t of the Gibbs sampler correspond to a mixture of Gaussians distributions, linear in $\mathbf{x}_*$, with parameters $\mathbf{S}^{(t)}, k^{(t)}, \theta^{(t)}, \gamma^{(t)}$. This density can be computed, using (15), for a dense grid of possible y_* values, obtaining a graphical representation of the mixture density. One can apply this procedure for a large number, T, of iterates after convergence of the Gibbs sampler, ending up with a trace of T predictive density distributions. From this trace, one can calculate the expected predictive density averaging over the large number of iterates. This remove also the conditioning on $\mathbf{S}, k, \theta, \gamma$. In practice, given $t = 1, \ldots, T$ we obtain the estimator:

$$\hat{f}(y_*|\mathbf{x}_*, \mathcal{D}, \alpha, \psi) = \frac{1}{T} \sum_{t=1}^{T} \left(\frac{\alpha}{\alpha + w_*(\psi)} \right) h_*(y_*|\mathbf{x}_*, \gamma^{(t)}) + \tag{16}$$

$$+ \left(\frac{1}{\alpha + w_*(\psi)} \right) \sum_{h=1}^{k} \bar{w}_{*,h}(\psi) \mathcal{N}(y_*|\mathbf{x}'_* \beta_h^{(t)}, \tau_h^{(t)\,-1}). \tag{17}$$

Furthermore, this pool of densities provide also an estimate of the variance of the predictive density. One can leverage this information to estimate the uncertainty underlying the abstraction procedure in a specific state $\mathbf{x}_* \in \mathcal{S}$.

Sampling from the Predictive Density. The same pool of T iterations can be used to sample from $f(\cdot|\mathbf{x}_*, \mathcal{D}, \alpha, \psi)$ in the following hierarchical way:

- randomly pick an iteration index $\hat{t} \in \{1, \ldots, T\}$,
- given the parameters at iteration $\hat{t}$, we sample a component $\hat{h}^{(\hat{t})}$ from the discrete vector of weights $[\alpha, \bar{w}_{*,1}, \ldots, \bar{w}_{*,k^{(\hat{t})}}]$, normalized by $(\alpha + w_*(\psi))$,
- from this $\hat{h}^{(\hat{t})}$ component, $\mathcal{N}(\cdot|\mathbf{x}'_* \beta_{\hat{h}^{(\hat{t})}}, \tau_{\hat{h}^{(\hat{t})}}^{-1})$ if $\hat{h}^{(\hat{t})} \neq 0$, we sample the desired value $\hat{y}$.

Iterating this procedure, we have an approximate simulation algorithm.

Consider a state of a d-dimensional system at time t, $\mathbf{x}_* = \eta_t$, and an abstract kernel trained on a given dataset and evaluated in $\mathbf{x}_*$. A sample from such kernel should return a full state of the system after a time Δt. In other words, the abstract kernel should approximate the joint probability distribution of the d variables composing the state space $\mathcal{S}$. Unfortunately, the proposed solution needs the response variable $y \in \mathcal{Y}$ to be one-dimensional, which means that we are actually approximating the marginal distribution of a single variable. Therefore, in order to simulate the full state of the system after a time Δt, i.e. $\eta_{t+\Delta t}$, we must sample from d different kernels, loosing the correlation between response variables. Nonetheless, the joint probability can be expressed in terms of chain of conditional probabilities. We are investigating a technique to approximate conditional distributions rather than marginals. The basic idea is to enlarge the input space by adding the response variables that condition the response of interest. Finally, the chain rule allows us to sample from the joint distribution while preserving the correlation between variables.

5   Experimental Results

We now validate the proposed Bayesian non-parametric approach on the two case studies introduced in Sect. 2: the SIR model and the Gene Regulatory Network (GRN) model.

Experimental Setting. Input data has been generated simulating the original CRN model by using both the direct and the τ-leaping SSA algorithms. The StochPy library (stochastic modeling in Python [12]) served for this purpose. The Gibbs sampling algorithm has been implemented in Python as well. All the computations were performed on a computer equipped with a CPU Intel x86 and 24 cores.

Setting the Hyperparameters. The proposed Bayesian method is nonparametric in the sense that it allows infinitely many parameters. However, some hyperparameters have to be fixed. As suggested in [6], we set, for both models, $\psi = n/25$, $a_\tau = 1$, $\beta_0 = 0$, $\Sigma_{\beta_0} = (\mathbf{X}'\mathbf{X})^{-1}$, $\nu_0 = d$, $V_0 = \mathbb{I}_d$, where d is the dimension of the state space, $a_0 = 1$ and $b_0 = 0.5$. In addition we set $\alpha = 0.5$ for SIR and, since we expect a larger number of cluster to be needed, we set $\alpha = 1$ for GRN. Data has been rescaled in order to have zero mean and variance one. This avoids sensitivity of the kernel function to different scales. Data has been scaled back after inference was performed, hence results are shown in the original scale. The Gibbs sampling algorithm performed $10,000$ iterations, with a burn-in period of $1,000$ iterations. The trace plots of different unknowns show that the convergence of the Gibbs sampler was rapid and mixing was good.

The training and validation set has been created as presented in Sect. 3. The time required to generate the dataset depends heavily on the length of the time step (Δt) considered and on the complexity of the model. For the SIR model, whose dynamic is rather simple, we fix $\Delta t = 0.1\,\text{s}$. The time required to simulate $10,000$ trajectories starting from a given state is $125\,\text{s}$, using the direct SSA method, and $23\,\text{s}$, using the τ-leaping method. For the GRN model, in order to observe a strong multimodality, we choose a much larger time interval: $\Delta t = 400\,\text{s}$. Here the computational effort required to generate the dataset is much higher. Generating $10,000$ trajectories takes around $40\,\text{h}$ with the direct SSA method, or around $20\,\text{h}$ with the τ-leaping approach. It is important to point out that the choice of Δt does not depend on the model itself, but rather on the intended use of abstraction. For instance, when using this abstraction in a multi-scale scenario (e.g. cells and tissues), Δt may be chosen as the integration step of the higher order model (e.g. the diffusion dynamics at the tissue level).

The training time of our approach increases with the number of training points. In order to sample from the predictive density and to compute the average predictive density we must fix the number T of Monte Carlo iterations that we are willing to consider. From a computational point of view, the sampling procedure is not affected by the dimension of T. However, computing the estimators for inter-densities mean and variance takes longer as T increases. Leveraging the multi-core hardware available, we ran, for the GRN model, 12 parallel Gibbs sampling for each response variable and we mixed the final traces by taking the last 200 iterations of each process. For the simple SIR model, we ran just 2 parallel processes. By doing so, a greater mixing can be achieved. Therefore, for the GRN model, $T = 1200$, whereas, for the SIR model, $T = 400$.

5.1 SIR Model

The constant parameters governing the transition rates are: $k_1 = 3$ and $k_2 = 1$. The model has been trained on a set of 500 observations. Experiments with more points showed similar results. Since we are considering a fixed population of size N, the model is two-dimensional ($d = 2$) and the number of recovered individuals is indeed $N - S - I$. We trained two separates models, one for each component/species. The first model predicts the number of susceptible individuals after Δt time units, whereas the second model predicts the number of infected individuals. The training of the two models has been performed in parallel and globally took 1.5 h. The method automatically fix the number of cluster needed. The average number of clusters is 7.2 for the response variable S and 22.1 for the response variable I. In Fig. 2, we can appreciate how a larger number of clusters is required as the intrinsic variance of the response I is larger with respect to the response variable S. Figure 2 shows the average density estimator, which is the average among T mixtures, against the empirical distribution obtained with 1,000 SSA trajectories. The shaded area denotes the 95% confidence interval, indicating the variance among the T mixtures. Given 1,000 samples from the true densities and 1,000 samples from the approximate densities, the average histogram distances, over 10 validation points and 100 bins, are 0.362 for response variable S and 0.235 for the response variable I. The self-distance bound is 0.357. The average estimator is a faithful model of the true density for all the validation points. However, sometimes the inter-densities variances are large, which results in potentially high variability while sampling. In fact, as we sample from a single normal component of a randomly picked mixture, the specific component may deviate considerably from the average estimator. This fact may introduce instabilities in trajectories simulated for more Δt time instants, resulting in an artificial increase of variance. We are currently exploring alternative simulation strategies based on the average estimator to ameliorate this problem. In any case, sampling 10,000 one-step trajectories from the abstract model takes on average 1.5 s, a considerable speed up of two orders of magnitude compared to the SSA algorithm.

5.2 GRN Model

The constants governing the transition rates are: $k_{prodP} = 350$, $k_{prodM} = 300$, $k_{degM} = 0.001$, $k_{degP} = 1.5$, $k_{deact} = 166$, $k_{act} = 1$. In contrast with the SIR model, which is unimodal, we set a long time interval, $\Delta t = 400$ s, in order to observe strong multimodality and test our approach in such extreme scenario.

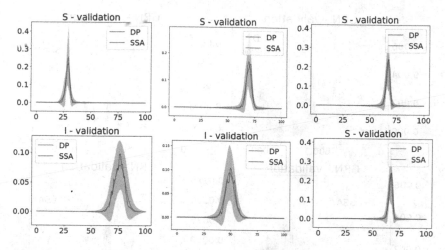

Fig. 2. SIR model: estimators of the average predictive densities and their confidence intervals, against the empirical simulated distributions for 3 validation points. Each column has a corresponding 2-dimensional value $\mathbf{x}_*$. The horizontal axis indicates the one-dimensional response space $\mathcal{Y}$. The first row shows the densities for the first component (susceptibles), while the second row shows results for the second component (infected). The grid on $\mathcal{Y}$, whose domain is $[0, 100] \subset \mathbb{N}$, contains 100 points. The empirical distribution is generated by 1,000 SSA trajectories.

Since the species G^{ON} and G^{OFF} are constrained, i.e. $G^{ON} + G^{OFF} = 1$, the model is three-dimensional ($d = 3$). We analyze the performance of our approach in predicting the protein outcomes, since it is the species with the multi-stable behaviour. The model has been trained on a set of 2,000 observations and training took around 20 h. The average number of clusters is 154.5. Figure 3 shows the average density estimator and the 95% confidence interval against the empirical distribution obtained through SSA simulations.

The true and the approximate distributions are clearly distinguishable, but the main qualitative characteristics of the system are captured. The DP approach manages to recognize the 3 modes associated with highest probability. When it does not recognize a mode, it shows tails with high-variance. The last picture of Fig. 3 shows the behaviour in situations where the new input point $\mathbf{x}_*$ has no neighbouring training points: it doesn't recognizes the modes but the variance is extremely large, reflecting the uncertainty in the reconstruction. Given 1,000 samples from the true densities and 1000 samples from the approximate densities, the average histogram distances, over 25 validation points and 200 bins, is 0.12 (the self-distance bound is 0.504).

The analyzed case studies coincide with the ones presented in [3], thus, results may be compared. The use of deep neural networks renders the abstract model more accurate and the training is faster with respect to the non-parametric approach, but provides no estimate of the uncertainty. However, once the training is over, the time required by the two methods to simulate an abstract trajectory is

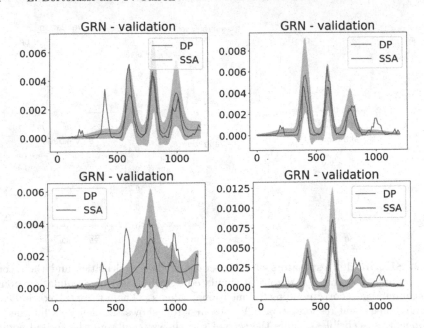

Fig. 3. GRN model: estimators of the average predictive densities and their confidence intervals, against the empirical simulated distributions for 4 validation points. The horizontal axis indicates the one-dimensional response space corresponding to the protein P. The grid on $\mathcal{Y}$ contains 100 points. The empirical distribution is generated by 1, 000 SSA trajectories. (Color figure online)

similar. It is reasonable to assume that the main reason behind the lower accuracy of our approach is due to the smaller training set, 2, 000 instead of 30, 000 data points. In fact, since the Monte Carlo method used is quite expensive, the Bayesian model is trained with relatively few points. The ability of the DP to recognize the modes is likely to increase as n grows. In order to significantly speed up the computations, we plan to develop a more efficient and compiled implementation of the algorithm. Finally, simulating 10, 000 one-step trajectories from the abstract model took on average 30 s, a tremendous speed-up compared to the SSA and τ-leaping algorithms.

6   Discussion

We presented a Bayesian approach to abstract the kernel of a Markov process. We used a mixture of Gaussian distributions with a mixing measure, a Dirichlet Process, flexible enough to vary accordingly with the state of the population. This work presents a first analysis on the performances of the proposed method, both in terms of accuracy and in terms of computational speed-up.

Results are encouraging, both in terms of the accuracy of the average estimator as a function of the input point and in terms of the computational gain. The variance of the reconstructed density is also a good indicator of the uncertainty in the reconstruction, and can be used to guide an active learning step to refine the abstraction improving in target points with low accuracy.

Another important issue is to improve the accuracy of the method when it comes to iterate the sampling of the kernel to generate trajectories longer than Δt. One possibility is to sample from the average estimator, rather than from a single component of the mixture. The drawback in this case is the increased cost of simulation per step, which may be tamed by learning a simplified model (e.g. a mixture of Gaussians with a fixed number of components) that interpolates the average estimator.

An additional improvement, also in terms of speedup may come from using either variational approaches for density regression [15]. Finally, we also plan to extend this approach to approximate more general classes of models, like time inhomogeneous systems (including time as a covariate), including non-Markovian models.

References

1. Barber, D.: Bayesian Reasoning and Machine Learning. Cambridge University Press, Cambridge (2012)
2. Bodei, C., Bortolussi, L., Chiarugi, D., Guerriero, M.L., Policriti, A., Romanel, A.: On the impact of discreteness and abstractions on modellingnoise in gene regulatory networks. Comput. Biol. Chem. **56**, 98–108 (2015)
3. Bortolussi, L., Palmieri, L.: Deep abstractions of chemical reaction networks. In: Češka, M., Šafránek, D. (eds.) CMSB 2018. LNCS, vol. 11095, pp. 21–38. Springer, Cham (2018). https://doi.org/10.1007/978-3-319-99429-1_2
4. Cao, Y., Petzold, L.: Accuracy limitations and the measurement of errors in the stochastic simulation of chemically reacting systems. J. Comput. Phys. **212**(1), 6–24 (2006)
5. Deisboeck, T.S., Wang, Z., Macklin, P., Cristini, V.: Multiscale cancer modeling. Annu. Rev. Biomed. Eng. **13**(1), 127–155 (2011). https://doi.org/10.1146/annurev-bioeng-071910-124729
6. Dunson, D.B.: Empirical bayes density regression. Statistica Sinica **17**(2), 481 (2007)
7. Dunson, D.B., Pillai, N., Park, J.H.: Bayesian density regression. J. Roy. Stat. Soc.: Ser. B (Stat. Methodol.) **69**(2), 163–183 (2007)
8. Ferguson, T.S.: A bayesian analysis of some nonparametric problems. Ann. Stat. **1**, 209–230 (1973)
9. Gelman, A., Stern, H.S., Carlin, J.B., Dunson, D.B., Vehtari, A., Rubin, D.B.: Bayesian Data Analysis. Chapman and Hall/CRC, Boca Raton (2013)
10. Gillespie, D.T., Petzold, L.: Numerical simulation for biochemical kinetics. In: Szallasi, Z., Stelling, J., Periwal, V. (eds.) Systems Modelling in Cellular Biology, pp. 331–354. MIT Press, Cambridge (2006)
11. Ishwaran, H., James, L.F.: Gibbs sampling methods for stick-breaking priors. J. Am. Stat. Assoc. **96**(453), 161–173 (2001)
12. Maarleveld, T.R., Olivier, B.G., Bruggeman, F.J.: Stochpy: a comprehensive, user-friendly tool for simulating stochastic biological processes. PLoS ONE **8**(11), e79345 (2013)

13. Michaelides, M., Hillston, J., Sanguinetti, G.: Statistical abstraction for multi-scale spatio-temporal systems. In: Bertrand, N., Bortolussi, L. (eds.) QEST 2017. LNCS, vol. 10503, pp. 243–258. Springer, Cham (2017). https://doi.org/10.1007/978-3-319-66335-7_15
14. Norris, J.R.: Markov Chains. Cambridge University Press, Cambridge (1998)
15. Nott, D.J., Tan, S.L., Villani, M., Kohn, R.: Regression density estimation with variational methods and stochastic approximation. J. Comput. Graph. Stat. 21(3), 797–820 (2012)
16. Palaniappan, S.K., Bertaux, F., Pichené, M., Fabre, E., Batt, G., Genest, B.: Abstracting the dynamics of biological pathways using information theory. Bioinformatics 33, 1980–1986 (2017)
17. Sethuraman, J.: A constructive definition of dirichlet priors. Statistica Sinica 4, 639–650 (1994)

UTOPIC: Under-Approximation Through Optimal Control

Josu Doncel[1], Nicolas Gast[2], Mirco Tribastone[3], Max Tschaikowski[4(✉)],
and Andrea Vandin[5]

[1] University of the Basque Country UPV/EHU, Leioa, Spain
[2] Inria, Univ. Grenoble Alpes, Saint-Martin-d'Hères, France
[3] IMT School for Advanced Studies Lucca, Lucca, Italy
[4] Technische Universität Wien, Wien, Austria
max.tschaikowski@tuwien.ac.at
[5] DTU Technical University of Denmark, Lyngby, Denmark

Abstract. We consider a class of nonlinear systems of differential equations with uncertainties, i.e., with lack of knowledge in some of the parameters that is represented by a time-varying unknown bounded functions. An under-approximation of such systems consists of a subset of its reachable set, for any value of the unknown parameters. By relying on optimal control theory through Pontryagin's principle, we provide an algorithm for the under-approximation of a linear combination of the state variables in terms of a fully automated tool-chain named UTOPIC. This allows to establish tight under-approximations of common benchmarks models with dimensions as large as sixty-five.

Keywords: Under-approximation · Uncertain nonlinear dynamics · Modeling

1 Introduction

Many safety-critical systems can be modeled as nonlinear ordinary differential equations (ODEs) which exhibit uncertain parameters due to finite precision measurements, lack of data, or noise [5]. In order to formally decide whether such a system is safe, it is necessary to determine whether the model can reach a bad state [9,14,36]. Since closed-form expressions for reachable sets of nonlinear ODE systems are not known in general [35], it becomes necessary to over- and under-approximate the reachable set. Over-approximation provides a superset of the reachable set, and thus can be used to prove that a model under study is safe. Instead, under-approximation gives a subset of the reachable set; therefore, it can be used to prove the presence of problems. While conceptually different, over- and under-approximation techniques are technically closely related.

In the case of linear dynamics with uncertain parameters, the reachable set can be shown to be convex. This has paved the way for efficient over- and under-approximation techniques for systems of linear ODEs, see for

© Springer Nature Switzerland AG 2019
D. Parker and V. Wolf (Eds.): QEST 2019, LNCS 11785, pp. 277–291, 2019.
https://doi.org/10.1007/978-3-030-30281-8_16

instance [4, 27, 28, 34] and references therein. Under-approximation techniques for nonlinear ODEs received less attention in the past, and often rely on over-approximation techniques for nonlinear systems. Indeed, by relying on the Hamilton-Jacobi equation, a special case of the Hamilton-Jacobi-Bellman equation used in over-approximation techniques [33], the recent work [38, 51] constructs a sequence of convex programs for polynomial ODE systems. The sequence is proven to converge to the actual reachable set but is truncated in practical computations to ensure a compromise between tightness and computational cost. [51] has been extended to cover time-varying uncertainties and non-singleton initial sets [52]. In [16, 53], instead, the main idea is to reverse time of the original ODE system and to over-approximate the so obtained backward flow while ensuring certain topological properties of the over-approximation. Similarly to [51], the work [53] was extended to cover time-varying uncertainties and non-singleton initial sets [54]. Another line of research is [29] and its extension [30] where over- and under-approximations are obtained via Taylor models.

The aforementioned approaches share the following features:

(i) the full reachable set (and not some projection of it) is approximated;
(ii) the approximation is given across the whole time course;
(iii) in addition to ODE parameters, the initial conditions may or must be uncertain.

Instead, the technique (presented in Sect. 2) and the tool (presented in Sect. 3)—UTOPIC—presented in this paper focus on the under-approximation of a one-dimensional projection of the reachable set at a given time point in the presence of time-varying uncertainties and fixed initial condition. The one-dimensional projection is given by a linear combination of state variables. Our restriction essentially amounts to computing an under-approximation of a one-dimensional linear projection of the reachable set; the whole time course can then be covered by performing the under-approximation for each grid point of a sufficiently fine discretization of the time interval.

The basic idea underlying our approach is to interpret uncertain parameters as controls and to minimize and maximize a given linear combination of state variables using Pontryagin's maximum principle, a well-established technique from optimal control theory [31]. The interval spanned by the so-obtained extrema can be shown to be an under-approximation of the reachable set of the linear combination. Moreover, thanks to the fact that Pontryagin's maximum principle is a necessary condition for optimality, the aforementioned interval is likely to be tight enough to cover the actual reachable set in many cases. This is confirmed by a numerical evaluation of UTOPIC in Sect. 4 on classical benchmark models from [15, 24], where the under-approximation is compared against over-approximations computed by state-of-the-art reachability analysis tools such as Flow* [15] and CORA [3]. Further, we show the scalability of the algorithm on larger scale models from biochemistry [13, 18] with 65 state variables.

Related Work. The most closely related approaches are [7,10] which propose the use of Pontryagin's maximum principle, but for over-approximation. Due to the fact that Pontryagin's maximum principle is only a necessary condition for optimality such a solution provides only a heuristic estimation which is not guaranteed to be valid in general [10]. The authors of [7] address this by restricting to bilinear systems for which Pontryagin's maximum principle can sometimes be shown to be a sufficient condition for optimality.

We next provide a brief account on over-approximation techniques for nonlinear ODEs. The abstraction approach locally approximates the nonlinear dynamics by multivariate polynomials or affine maps, see [2,6,15,16,18]. While abstraction techniques can cover many practical models, they are prone to the curse of dimensionality, i.e., their run-time may be exponential in the size of the ODE system, as discussed in [2,22]. A similar remark applies to the worst-case complexities of techniques relying on the Hamilton-Jacobi-Bellman equation [33] and Satisfiability Modulo Theories (SMT) solvers [32], even though certain combinations with model reductions exist [47]. Another classic approach is based on Lyapunov-like functions known from the stability theory of ODEs [21,44,48,55]. Unfortunately, for nonlinear systems the automatic computation of Lyapunov-like functions remains a challenging task. Restricting to special classes of such functions (e.g., sum-of-squares polynomials [26]) leads to efficient construction algorithms which may provide tight bounds, but existence is not guaranteed. Approximations with differential inequalities [45,49] and interval arithmetics [40,46], on the other hand, can be computed efficiently, but are loose in general [10].

To the best of our knowledge, C2E2 [24] is the only tool that supports under-approximation, but only for linear systems with time-varying uncertain parameters [22]. Instead, over-approximation is supported for nonlinear systems with time-invariant uncertain parameters [23]. As UTOPIC covers the under-approximation of nonlinear systems with time-varying uncertain parameters (and C2E2 relies on closed-form expressions in the case of linear systems that will outperform any known over- and under-approximation technique for nonlinear systems), Flow* and CORA are the most related tools to our work because they both support nonlinear systems with time-varying uncertain parameters. That is why we chose them for our numerical tests. In a broader context, we mention here also the tools Breach [20] and S-Taliro [42] which can be used to falsify given properties by relying on approximate sensitivity analysis and randomized testing, respectively.

2   Theoretical Background

Problem Statement. We consider a system of ODEs $\dot{x} = f(x, u)$ over state variables $x_1, \ldots, x_n$, and parameter variables $u_1, \ldots, u_m$. We assume that our dynamical system is affine in controls [39], i.e., it holds that

$$f(x, u) = g_0(x) + \sum_{j=1}^{m} u_j g_j(x),$$

for some differentiable functions $g_0, g_1, \ldots, g_m \colon \mathbb{R}^n \to \mathbb{R}^n$.

Given an initial condition $x(0) \in \mathbb{R}^n$, a finite time horizon $T > 0$, and a weight vector $\alpha = (\alpha_1, \ldots, \alpha_n) \in \mathbb{R}^n$, our goal is to compute an under-approximation at time T of the one-dimensional projection

$$\mathcal{P}_\alpha(T) = \Big\{ \sum_{i=1}^n \alpha_i x_i(T) \mid \dot{x} = f(x, u), u \in \mathcal{U}(\underline{\beta}, \overline{\beta}) \Big\},$$

where the set $\mathcal{U}(\underline{\beta}, \overline{\beta})$, defined as

$$\mathcal{U}(\underline{\beta}, \overline{\beta}) = \{ u : [0; T] \to \prod_{j=1}^m [\underline{\beta}_j; \overline{\beta}_j] \mid u \text{ measurable} \},$$

describes the admissible parameter functions. We assume that $\dot{x} = f(x, u)$ has a solution on $[0; T]$ for any admissible function u, thus excluding in particular finite explosion times.

We seek to compute parameter functions $\underline{u}, \overline{u} \in \mathcal{U}(\underline{\beta}, \overline{\beta})$ such that

$$\underline{u} \in \arg\min \Big\{ \sum_{i=1}^n \alpha_i x_i(T) \mid u \in \mathcal{U}(\underline{\beta}, \overline{\beta}), \dot{x} = f(x, u) \Big\}$$

$$\overline{u} \in \arg\max \Big\{ \sum_{i=1}^n \alpha_i x_i(T) \mid u \in \mathcal{U}(\underline{\beta}, \overline{\beta}), \dot{x} = f(x, u) \Big\} \tag{1}$$

Thanks to the fact that our dynamical system is affine in controls, the above optimization problems have solutions [39].

Problem Solution. We tackle the optimization problems (1) by observing that Pontryagin's principle implies that any optimal parameter function $\underline{u}$ (resp. $\overline{u}$) solves, almost everywhere on $[0; T]$, the differential inclusion

$$\dot{x}(t) = f(x(t), u(t)) \tag{2}$$

$$\dot{p}(t) = -(\partial_x H)(x(t), u(t), p(t)) \tag{3}$$

$$u(t) \in \arg\max_{\underline{\beta} \le u \le \overline{\beta}} H(x(t), u(t), p(t)) \tag{4}$$

where $H(x, u, p) = \langle p, f(x, u) \rangle$ denotes the Hamiltonian, $\langle \cdot, \cdot \rangle$ refers to the scalar product on $\mathbb{R}^n$, and we set $p_i(T) = -\alpha_i$ (resp., $p_i(T) = \alpha_i$) for all $1 \le i \le n$ in the case of minimization (resp., maximization); in (4), the notation $\underline{\beta} \le u \le \overline{\beta}$ is shorthand for $\underline{\beta}_j \le u_j \le \overline{\beta}_j$ for all $1 \le j \le m$.

It is possible to approximate an optimal control satisfying (2–4) by relying on a gradient projection algorithm [31, Section 6.2]. To this end, we consider the cost function

$$J : \mathcal{U}(\underline{\beta}, \overline{\beta}) \to \mathbb{R}, \quad u \mapsto \sum_{i=1}^n \alpha_i x_i(T),$$

where x is a solution of $\dot{x}(t) = f(x(t), u(t))$. The main idea to lift the common gradient descent algorithm over reals to the space of functions. More specifically, one interprets J as a functional on the Hilbert space $\mathcal{H} = L^\infty([0;T])$, where

$$L^\infty([0;T]) = \{h : [0;T] \to \mathbb{R}^m \mid h \text{ is measurable and}$$
$$\|h(t)\|_\infty \leq C \text{ for almost all } t \text{ for some finite } C > 0\}$$

endowed with the scalar product $\langle g, h \rangle = \int_0^T g(s)h(s)ds$. With this, it is possible to prove (see [43]) that J is Fréchet differentiable at any function vector $u \in \mathcal{U}(\underline{\beta}, \overline{\beta}) \subseteq \mathcal{H}$ and that its derivative is given by the function vector

$$(\partial J)(u) : t \mapsto -(\partial_u H)(x(t), u(t), p(t))$$

That is, the value of the derivative at the function vector $t \mapsto u(t)$ is given by the function vector $t \mapsto -(\partial_u H)(x(t), u(t), p(t))$. (Note the similarity to the common gradient descent method over the reals where the value of the derivative at a point vector is given by another point vector.)

In order to compute $(\partial J)(u)$ for a given u, we first solve (2); afterwards, we compute p from (3) which depends on u and x. Similarly to the gradient projection algorithm in finite dimensions, a given $u \in \mathcal{U}(\underline{\beta}, \overline{\beta})$ is either stationary or there is some $\gamma > 0$ such that

$$J\big(\mathfrak{P}(u + \gamma \cdot (\partial J)(u))\big) < J(u),$$

where $\mathfrak{P} : \mathcal{H} \to \mathcal{U}(\underline{\beta}, \overline{\beta})$ is the projection from $\mathcal{H}$ onto $\mathcal{U}(\underline{\beta}, \overline{\beta})$. Following [43] and references therein, $\mathfrak{P}(v) \in \mathcal{U}(\underline{\beta}, \overline{\beta})$, where $v \in \mathcal{H}$, is given by

$$(\mathfrak{P}(v))(t) = \begin{cases} v_i(t) & \text{if } \underline{\beta}_i \leq v_i(t) \leq \overline{\beta}_i \\ \overline{\beta}_i & \text{if } v_i(t) > \overline{\beta}_i \\ \underline{\beta}_i & \text{if } v_i(t) < \underline{\beta}_i \end{cases}$$

We are now in a position to perform a gradient (descent) projection algorithm that computes an approximation of an optimal control u: Algorithm 1 takes as inputs the maximal number of steps ν, the positive step sizes $\gamma^1, \gamma^2, \ldots, \gamma^\nu$ and the numerical threshold $\varepsilon \geq 0$ which triggers a termination based on a relative convergence criterion when $|J(u^k) - J(u^{k-1})| < \varepsilon$, where u^k denotes the approximation of the optimal control u obtained after k steps.

The following result can be proven.

Theorem 1. *For a sufficiently small step $\gamma_0 > 0$, Algorithm 1 induces, if applied to $\nu = \infty$, $\varepsilon = 0$ and the constant sequence $(\gamma_0, \gamma_0, \ldots)$, a sequence of controls $(u^k)_k$ which, in turn, induces a decreasing sequence $(J(u^k))_k$ which converges to a stationary control of the cost functional J.*

Proof. In the case of unbounded controls, the proof is at the heart of the calculus of variations, see [39, Section 3.4]. The case of bounded controls, instead, requires a refined reasoning, see [39, Chapter 5] and [43]. $\qquad\square$

```
Gradient(ẋ = f(x,u), x(0), α, β, γ, T, ε, ν)
  set k = 1
  choose some initial parameter function uᵏ
  while(true)
    solve ẋᵏ = f(xᵏ,uᵏ) forward in time using uᵏ and x(0)
    solve ṗᵏ = -(∂ₓH)(xᵏ,uᵏ,pᵏ) backward in time using uᵏ, xᵏ, p(T)
    compute Jᵏ = ∑ⁿᵢ₌₁ αᵢxᵏ(T)
    compute uᵏ⁺¹(·) = 𝔓(uᵏ(·) - γᵏ · (∂ᵤH)(xᵏ(·),uᵏ(·),pᵏ(·)))
    if (k ≥ ν ∨ (k ≥ 2 ∧ |Jᵏ - Jᵏ⁻¹| < ε))
      break
    else
      set k = k + 1
    end
  end
  return uᵏ⁺¹
```

Algorithm 1. Gradient projection algorithm for solving (1). Inputs are: the ODE system $\dot{x} = f(x, u)$, the initial values $x(0)$, the weight vector α, the boundary vector β, the time horizon T, a sequence of positive step sizes γ, the numerical threshold $\varepsilon \geq 0$, and the maximal number of steps $\nu \geq 0$.

We remark that Algorithm 1 requires the user to provide a sequence of step sizes $(\gamma^k)_k$. Those are not known a priori and have to be either inferred through trial and error or estimated by some other means. In the case of unconstrained optimization, estimations can be obtained via backtracking. For constrained optimization such as here, instead, the basic line search algorithm can be used. For more information on step size estimation, see [19]. Another point worth noticing is that $(J(u^k))_k$ may converge to a saddle, instead of an extreme, control. In practice, this problem can be tackled by introducing a random noise term. A formal proof of this heuristics has been recently established in the finite dimensional case [37].

We invoke Algorithm 1 for the computation of optimal control candidates $\underline{u}'$ and $\overline{u}'$ for $\underline{u}$ and $\overline{u}$, respectively, see (1). As stated next, $\underline{u}'$ and $\overline{u}'$ induce solution points $\underline{x}'(T)$ and $\overline{x}'(T)$ whose projections span a subinterval of $\mathcal{P}_\alpha(T) \subseteq \mathbb{R}$.

Theorem 2. *Assume that $\underline{u}'$ and $\overline{u}'$ are candidates for $\underline{u}$ and $\overline{u}$ underlying $p(T) = -\alpha$ and $p(T) = \alpha$, respectively. Then, if $\underline{x}'$ and $\overline{x}'$ denotes the solution underlying $\underline{u}'$ and $\overline{u}'$, respectively, it holds that*

$$\Big[\sum_{i=1}^n \alpha_i \underline{x}'_i(T); \sum_{i=1}^n \alpha_i \overline{x}'_i(T)\Big] \subseteq \mathcal{P}_\alpha(T)$$

when $\sum_{i=1}^n \alpha_i \underline{x}'_i(T) \leq \sum_{i=1}^n \alpha_i \overline{x}'_i(T)$.

Proof. Given two arbitrary piecewise continuous functions $\underline{u}'$ and $\overline{u}'$, we define the function $u'_\lambda = \lambda \underline{u}' + (1 - \lambda)\overline{u}'$ for all $0 \leq \lambda \leq 1$ and note that $u'_\lambda \in \mathcal{U}(\underline{\beta}, \overline{\beta})$ for all $0 \leq \lambda \leq 1$. Using this, we further define $\Phi(\lambda) = \sum_{i=1}^n \alpha_i x_i(T)$, where x satisfies $\dot{x} = f(x, u'_\lambda)$. Thanks to the intermediate value theorem, it thus suffices to show that Φ is continuous. Since for any $C > 0$ and $\varepsilon > 0$ there exists a $\delta > 0$ such that $|\lambda - \lambda'| < \delta$ implies $\|f(x, u'_\lambda(t)) - f(x, u'_{\lambda'}(t))\|_\infty \le \varepsilon$ for all $x \in [-C; C]^n$ and $0 \le t \le T$, Gronwall's inequality yields the claim. $\qquad\square$

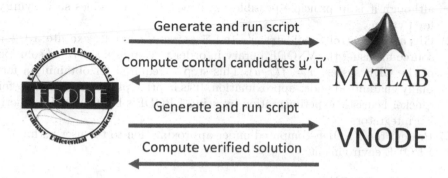

Fig. 1. Pictorial representation of UTOPIC's workflow.

3   UTOPIC

Architecture. Our approach is implemented as a fully automated workflow named UTOPIC—under-approximation through optimal control. UTOPIC involves the following tools: ERODE, MATLAB, and VNODE [41], where:

- ERODE [12], a Java-based cross-platform tool for the specification, evaluation, and reduction of nonlinear ODEs, providing an integrated development environment.
- MATLAB, and in particular the *Symbolic Math Toolbox* to symbolic computation.
- VNODE [41], a C++-based validated solver for computing bounds on the solution of ODEs. Traditional ODE solvers compute approximate solutions. Instead, VNODE first proves that a unique solution exists, and then computes formal bounds that are guaranteed to contain it.

Workflow. UTOPIC is implemented within the ERODE environment by means of the dedicated `utopic` command. The input is a specification of a nonlinear ODE system. Currently, this consists of the differentiable fragment of the IDOL language [11], which corresponds to nonlinear ODEs with derivatives that are multivariate rational expressions over the system's variables. The execution of the ERODE specification triggers a fully automated orchestration, as depicted in Fig. 1. In particular ERODE performs the following actions.

(1) Generate and execute a MATLAB specification implementing Algorithm 1 for the ODE under study, in order to compute the control candidates $\underline{u}'$, and $\overline{u}'$. This activity is delegated to an external tool to leverage an already available symbolic engine in order to perform the differentiations in (2)–(4). We remark that MATLAB is the tool of choice in this version of UTOPIC, although it is in principle possible to allow for other engines such SymPy for Python.

(2) Given the control candidates $\underline{u}'$, and $\overline{u}'$, create, compile, execute a C++ source file using the VNODE library in order to compute a *verified* solution of the ODE system $\dot{x} = f(x, u)$. This step is required for obtaining a formally guaranteed under-approximation. It is in principle possible to allow for quicker heuristic experimentation by using ERODE's built-in ODE numerical integrators.

(3) Present the VNODE-computed under-approximation to the user within the ERODE environment.

Deployment. ERODE does not require any installation process, and it is available, together with a manual, at http://sysma.imtlucca.it/tools/erode. Instead, http://www.cas.mcmaster.ca/~nedialk/ provides installation notes on VNODE.

Example. Figure 2 shows a screenshot of the standard Brusselator model [15, 24] defined in the ERODE editor, together with an example of the utopic command which triggers the analysis workflow. In particular, the file defines the ODEs

$$\dot{x} = x^2 y - bx + a$$
$$\dot{y} = -x^2 y + cx \tag{5}$$

with ODE variables x, y and parameters a, b and c. The initial conditions are chosen to be $x(0) = 1.0$ and $y(0) = 1.2$. Instead, the uncertain parameters u are specified using the paramsToPerturb argument, while the weight vector α specifying the query is given with the coefficients argument. Finally, delta specifies the numerical error threshold ε of Algorithm 1, while step is used to describe a sequence of uniform step sizes $\gamma = (\text{step}, \text{step}, \ldots)$. The maximal number of iterations performed by the algorithm is given by kMax. Instead, integrationStep specifies the integration time step used in the algorithm.

In the case of the Brusselator model, setting coefficient = { x : 1.0 } and tEnd = 1.0 ensures that utopic computes an under-approximation of the reachable set

$$\left\{ 1.0 \cdot x(1.0) \mid (\dot{x}(t), \dot{y}(t)) = f(x(t), y(t), a(t), b(t), c(t)) \right.$$

$$\left. \text{where } a(\cdot) \in [0.9; 1.1], b(\cdot) \in [2.3; 2.7], c(\cdot) \in [1.1; 1.6] \text{ and } t \in [0; 1.0] \right\},$$

where f is as in (5). Note that paramsToPerturb overwrites the parameter choice $a = 1.0$, $b = 2.5$ and $c = 1.5$ at the beginning of the file; we require the specification of all parameters because only a subset of parameters is subject to

Fig. 2. Graphical-user interface of ERODE in the case of the standard Brusselator model [15,24]. The ERODE file defines an ODE system with two ODE variables (x and y) and three parameters (a, b and c). In the visualized example, all three parameters are subject to uncertainty, thus turning them in time-varying uncertain parameter functions. For instance, $a(t) \in [0.9; 1.1]$ for all $t \in [0; 1.0]$.

uncertainty in general. The field `coefficient` can be used to describe arbitrary weighted sums of state variables, e.g., `coefficient = { x : 0.5, y : 2.0 }` leads to the under-approximation of the set

$$\left\{ 0.5 \cdot x(1.0) + 2.0 \cdot y(1.0) \mid (\dot{x}(t), \dot{y}(t)) = f(x(t), y(t), a(t), b(t), c(t)) \right.$$

$$\left. \text{where } a(\cdot) \in [0.9; 1.1], b(\cdot) \in [2.3; 2.7], c(\cdot) \in [1.1; 1.6] \text{ and } t \in [0; 1.0] \right\}.$$

Executing the model in Fig. 2 presents the under-approximation results as

```
Maximum reachable point enclosure at t = [1.0, 1.0]
[1.17, 1.17]
Minimum reachable point enclosure at t = [1.0, 1.0]
[0.578, 0.578]
```

This is the processed output of VNODE execution. Being formally an over-approximation technique that is based on interval arithmetics [40,46], VNODE returns intervals instead of points. This is because any numerical integration scheme introduces truncations due to finite-precision arithmetics. Thus, already after one time step, the solution of a verified ODE solver has to be represented by a (very small) set instead of a point. Algorithmically, verified integration there-fore corresponds to over-approximation in the special case where the initial set is a singleton. This greatly simplifies the task and often leads to over-approximation sets with negligibly small diameters. In the above example, for instance, VNODE returns $[1.17, 1.17]$ and $[0.578, 0.578]$.

Complexity Analysis. The complexity of Algorithm 1 depends on the number of state and parameter variables n and m, respectively, the step sizes $(\gamma_1, \gamma_2, \ldots)$, the numerical threshold ε and the maximal number of iterations ν. Assuming that ν is finite, the approach requires one to solve a sequence of n dimensional ODE systems whose length is polynomial in ν and m. Heuristic ODE solvers like ode45 of MATLAB enjoy a polynomial time and space complexity in n and m. Instead, the complexity of the two VNODE invocations in the last step of the workflow, see Fig. 1, may be computationally expensive and depend, in general, on factors such as the stiffness of the ODE system, the time discretization step and the order of the integration scheme. This notwithstanding we stress that many practical models can be covered efficiently because the initial set is a singleton, as outlined in the preceding paragraph.

4   Evaluation

Set-Up. In this section we evaluate UTOPIC on benchmark models from the literature. The main measure of performance is the scalability with respect to the ODE system size. In addition, we investigated the tightness of the under-approximation by comparing it against an over-approximation, since the exact reachable sets are not known for these models. For this comparison we used the tools Flow* [15] and CORA [3]. Ariadne [8] or HySAT/iSAT [25] were not consid-ered because they have been compared to Flow* already [15]; similarly, C2E2 [24] was not considered because it does not support time-varying uncertainties.

In all experiments, the error threshold ε of Algorithm 1 was set to zero (thus enforcing ν iterations). The parameters γ and ν were found by trail and error by looking for values that yielded a bang-bang control; this was motivated by the fact that the vast majority of optimal controls can be expected to belong to this class if the input sets are compact [39]. To avoid that the algorithm does not "overshoot" due to too large γ values, γ was initialized with a small number and increased iteratively, while comparing the control plots $u_1, \ldots, u_m$. The expert settings of CORA have been chosen as in the nonlinear tank model in the CORA manual [1]. A similar approach was taken in the case of Flow*.

Benchmarks Description. The benchmark models are listed in the first column of Table 1. In addition to models taken from the literature, we considered a *binding model*, a biochemical model that can be parameterized in order to generate instances of increasing size.

The binding model is a simple protein-interaction network, taken from [13], where an enzyme, E, can form a complex with a substrate, P, according to a

Table 1. Numerical evaluation (with termination denoted by ✓). Unless specified, the benchmark models are taken from https://publish.illinois.edu/c2e2-tool/example/ (hybrid models were omitted, since Algorithm 1 applies to continuous systems only). TO: time-out (>10 h) during the symbolic computation of the Jacobian matrices; (a) Divide-by-zero error; (b) "Cannot compute flowpipes" error. The number of binding sites is shown within parenthesis alongside the model name.

Model	Size	UTOPIC					CORA			Flow*		
		Min	Max	γ	ν	Term.	Min	Max	Term.	Min	Max	Term.
Brusselator	2	0.5780	1.1700	8	20	✓	−0.3931	2.3485	✓	−1.3139	3.1541	✓
Buckling	2	−1.1829	−0.9271	8	20	✓	−1.2039	−0.8817	✓	−1.4706	−0.6234	✓
JetEngine	2	−0.8717	−0.5813	100	50	✓	−1.0332	−0.3968	✓	−1.6446	−0.7168	✓
Van der Pol	2	0.9330	1.2300	10	100	✓	0.8690	1.3530	✓	0.7414	1.4785	✓
Robot Arm	4	0.597	0.738	1000	200	✓	0.5453	0.7767	✓	0.3719	0.9565	✓
Enzyme [17]	7	1.4773	1.7592	1000	50	✓	1.2473	1.9398	✓	−0.2956	3.4850	✓
Oscillator [50]	9	0.1242	0.1393	5	50	✓	0.1216	0.1422	✓	0.0910	0.1730	✓
Helicopter	28	2.3137	3.3600	2000	25	✓	2.1429	3.5551	✓	—		—b
Binding (2)	5	0.0454	0.0467	5	20	✓	0.0452	0.0482	✓	0.0428	0.0491	✓
Binding (3)	9	0.0242	0.0249	5	20	✓	0.0239	0.0265	✓	−0.0280	0.3149	✓
Binding (4)	17	0.0114	0.0117	10	20	✓	0.0112	0.0123	✓	−0.1816	0.3997	✓
Binding (5)	33	0.0104	0.0109	50	20	✓	—		TO	—		—b
Binding (6)	65	0.0100	0.0108	100	50	✓	—		TO	—		—b

reversible reaction occurring at any of the n independent and identical binding sites of P. The model can be specified using the chemical reaction network

$$P_{(s_1,\dots,s_i,0,s_{i+1},\dots,s_n)} + E \xrightarrow{u_{a_i}} P_{(s_1,\dots,s_i,1,s_{i+1},\dots,s_n)}$$

$$P_{(s_1,\dots,s_i,E,s_{i+1},\dots,s_n)} \xrightarrow{u_{d_i}} P_{(s_1,\dots,s_i,0,s_{i+1},\dots,s_n)} + E,$$

where the first reaction models the binding of E to to the i-th binding site of P; instead, the second reaction refers to the unbinding. The parameters u_{a_i} and u_{d_i}, for $i = 1,\dots,n$, give the association and dissociation constants for this model, respectively. The dynamics is visualized in Fig. 3, while the underlying ODEs are given, for any $n \geq 1$, by

$$\dot{x}_{P_s}(t) = - \sum_{i:s(i)=1} u_{d_i}(t)x_{P_s}(t) - \sum_{i:s(i)=0} u_{a_i}(t)x_{P_s}(t)x_E(t)$$

$$+ \sum_{i:s(i)=0} u_{d_i}(t)x_{A_{s+e^i}}(t) + \sum_{i:s(i)=1} u_{a_i}(t)x_{A_{s-e^i}}(t)x_E(t)$$

$$\dot{x}_E(t) = \sum_s \Big[\sum_{i:s(i)=1} u_{d_i}(t)x_{P_s}(t) - \sum_{i:s(i)=0} u_{a_i}(t)x_{P_s}(t)x_E(t) \Big]$$

In above ODEs, $s = (s_1,\dots,s_n) \in \{0,1\}^n$ and $e^i \in \{0,1\}^n$ is such that $e^i_j = 1$ if $i = j$ and $e^i_j = 0$ otherwise. The ODE system of dimension $2^n + 1$ is obtained by interpreting the chemical reactions according to common law of mass action.

Results. The results reported in Table 1 confirm that the under-approximated reachable intervals are contained in the respective over-approximations.

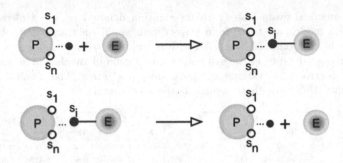

Fig. 3. Visualization of the binding unbinding pattern appearing in biochemical models [13]. The protein molecule P has n binding sites $s_1, \ldots, s_n$ that can be either empty or occupied by the molecule E.

In particular, in the models where all techniques terminated (we fixed a time-out of 10 h on a machine equipped with a 2.7 GHz CPU core and 8 GB RAM), the under-approximation computed by UTOPIC is close to the tightest over-approximation. Since the binding models could neither be analyzed by CORA nor Flow* beyond the four-domain binding example (corresponding to 17 ODEs), larger model instances feature only results of UTOPIC.

We did not compare the runtimes because a fair comparison would be difficult for the following reasons:

(i) CORA and Flow* are over-approximation techniques, while UTOPIC is an under-approximation technique.
(ii) The runtimes may be sensibly affected by the choice of parameters.
(iii) CORA and Flow* approximate the whole reachable set across the whole time interval, while UTOPIC considers a one-dimensional linear projection of the state space at a given time point.

However, we report that the largest runtime of our algorithm was around ten minutes (of which eight were due to VNODE) for the binding model with 6 sites, indicating good scalability with increasing system sizes.

5   Conclusion

We presented a tool for the under-approximation of reachable sets of nonlinear ODEs with time-varying uncertainties. The approach was implemented in UTOPIC, a fully automated tool-chain involving ERODE [12], MATLAB and VNODE [41]. An extensive evaluation demonstrated that UTOPIC provides tight under-approximations in benchmark models and scales to nonlinear systems with sizes as large as sixty-five. Overall, the results suggest that our algorithm can be a useful tool in the analysis of uncertain nonlinear systems.

Acknowledgement. Josu Doncel is supported by the Marie Sklodowska-Curie grant No. 777778, the Basque Government, Spain, Consolidated Research Group Grant IT1294-19, & the Spanish Ministry of Economy and Competitiveness project MTM2016-76329-R. Max Tschaikowski is supported by a Lise Meitner Fellowship funded by the Austrian Science Fund (FWF) under grant No. M-2393-N32 (COCO).

References

1. Althoff, M.: Cora 2016 manual
2. Althoff, M.: Reachability analysis of nonlinear systems using conservative polynomialization and non-convex sets. In: HSCC, pp. 173–182 (2013)
3. Althoff, M.: An introduction to CORA 2015. In: Proceedings of the Workshop on Applied Verification for Continuous and Hybrid Systems (2015)
4. Althoff, M., Le Guernic, C., Krogh, B.H.: Reachable set computation for uncertain time-varying linear systems. In: HSCC pp. 93–102 (2011)
5. Althoff, M., Stursberg, O., Buss, M.: Reachability analysis of nonlinear systems with uncertain parameters using conservative linearization. In: CDC, pp. 4042–4048 (2008)
6. Asarin, E., Dang, T., Girard, A.: Reachability analysis of nonlinear systems using conservative approximation. In: Maler, O., Pnueli, A. (eds.) HSCC 2003. LNCS, vol. 2623, pp. 20–35. Springer, Heidelberg (2003). https://doi.org/10.1007/3-540-36580-X_5
7. Asarin, E., Dang, T.: Abstraction by projection and application to multi-affine systems. In: Alur, R., Pappas, G.J. (eds.) HSCC 2004. LNCS, vol. 2993, pp. 32–47. Springer, Heidelberg (2004). https://doi.org/10.1007/978-3-540-24743-2_3
8. Benvenuti, L., et al.: Reachability computation for hybrid systems with ariadne. In: Proceedings of the 17th IFAC World Congress, vol. 41, pp. 8960–8965 (2008)
9. Bisgaard, M., Gerhardt, D., Hermanns, H., Krcál, J., Nies, G., Stenger, M.: Battery-aware scheduling in low orbit: the GomX-3 case. In: FM, pp. 559–576 (2016)
10. Bortolussi, L., Gast, N.: Mean field approximation of uncertain stochastic models. In: DSN (2016)
11. Cardelli, L., Tribastone, M., Tschaikowski, M., Vandin, A.: Symbolic computation of differential equivalences. In: POPL (2016)
12. Cardelli, L., Tribastone, M., Tschaikowski, M., Vandin, A.: ERODE: a tool for the evaluation and reduction of ordinary differential equations. In: Legay, A., Margaria, T. (eds.) TACAS 2017. LNCS, vol. 10206, pp. 310–328. Springer, Heidelberg (2017). https://doi.org/10.1007/978-3-662-54580-5_19
13. Cardelli, L., Tribastone, M., Tschaikowski, M., Vandin, A.: Maximal aggregation of polynomial dynamical systems. PNAS **114**(38), 10029–10034 (2017)
14. Chen, T., Forejt, V., Kwiatkowska, M.Z., Parker, D., Simaitis, A.: Automatic verification of competitive stochastic systems. In: TACAS, pp. 315–330 (2012)
15. Chen, X., Ábrahám, E., Sankaranarayanan, S.: Flow*: an analyzer for non-linear hybrid systems. In: Sharygina, N., Veith, H. (eds.) CAV 2013. LNCS, vol. 8044, pp. 258–263. Springer, Heidelberg (2013). https://doi.org/10.1007/978-3-642-39799-8_18
16. Chen, X., Sankaranarayanan, S., Ábrahám, E.: Under-approximate flowpipes for non-linear continuous systems. In: FMCAD, pp. 59–66 (2014)
17. Craciun, G., Tang, Y., Feinberg, M.: Understanding bistability in complex enzyme-driven reaction networks. PNAS **103**(23), 8697–8702 (2006)

18. Dang, T., Guernic, C.L., Maler, O.: Computing reachable states for nonlinear biological models. TCS **412**(21), 2095–2107 (2011)
19. Dennis Jr., J.E., Schnabel, R.B.: Numerical Methods for Unconstrained Optimization and Nonlinear Equations. SIAM (1996)
20. Donzé, A.: Breach, a toolbox for verification and parameter synthesis of hybrid systems. In: Touili, T., Cook, B., Jackson, P. (eds.) CAV 2010. LNCS, vol. 6174, pp. 167–170. Springer, Heidelberg (2010). https://doi.org/10.1007/978-3-642-14295-6_17
21. Duggirala, P.S., Mitra, S., Viswanathan, M.: Verification of annotated models from executions. In: EMSOFT, pp. 26:1–26:10 (2013)
22. Duggirala, P.S., Viswanathan, M.: Parsimonious, simulation based verification of linear systems. In: Chaudhuri, S., Farzan, A. (eds.) CAV 2016. LNCS, vol. 9779, pp. 477–494. Springer, Cham (2016). https://doi.org/10.1007/978-3-319-41528-4_26
23. Fan, C., Mitra, S.: Bounded verification with on-the-fly discrepancy computation. In: Finkbeiner, B., Pu, G., Zhang, L. (eds.) ATVA 2015. LNCS, vol. 9364, pp. 446–463. Springer, Cham (2015). https://doi.org/10.1007/978-3-319-24953-7_32
24. Fan, C., Qi, B., Mitra, S., Viswanathan, M., Duggirala, P.S.: Automatic reachability analysis for nonlinear hybrid models with C2E2. In: Chaudhuri, S., Farzan, A. (eds.) CAV 2016. LNCS, vol. 9779, pp. 531–538. Springer, Cham (2016). https://doi.org/10.1007/978-3-319-41528-4_29
25. Fränzle, M., Herde, C., Teige, T., Ratschan, S., Schubert, T.: Efficient solving of large non-linear arithmetic constraint systems with complex boolean structure. J. Satisf. Boolean Model. Computation **1**, 209–236 (2007)
26. Girard, A., Pappas, G.J.: Approximate bisimulations for nonlinear dynamical systems. In: CDC, pp. 684–689 (2005)
27. Girard, A., Le Guernic, C.: Efficient reachability analysis for linear systems using support functions. In: Proceedings of the 17th IFAC World Congress, vol. 41, pp. 8966–8971 (2008)
28. Girard, A., Le Guernic, C., Maler, O.: Efficient computation of reachable sets of linear time-invariant systems with inputs. In: Hespanha, J.P., Tiwari, A. (eds.) HSCC 2006. LNCS, vol. 3927, pp. 257–271. Springer, Heidelberg (2006). https://doi.org/10.1007/11730637_21
29. Goubault, E., Putot, S.: Forward inner-approximated reachability of non-linear continuous systems. In: HSCC, pp. 1–10 (2017)
30. Goubault, E., Putot, S.: Inner and outer reachability for the verification of control systems. In: HSCC, pp. 11–22 (2019)
31. Kirk, D.E.: Optimal Control Theory: An Introduction. Dover Publications, Mineola (1970)
32. Kong, S., Gao, S., Chen, W., Clarke, E.: dReach: δ-reachability analysis for hybrid systems. In: Baier, C., Tinelli, C. (eds.) TACAS 2015. LNCS, vol. 9035, pp. 200–205. Springer, Heidelberg (2015). https://doi.org/10.1007/978-3-662-46681-0_15
33. Kurzhanski, A.B., Varaiya, P.: Dynamic optimization for reachability problems. J. Optim. Theory Appl. **108**(2), 227–251 (2001)
34. Kurzhanski, A.B., Varaiya, P.: Ellipsoidal techniques for reachability analysis: internal approximation. Syst. Control Lett. **41**(3), 201–211 (2000)
35. Lafferriere, G., Pappas, G.J., Yovine, S.: A new class of decidable hybrid systems. In: Vaandrager, F.W., van Schuppen, J.H. (eds.) HSCC 1999. LNCS, vol. 1569, pp. 137–151. Springer, Heidelberg (1999). https://doi.org/10.1007/3-540-48983-5_15
36. Larsen, K.G.: Validation, synthesis and optimization for cyber-physical systems. In: Legay, A., Margaria, T. (eds.) TACAS 2017. LNCS, vol. 10205, pp. 3–20. Springer, Heidelberg (2017). https://doi.org/10.1007/978-3-662-54577-5_1

37. Lee, J.D., Simchowitz, M., Jordan, M.I., Recht, B.: Gradient descent only converges to minimizers. In: COLT, pp. 1246–1257 (2016)
38. Li, M., Mosaad, P.N., Fränzle, M., She, Z., Xue, B.: Safe over- and under-approximation of reachable sets for autonomous dynamical systems. In: Jansen, D.N., Prabhakar, P. (eds.) FORMATS 2018. LNCS, vol. 11022, pp. 252–270. Springer, Cham (2018). https://doi.org/10.1007/978-3-030-00151-3_15
39. Liberzon, D.: Calculus of Variations and Optimal Control Theory: A Concise Introduction. Princeton University Press, Princeton (2011)
40. Ramon, E., Moore, R., Kearfott, B., Cloud, M.J.: Introduction to Interval Analysis. SIAM (2009)
41. Nedialkov, N.S.: Implementing a rigorous ODE solver through literate programming. In: Rauh, A., Auer, E. (eds.) Modeling, Design, and Simulation of Systems with Uncertainties. Mathematical Engineering, vol. 3, pp. 3–19. Springer, Heidelberg (2011). https://doi.org/10.1007/978-3-642-15956-5_1
42. Nghiem, T., Sankaranarayanan, S., Fainekos, G.E., Ivancic, F., Gupta, A., Pappas, G. J.: Monte-carlo techniques for falsification of temporal properties of non-linear hybrid systems. In: HSCC, pp. 211–220 (2010)
43. Nikolskii, M.S.: Convergence of the gradient projection method in optimal control-problems. Comput. Math. Model. **18**, 148–156 (2007)
44. Prajna, S.: Barrier certificates for nonlinear model validation. Automatica **42**(1), 117–126 (2006)
45. Ramdani, N., Meslem, N., Candau, Y.: Computing reachable sets for uncertain nonlinear monotone systems. Nonlinear Anal.: Hybrid Syst. **4**(2), 263–278 (2010). IFAC World Congress 2008
46. Scott, J.K., Barton, P.I.: Bounds on the reachable sets of nonlinear control systems. Automatica **49**(1), 93–100 (2013)
47. Shoukry, Y., et al.: Scalable lazy SMT-based motion planning. In: CDC, pp. 6683–6688 (2016)
48. Tkachev, I., Abate, A.: A control Lyapunov function approach for the computation of the infinite-horizon stochastic reach-avoid problem. In: CDC, pp. 3211–3216 (2013)
49. Tschaikowski, M., Tribastone, M.: Approximate reduction of heterogenous nonlinear models with differential hulls. IEEE Trans. Autom. Control **61**(4), 1099–1104 (2016)
50. Vilar, J.M.G., Kueh, H.Y., Barkai, N., Leibler, S.: Mechanisms of noise-resistance in genetic oscillators. PNAS **99**(9), 5988–5992 (2002)
51. Xue, B., Fränzle, M., Zhan, N.: Under-approximating reach sets for polynomial continuous systems. In: HSCC, pp. 51–60 (2018)
52. ue, B., Fränzle, M., Zhan, N.: Inner-approximating reachable sets for polynomial systems with time-varying uncertainties. CoRR (2018)
53. Xue, B., She, Z., Easwaran, A.: Under-approximating backward reachable sets by polytopes. In: Chaudhuri, S., Farzan, A. (eds.) CAV 2016. LNCS, vol. 9779, pp. 457–476. Springer, Cham (2016). https://doi.org/10.1007/978-3-319-41528-4_25
54. Xue, B., Wang, Q., Feng, S., Zhan, N.: Over- and under-approximating reachable sets for perturbed delay differential equations. CoRR (2019)
55. Zamani, M., Majumdar, R.: A Lyapunov approach in incremental stability. In: CDC (2011)

Reducing Spreading Processes on Networks to Markov Population Models

Gerrit Großmann[1]([⊠]) [ID] and Luca Bortolussi[1,2] [ID]

[1] Saarland University, 66123 Saarbrücken, Germany
gerrit.grossmann@uni-saarland.de
[2] University of Trieste, Trieste, Italy
lbortolussi@units.it

Abstract. Stochastic processes on complex networks, where each node is in one of several compartments, and neighboring nodes interact with each other, can be used to describe a variety of real-world spreading phenomena. However, computational analysis of such processes is hindered by the enormous size of their underlying state space.

In this work, we demonstrate that lumping can be used to reduce any epidemic model to a Markov Population Model (MPM). Therefore, we propose a novel lumping scheme based on a partitioning of the nodes. By imposing different types of counting abstractions, we obtain coarse-grained Markov models with a natural MPM representation that approximate the original systems. This makes it possible to transfer the rich pool of approximation techniques developed for MPMs to the computational analysis of complex networks' dynamics.

We present numerical examples to investigate the relationship between the accuracy of the MPMs, the size of the lumped state space, and the type of counting abstraction.

Keywords: Epidemic modeling · Markov Population Model · Lumping · Model reduction · Spreading process · SIS model · Complex networks

1 Introduction

Computational modeling and analysis of dynamic processes on networked systems is a wide-spread and thriving research area. In particular, much effort has been put into the study of spreading phenomena [2,16,28,38]. Arguably, the most common formalism for spreading processes is the so-called Susceptible-Infected-Susceptible (SIS) model with its variations [28,38,39].

In the SIS model, each node is either *infected* (I) or *susceptible* (S). Infected nodes propagate their infection to neighboring susceptible nodes and become susceptible again after a random waiting time. Naturally, one can extend the number of possible node states (or compartments) of a node. For instance, the SIR model introduces an additional *recovered* state in which nodes are immune to the infection.

© Springer Nature Switzerland AG 2019
D. Parker and V. Wolf (Eds.): QEST 2019, LNCS 11785, pp. 292–309, 2019.
https://doi.org/10.1007/978-3-030-30281-8_17

SIS-type models are remarkable because—despite their simplicity—they allow the emergence of complex macroscopic phenomena guided by the topological properties of the network. There exists a wide variety of scenarios which can be described using the SIS-type formalism. For instance, the SIS model has been successfully used to study the spread of many different pathogens like influenza [26], dengue fever [40], and SARS [36]. Likewise, SIS-type models have shown to be extremely useful for analyzing and predicting the spread of opinions [29,49], rumors [52,53], and memes [51] in online social networks. Other areas of applications include the modeling of neural activity [15], the spread of computer viruses [11] as well as blackouts in financial institutions [34].

The semantics of SIS-type processes can be described using a continuous-time Markov chain (CTMC) [28,47] (cf. Sect. 3 for details). Each possible assignment of nodes to the two node states S and I constitutes an individual state in the CTMC (here referred to as *network state* to avoid confusion[1]). Hence, the CTMC state space grows exponentially with the number of nodes, which renders the numeral solution of the CTMC infeasible for most realistic contact networks.

This work investigates an aggregation scheme that *lumps* similar network states together and thereby reduces the size of the state space. More precisely, we first partition the nodes of the contact network. After which, we impose a counting abstraction on each partition. We only lump two networks states together when their corresponding counting abstractions coincide on each partition.

As we will see, the counting abstraction induces a natural representation of the lumped CTMC as a Markov Population Model (MPM). In an MPM, the CTMC states are vectors which, for different types of species, count the number of entities of each species. The dynamics can elegantly be represented as species interactions. More importantly, a very rich pool of approximation techniques has been developed on the basis of MPMs, which can now be applied to the lumped model. These include efficient simulation techniques [1,7], dynamic state space truncation [24,33], moment-closure approximations [19,44], linear noise approximation [18,46], and hybrid approaches [4,43].

The remainder of this work is organized as follows: Sect. 2 shortly revises related work, Sect. 3 formalized SIS-type models and their CTMC semantics. Our lumping scheme is developed in Sect. 4. In Sect. 5, we show that the lumped CTMCs have a natural MPM representation. Numerical results are demonstrated in Sect. 6 and some conclusions in Sect. 7 complete the paper and identify open research problems.

2 Related Work

The general idea behind *lumping* is to reduce the complexity of a system by aggregating (i.e., lumping) individual components of the system together. Lumping is a popular model reduction technique which has been used to reduce the number of equations in a system of ODEs and the number of states in a Markov

---

[1] In the following, we will use the term CTMC state and network state interchangeably.

chain, in particular in the context of biochemical reaction networks [6,8,31,50]. Generally speaking, one can distinguish between *exact* and *approximate* lumping [6,31].

Most work on the lumpability of epidemic models has been done in the context of exact lumping [28,42,48]. The general idea is typically to reduce the state space by identifying symmetries in the CTMC which themselves can be found using symmetries (i.e., automorphisms) in the contact network. Those methods, however, are limited in scope because these symmetries are infeasible to find in real-world networks and the state space reduction is not sufficient to make realistic models small enough to be solvable.

This work proposes an approximate lumping scheme. Approximate lumping has been shown to be useful when applied to mean-field approximation approaches of epidemic models like the degree-based mean-field and pair approximation equations [30], as well as the approximate master equation [14,21]. However, mean-field equations are essentially inflexible as they do not take topological properties into account or make unrealistic independence assumptions between neighboring nodes.

Moreover, [27] proposed using local symmetries in the contact network instead of automorphisms to construct a lumped Markov chain. This scheme seems promising, in particular on larger graphs where automorphisms often do not even exist, however, the limitations for real-world networks due to a limited amount of state space reduction and high computational costs seem to persist.

Conceptually similar to this work is also the *unified mean-field framework* (UMFF) proposed by Devriendt et al. in [10]. Devriendt et al. also partition the nodes of the contact network but directly derive a mean-field equation from it. In contrast, this work focuses on the analysis of the lumped CTMC and its relation to MPMs. Moreover, we investigate different types of counting abstractions, not only node based ones. The relationship between population dynamics and networks has also been investigated with regard to Markovian agents [3].

3   Spreading Processes

Let $\mathcal{G} = (\mathcal{N}, \mathcal{E})$ be a an undirected graph without self-loops. At each time point $t \in \mathbb{R}_{\geq 0}$ each node occupies one of m different node states, denoted by $\mathcal{S} = \{s_1, s_2, \ldots, s_m\}$ (typically, $\mathcal{S} = \{\texttt{S}, \texttt{I}\}$). Consequently, the network state is given by a labeling $x : \mathcal{N} \to \mathcal{S}$. We use

$$\mathcal{X} = \{x \mid x : \mathcal{N} \to \mathcal{S}\}$$

to denote all possible labelings. $\mathcal{X}$ is also the state space of the underlying CTMC. As each of the $|\mathcal{N}|$ nodes occupies one of m states, we find that $|\mathcal{X}| = |\mathcal{S}|^{|\mathcal{N}|}$.

A set of stochastic rules determines the particular way in which nodes change their corresponding node states. Whether a rule can be applied to a node depends on the state of the node and of its immediate neighborhood.

The neighborhood of a node is modeled as a vector $\mathbf{m} \in \mathbb{Z}_{\geq 0}^{|\mathcal{S}|}$ where $\mathbf{m}[s]$ denotes the number of neighbors in state $s \in \mathcal{S}$ (we assume an implicit enumeration of states). Thus, the degree (number of neighbors, denoted by k) of

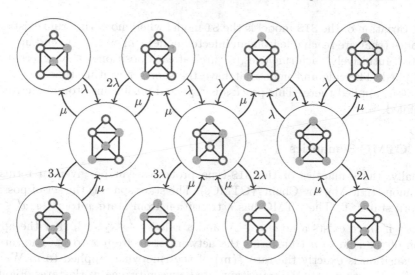

Fig. 1. The CTMC induced by the SIS model (S: *blue*, I: *magenta, filled*) on a toy graph. Only a subset of the CTMC spate space (11 out of $2^6 = 64$ network states) is shown. (Color figure online)

a node is equal to the sum over its associated neighborhood vector, that is, $k = \sum_{s \in \mathcal{S}} \mathbf{m}[s]$. The set of possible neighborhood vectors is denoted as

$$\mathcal{M} = \left\{ \mathbf{m} \in \mathbb{Z}_{\geq 0}^{|\mathcal{S}|} \, \middle| \, \sum_{s \in \mathcal{S}} \mathbf{m}[s] \leq k_{\max} \right\},$$

where $k_{\max}$ denotes the maximal degree in a given network.

Each rule is a triplet $s_1 \xrightarrow{f} s_2$ $(s_1, s_2 \in \mathcal{S}, s_1 \neq s_2)$, which can be applied to each node in state s_1. When the rule "fires" it transforms the node from s_1 into s_2. The rate at which a rule "fires" is specified by the rate function $f : \mathcal{M} \to \mathbb{R}_{\geq 0}$ and depends on the node's neighborhood vector. The time delay until the rule is applied to the network state is drawn from an exponential distribution with rate $f(\mathbf{m})$. Hence, higher rates correspond to shorter waiting times. For the sake of simplicity and without loss of generality, we assume that for each pair of states s_1, s_2 there exists at most one rule that transforms s_1 to s_2.

In the well-known SIS model, infected nodes propagate their infection to susceptible neighbors. Thus, the rate at which a susceptible node becomes infected is proportional to its number of infected neighbors:

$$\mathtt{S} \xrightarrow{f} \mathtt{I} \quad \text{with} \quad f(\mathbf{m}) = \lambda \cdot \mathbf{m}[\mathtt{I}],$$

where $\lambda \in \mathbb{R}_{\geq 0}$ is a rule-specific rate constant (called *infection rate*) and $\mathbf{m}[\mathtt{I}]$ denotes the number of infected neighbors. Furthermore, a recovery rule transforms infected nodes back to being susceptible:

$$\mathtt{I} \xrightarrow{f} \mathtt{S} \quad \text{with} \quad f(\mathbf{m}) = \mu,$$

where $\mu \in \mathbb{R}_{\geq 0}$ is a rule-specific rate constant called *recovery rate*.

A variation of the SIS model is the SI model where no curing rule exists and all nodes (that are reachable from an infected node) will eventually end up being infected. Intuitively, each rule tries to "fire" at each position $n \in \mathcal{N}$ where it can be applied. The rule and node that have the shortest waiting time "win" and the rule is applied there. This process is repeated until some stopping criterion is fulfilled.

3.1 CTMC Semantics

Formally, the semantics of the SIS-type processes can be given in terms of continuous-time Markov Chains (CTMCs). The state space is the set of possible network states $\mathcal{X}$. The CTMC has a transition from state x to x' ($x, x' \in \mathcal{X}$, $x \neq x'$) if there exists a node $n \in \mathcal{N}$ and a rule $s_1 \xrightarrow{f} s_2$ such that the application of the rule to n transforms the network state from x to x'. The rate of the transition is exactly the rate $f(\mathbf{m})$ of the rule when applied to n. We use $q(x, x') \in \mathbb{R}_{\geq 0}$ to denote the transition rate between two network states. Figure 1 illustrates the CTMC corresponding to an SIS process on a small toy network.

Explicitly computing the evolution of the probability of $x \in \mathcal{X}$ over time with an ODE solver, using numerical integration, is only possible for very small contact networks, since the state space grows exponentially with the number of nodes. Alternative approaches include sampling the CTMC, which can be done reasonably efficiently even for comparably large networks [9,22,45] but is subject to statistical inaccuracies and is mostly used to estimate global properties.

4   Approximate Lumping

Our lumping scheme is composed of three basic ingredients:
Node Partitioning: The partitioning over the nodes $\mathcal{N}$ that is explicitly provided.
Counting Pattern: The type of features we are counting, i.e., nodes or edges.
Implicit State Space Partitioning: The CTMC state space is implicitly partitioned by counting the nodes or edges on each node partition.

We will start our presentation discussing the partitioning of the state space, then showing how to obtain it from a given node partitioning and counting pattern. To this end, we use $\mathcal{Y}$ to denote the new *lumped* state space and assume that there is a surjective[2] lumping function

$$\mathcal{L} : \mathcal{X} \to \mathcal{Y}$$

that defines which network states will be lumped together. Note that the lumped state space is the image of the lumping function and that all network states $x \in \mathcal{X}$ which are mapped to the same $y \in \mathcal{Y}$ will be aggregated.

---

[2] If $\mathcal{L}$ is not surjective, we consider only the image of $\mathcal{L}$ to be the lumped state space.

Later in this section, we will discuss concrete realizations of $\mathcal{L}$. In particular, we will construct $\mathcal{L}$ based on a node partitioning and a counting abstraction of our choice. Next, we define the transition rates $q(y, y')$ (where $y, y' \in \mathcal{Y}$, $y \neq y'$) between the states of the lumped Markov chain:

$$q(y, y') = \frac{1}{|\mathcal{L}^{-1}(y)|} \sum_{x \in \mathcal{L}^{-1}(y)} \sum_{x' \in \mathcal{L}^{-1}(y')} q(x, x'). \tag{1}$$

This is simply the mean transition rate at which an original state from x goes to some $x' \in \mathcal{L}^{-1}(y')$. Technically, Eq. (1) corresponds to the following *lumping assumption*: we assume that at each point in time all network states belonging to a lumped state y are equally likely.

4.1   Partition-Based Lumping

Next, we construct the lumping function $\mathcal{L}$. Because we want to make our lumping aware of the contact network's topology, we assume a given partitioning $\mathcal{P}$ over the nodes $\mathcal{N}$ of the contact network. That is, $\mathcal{P} \subset 2^{\mathcal{N}}$ and $\bigcup_{P \in \mathcal{P}} P = \mathcal{N}$ and all $P \in \mathcal{P}$ are disjoint and non-empty. Based on the node partitioning, we can now impose different kinds of counting abstractions on the network state. This work considers two types: counting nodes and counting edges. The counting abstractions are visualized in Fig. 3. A full example of how a lumped CTMC of an SI model is constructed using the node-based counting abstraction is given in Fig. 2.

Node-Based Counting Abstraction. We count the number of nodes in each state and partition. Thus, for a given network state $x \in \mathcal{X}$, we use $y(s, P)$ to denote the number of nodes in state $s \in \mathcal{S}$ in partition $P \in \mathcal{P}$. The lumping function $\mathcal{L}$ projects x to the corresponding counting abstraction. Formally:

$$\mathcal{Y} = \{y \mid y : \mathcal{S} \times \mathcal{P} \to \mathbb{Z}_{\geq 0}\}$$

$$\mathcal{L}(x) = y$$

with: $y(s, P) = |\{n \in \mathcal{N} \mid X(n) = s, n \in P\}|.$

Edge-Based Counting Abstraction. Again, we assume that a network state x and a node partitioning $\mathcal{P}$ are given. Now we count the edges, that is for each pair of states $s, s' \in \mathcal{S}$ and each pair of partitions $P, P' \in \mathcal{P}$, we count $y(s, P, s', P')$ which is the number of edges $(n, n') \in \mathcal{E}$ where $x(n) = s$, $n \in P$, $x(n') = s'$, $n' \in P'$. Note that this includes cases where $P = P'$ and $s = s'$. However, only counting the edges does not determine how many nodes there are in each state (see Fig. 3 for an example).

In order to still have this information encoded in each lumped state, we slightly modify the network structure by adding a new dummy node n_* and connecting each node to it . The dummy node has a dummy state denoted by

Graph Partition   SI Rule

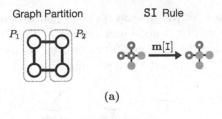

(a)

Original Markov Model

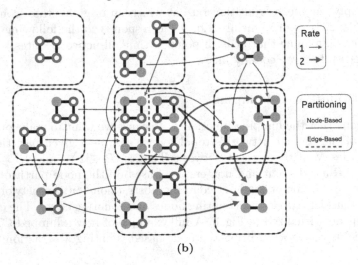

(b)

Lumped Markov Model

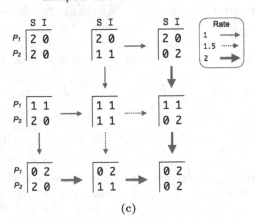

(c)

Fig. 2. Illustration of the lumping process. (a): Model. A basic SI-Process where infected nodes (magenta, filled) infect susceptible neighbors (blue) with rate infection $\lambda = 1$. The contact graph is divided into two partitions. (b): The underlying CTMC with $2^4 = 16$ states. The graph partition induces the edge-based and node-based lumping. The edge-based lumping refines the node-based lumping and generates one partition more (vertical line in the central partition). (c): The lumped CTMC using node-based counting abstraction with only 9 states. The rates are the averaged rates from the full CTMC. (Color figure online)

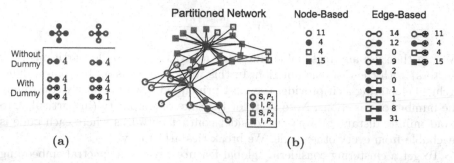

(a)   (b)

Fig. 3. (a) By adding the dummy-node, the edge-based abstraction is able to differ-entiate the two graphs. Adding the dummy-node ensures that the nodes in each state are counted in the edge-based abstraction. (b) Left: A partitioned network (Zachary's Karate Club graph from [12]) (S: *blue,* I: *magenta, filled*). The network is partitioned into P_1 (○-nodes) and P_2 (□-nodes). Right: The corresponding counting abstractions. (Color figure online)

$\star$ which never changes, and it can be assigned to a new dummy partition $P_\star$. Formally,

$$\mathcal{N} := \mathcal{N} \cup \{n_\star\} \quad \mathcal{S} := \mathcal{S} \cup \{\star\} \quad L(n_\star) = \star \quad \mathcal{P} := \mathcal{P} \cup \{P_\star\}$$
$$\mathcal{E} := \mathcal{E} \cup \{(n, n_\star) \mid n \in \mathcal{N}, n \neq n_\star\}.$$

Note that the rate function f ignores the dummy node. The lumped representation is then given as:

$$\mathcal{Y} = \{y \mid y : \mathcal{S} \times \mathcal{P} \times \mathcal{S} \times \mathcal{P} \to \mathbb{Z}_{\geq 0}\}$$
$$\mathcal{L}(x) = y$$

with: $y(s, P, s', P') = |\{(n, n') \in \mathcal{E} \mid x(n) = s, n \in P, x(n') = s', n' \in P'\}|$

Example. Figure 2 illustrates how a given partitioning and the node-based counting approach induces a lumped CTMC. The partitions induced by the edge-based counting abstracting are also shown. In this example, the edge-based lumping aggregates only isomorphic network states.

4.2   Graph Partitioning

Broadly speaking, we have three options to partition the nodes based on local features (e.g., its degree) or global features (e.g., communities in the graph) or randomly. As a baseline, we use a random node partitioning. Therefore, we fix the number of partitions and randomly assign each node to a partition while enforcing that all partitions have, as far as possible, the same number of elements.

Moreover, we investigate a degree-based partitioning, where we define the distance between to nodes n, n' as their relative degree difference (similar to [30]):

$$d_{\mathrm{k}}(n, n') = \frac{|k_n - k_{n'}|}{\max(k_n, k_{n'})} .$$

We can then use any reasonable clustering algorithm and build partitions (i.e., clusters) with the distance function. In this work, we focus on bottom-up hierarchical clustering as it provides the most principled way of precisely controlling the number of partitions. Note that, for the sake of simplicity (in particular, to avoid infinite distances), we only consider contact networks where each node is reachable from every other node. We break ties arbitrarily.

To get a clustering considering global features we use a spectral embedding of the contract network. Specifically, we use the `spectral_layout` function from the `NetworkX` Python-package [23] with three dimensions and perform hierarchical clustering on the embedding. In future research, it would be interesting to compute node distances based on more sophisticated graph embedding as the ones proposed in [17]. Note that in the border cases $|\mathcal{P}| = 1$ and $|\mathcal{P}| = |\mathcal{N}|$ all methods yield the same partitioning.

5   Markov Population Models

Markov Population Models (MPMs) are a special form of CTMCs where each CTMC state is a population vector over a set of species. We use $\mathcal{Z}$ to denote the finite set of species (again, with an implicit enumeration) and $\mathbf{y} \in \mathbb{Z}_{\geq 0}^{|\mathcal{Z}|}$ to denote the population vector. Hence, $\mathbf{y}[z]$ identifies the number of entities of species z. The stochastic dynamics of MPMs is typically expressed as a set of reactions $\mathcal{R}$, each reaction, $(\alpha, \mathbf{b}) \in \mathcal{R}$, is comprised of a propensity function $\alpha : \mathbb{Z}_{\geq 0}^{|\mathcal{Z}|} \to \mathbb{R}_{\geq 0}$ and a change vector $\mathbf{b} \in \mathbb{Z}^{|\mathcal{Z}|}$. When reaction $(\alpha, \mathbf{b})$ is applied, the system moves from state $\mathbf{y}$ to state $\mathbf{y} + \mathbf{b}$. The corresponding rate is given by the propensity function. Therefore, we can rewrite the transition matrix of the CTMC as[3]:

$$q(\mathbf{y}, \mathbf{y}') = \begin{cases} \alpha(\mathbf{y}) & \text{if } \exists (\alpha, \mathbf{b}) \in \mathcal{R}, \mathbf{y}' = \mathbf{y} + \mathbf{b} \\ 0 & \text{otherwise} \end{cases} .$$

Next, we show that our counting abstractions have a natural interpretation as MPMs.

5.1   Node-Based Abstraction

First, we define the set of species $\mathcal{Z}$. Conceptually, species are node states which are aware of their partition:

$$\mathcal{Z} = \{(s, P) \mid s \in \mathcal{S}, P \in \mathcal{P}\} .$$

---

[3] Without loss of generality, we assume that different reactions have different change vectors. If this is not the case, we can merge reactions with the same update by summing their corresponding rate functions.

Again, we assume an implicit enumeration of $\mathcal{Z}$. We use $z.s$ and $z.P$ to denote the components of a give species z.

We can now represent the lumped CTMC state as a single population vector $\mathbf{y} \in \mathbb{Z}_{\geq 0}^{|\mathcal{Z}|}$, where $\mathbf{y}[z]$ the number of nodes belonging to species z (i.e., which are in state $z.s$ and partition $z.P$). The image of the lumping function $\mathcal{L}$, i.e. the lumped state space $\mathcal{Y}$, is now a subset of non-negative integer vectors: $\mathcal{Y} \subset \mathbb{Z}_{\geq 0}^{|\mathcal{Z}|}$.

Next, we express the dynamics by a set of reactions. For each rule $r = s_1 \xrightarrow{f} s_2$ and each partition $P \in \mathcal{P}$, we define a reaction $(\alpha_{r,P}, \mathbf{b}_{r,P})$ with propensity function as:

$$\alpha_{r,P} : \mathcal{Y} \to \mathbb{R}_{\geq 0}$$

$$\alpha_{r,P}(\mathbf{y}) = \frac{1}{\mathcal{L}^{-1}(\mathbf{y})} \sum_{x \in \mathcal{L}^{-1}(\mathbf{y})} \sum_{n \in P} f(\mathbf{m}_{x,n}) \mathbb{1}_{x(n)=s_1},$$

where $\mathbf{m}_{x,n}$ denotes the neighborhood vector of n in network state x. Note that this is just the instantiation of Eq. 1 to the MPM framework.

The change vector $\mathbf{b}_{r,P} \in \mathbb{Z}^{|\mathcal{Z}|}$ is defined element-wise as:

$$\mathbf{b}_{r,P}[z] = \begin{cases} 1 & \text{if } z.s = s_2, P = z.P \\ -1 & \text{if } z.s = s_1, P = z.P \\ 0 & \text{otherwise} \end{cases}.$$

Note that s_1, s_2 refer to the current rule and $z.s$ to the entry of $\mathbf{b}_{r,P}$.

5.2 Edge-Based Counting Abstraction

We start by defining a *species neighborhood*. The species neighborhood of a node n is a vector $\mathbf{v} \in \mathbb{Z}_{\geq 0}^{|\mathcal{Z}|}$, where $\mathbf{v}[z]$ denotes the number of neighbors of species z. We define $\mathcal{V}_n$ to be the set of possible species neighborhoods for a node n, given a fixed contact network and partitioning. Note that we still assume that a dummy node is used to encode the number of states in each partition.

Assuming an arbitrary ordering of pairs of states and partitions, we define

$$\mathcal{Z} = \big\{ (s_{source}, P_{source}, s_{target}, P_{target}) \,|\, s_{source}, s_{target} \in \mathcal{S}, P_{source}, P_{target} \in \mathcal{P},$$
$$(s_{source}, P_{source}) \leq (s_{target}, P_{target}) \big\}.$$

Let us define $\mathcal{V}_P$ to be the set of partition neighborhoods all nodes in P can have:

$$\mathcal{V}_P = \bigcup_{n \in P} \mathcal{V}_n.$$

For each rule $r = s_1 \xrightarrow{f} s_2$, and each partition $P \in \mathcal{P}$, and each $\mathbf{v} \in \mathcal{V}_P$, we define a propensity function $\alpha_{r,P,\mathbf{v}}$ with:

$$\alpha_{r,P,\mathbf{v}} : \mathcal{Y} \to \mathbb{R}_{\geq 0}$$

$$\alpha_{r,P,\mathbf{v}}(\mathbf{y}) = \frac{1}{\mathcal{L}^{-1}(\mathbf{y})} \sum_{x \in \mathcal{L}^{-1}(\mathbf{y})} \sum_{n \in P} f(\mathbf{m}_{x,n}) \mathbb{1}_{x(n)=s_1, V(n)=\mathbf{v}}.$$

Note that the propensity does not actually depend on $\mathbf{v}$, it is simply individually defined for each $\mathbf{v}$. The reason for this is that the change vector depends on the a node's species neighborhood. To see this, consider a species $z = (s_{source}, P_{source}, s_{target}, P_{target})$, corresponding to edges connecting a node in state s_{source} and partition P_{source} to a node in state s_{target} and partition P_{target}. There are two scenarios in which the corresponding counting variable has to change: (a) when the node changing state due to an application of rule r is the source node, and (b) when it is the target node. Consider case (a); we need to know how many edges are connecting the updated node (which was in state s_1 and partition P) to a node in state s_{target} and partition P_{target}. This information is stored in the vector $\mathbf{v}$, specifically in position $\mathbf{v}[s_{target}, P_{target}]$. The case in which the updated node is the target one is treated symmetrically. This gives rise to the following definition:

$$
\mathbf{b}_{r,P,\mathbf{v}}[z] = \begin{cases}
\mathbf{v}[z.s_{target}, z.P_{target}] & \text{if } s_2 = z.s_{source}, P = z.P_{source} \\
-\mathbf{v}[z.s_{target}, z.P_{target}] & \text{if } s_1 = z.s_{source}, P = z.P_{source} \\
\mathbf{v}[z.s_{source}, z.P_{source}] & \text{if } s_2 = z.s_{target}, P = z.P_{target} \\
-\mathbf{v}[z.s_{source}, z.P_{source}] & \text{if } s_1 = z.s_{target}, P = z.P_{target} \\
0 & \text{otherwise}
\end{cases}
$$

The first two lines of the definition handle cases in which the node changing state is the source node, while the following two lines deal with the case in which the node changing state appears as target.

Figure 4 illustrates how a lumped network state is influenced by the application of an infection rule.

5.3   Direct Construction of the MPM

Approximating the solution of an SIS-type process on a contact network by lumping the CTMC first, already reduces the computational costs by many orders of magnitude. However, this scheme is still only applicable when it is possible to construct the full CTMC in the first place. Recall that the number of network states is exponential in the number of nodes of the contact network, that is, $|\mathcal{X}| = |\mathcal{S}|^{|\mathcal{N}|}$.

However, in recent years, substantial effort was dedicated to the analysis of very small networks [25, 32, 35, 37, 48]. One reason is that when the size of a network increases, the (macro-scale) dynamics becomes more deterministic because stochastic effects tend to cancel out. For small contact networks, however, methods which capture the full stochastic dynamics of the system, and not only the mean behavior, are of particular importance.

A substantial advantage of the reduction to MPM is the possibility of constructing the lumped CTMC without building the full CTMC first. In particular, this can be done exactly for the node counting abstraction. On the other hand, for the edge counting we need to introduce an extra approximation in the definition of the rate function, roughly speaking introducing an approximate probability

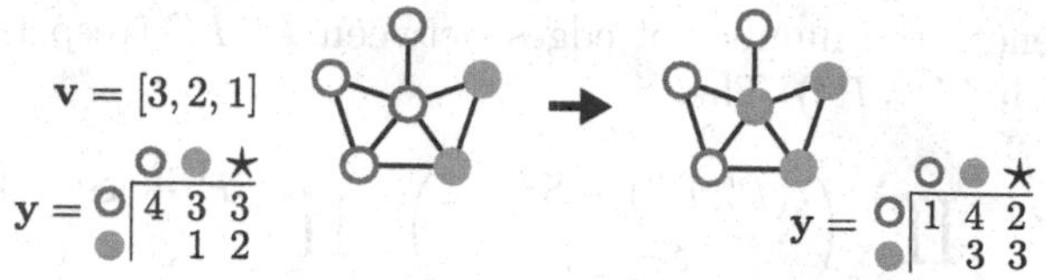

$$\mathbf{v} = [3, 2, 1]$$

$$\mathbf{y} = \begin{array}{c|ccc} & \text{O} & \bullet & \star \\ \hline \text{O} & 4 & 3 & 3 \\ \bullet & & 1 & 2 \end{array}$$

$$\mathbf{y} = \begin{array}{c|ccc} & \text{O} & \bullet & \star \\ \hline \text{O} & 1 & 4 & 2 \\ \bullet & & 3 & 3 \end{array}$$

Fig. 4. Example of how the neighborhood $\mathbf{v}$ influences the update in the edge-based counting abstraction on an example graph. Here, all nodes belong to the same partition (thus, nodes states and species are conceptually the same) and the node states are ordered $[\text{S}, \text{I}, \star]$. The population vector $\mathbf{y}$ is given in matrix form for the ease of presentation.

distribution over neighboring vectors, as knowing how many nodes have a specific neighboring vector requires us full knowledge of the original CTMC. We present full details of such direct construction in the Appendix of [20].

5.4   Complexity of the MPM

The size of the lumped MPM is critical for our method, as it determines which solution techniques are computationally tractable and provides guidelines on how many partitions to choose. There are two notions of size to consider: (a) the number of population variables and (b) the number of states of the underlying CTMC. While the latter governs the applicability of numerical solutions for CTMCs, the former controls the complexity of a large number of approximate techniques for MPMs, like mean field or moment closure.

Node-Based Abstraction. In this abstraction, the population vector is of length $|\mathcal{S}| \cdot |\mathcal{P}|$, i.e. there is a variable for each node state and each partition.

Note that the sum of the population variables for each partition P is $|P|$, the number of nodes in the partition. This allows us to count easily the number of states of the CTMC of the population model: for each partition, we need to subdivide $|P|$ different nodes into $|\mathcal{S}|$ different classes, which can be done in $\binom{|P|+|\mathcal{S}|-1}{|\mathcal{S}|-1}$ ways, giving a number of CTMC states exponential in the number $|\mathcal{S}|$ of node states and $|\mathcal{P}|$ of partitions, but polynomial in the number of nodes:

$$|\mathcal{Y}| = \prod_{P \in \mathcal{P}} \binom{|P| + |\mathcal{S}| - 1}{|\mathcal{S}| - 1}.$$

Edge-Based Abstraction. The number of population variables, in this case, is one for each edge connecting two different partitions, plus those counting the number of nodes in each partition and each node state, due to the presence of the dummy state. In total, we have $\frac{q(q-1)}{2} + q$ population variables, with $q = |\mathcal{S}| \cdot |\mathcal{P}|$.

In order to count the number of states of the CTMC in this abstraction, we start by observing that the sum of all variables for a given pair of partitions P', P'' is the number of edges connecting such partitions in the graph. We use

$\epsilon(P', P'')$ to denote the number of edges between P', P'' (resp. the number of edges inside P' if $P' = P''$). Thus,

$$|\mathcal{Y}| \le \prod_{\substack{P',P'' \in \mathcal{P}^2 \\ P' \le P''}} \binom{\epsilon(P', P'') + \mathcal{S}^2 - 1}{\mathcal{S}^2 - 1} \cdot \prod_{P \in \mathcal{P}} \binom{|P| + |\mathcal{S}| - 1}{|\mathcal{S}| - 1}.$$

This is an over-approximation, because not all combinations are consistent with the graph topology. For example, a high number of infected nodes in a partition might not be consistent with a small number of I − I-edges inside the partition. Note that also this upper bound is exponential in $|\mathcal{S}|$ and $|\mathcal{P}|$ but still polynomial in the number of nodes N, differently from the original network model, whose state space is exponential in N.

The exponential dependency on the number of species (i.e., dimensions of the population vector) makes the explicit construction of the lumped state space viable only for very small networks with a small number of node states. However, this is typically the case for spreading models like SIS or SIR. Yet, also the number of partitions has to be kept small, particularly in realistic models. We expect that the partitioning is especially useful for networks showing a small number of large-scale homogeneous structures, as happens in many real-world networks [12].

An alternative strategy for analysis is to derive mean-field [5] or moment closure equations [41] for MPMs, which can be done without explicitly constructing the lumped (and the original) state space. These are sets of ordinary differential equation (ODE) describing the evolution of (moments of) the population variables. We refer the reader to [10] for a similar approach regarding the node-based abstraction.

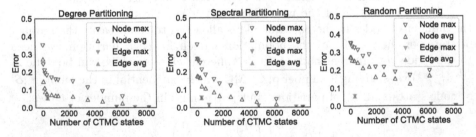

Fig. 5. Trade of between accuracy and state space size for the node-based (blue) and edge-based (magenta, filled) counting abstraction. Results are shown for node partitions based on the degree (l.), spectral embedding (c.), and random partitioning (r.). The accuracy is measured as the mean (△) and maximal (▽) difference between the original and lumped solution over all timepoints. (Color figure online)

6 Numerical Results

In this section, we compare the numerical solution of the original model—referred to as baseline model—with different lumped MPMs. The goal of this comparison is to provide evidence supporting the claim that the lumping preserves the dynamics of the original system, with an accuracy increasing with the resolution of the MPM. We will perform the comparison by solving numerically the ground and the lumped system, thus comparing the probability of each state in each point in time. In practical applications of our method, exact transient or steady state solutions may not be feasible, but in this case we can still rely to approximation methods for MPM [5,41]. Determining which of those techniques performs best in this context is a direction of future exploration.

A limit of the comparison based on numerical solution of the CTMC is that the state space of the original model has $|\mathcal{S}|^{|\mathcal{N}|}$ states, which limits the size of the contact network strongly[4].

Let $P(X(t) = x)$ denote the probability that the baseline CTMC occupies network state $x \in \mathcal{X}$ at time $t \geq 0$. Furthermore, let $P(Y(t) = y)$ for $t \geq 0$ and $y \in \mathcal{Y}$ denote the same probability for a lumped MPM (corresponding to a specific partitioning and counting abstraction). To measure their difference, we first approximate the probability distribution of the original model using the lumped solution, invoking the lumping assumption which states that all network states which are lumped together have the same probability mass. We use P_L to denote the *lifted* probability distribution over the original state space given a lumped solution. Formally,

$$P_L\big(Y(t) = x\big) = \frac{P\big(Y(t) = y\big)}{|\mathcal{L}^{-1}(y)|} \quad \text{where } y \text{ is s.t. } L(x) = y.$$

We measure the difference between the baseline and a lumped solution at a specific time point by summing up the difference in probability mass of each state, then take the maximum error in time:

$$d(P, P_L) = \max_t \sum_{x \in \mathcal{X}} \Big| P_L\big(Y(t) = x\big) - P\big(X(t) = x\big) \Big|.$$

In our experiments, we used a small toy network with 13 nodes and 2 states ($2^{13} = 8192$ network states). We generated a synthetic contact network following the Erdős–Rényi graph model with a connection probability of 0.5. We use a SIS model with an infection rate of $\lambda = 1.0$ and a recovery rate of $\mu = 1.3$. Initially, we assign an equal amount of probability mass to all network states.

Figure 5 shows the relationship between the error of the lumped MPM, the type of counting abstraction and the method used for node partitioning. We also report the mean difference together with the maximal difference over time.

From our results, we conclude that the edge-based counting abstraction yields a significantly better trade-off between state space size and accuracy. However,

---

[4] Code is available at github.com/gerritgr/Reducing-Spreading-Processes.

it generates larger MPM models than the node-based abstraction when adding a new partition. We also find that spectral and degree-based partitioning yield similar results for the same number of CTMC states and that random partitioning performed noticeably worse, for both edge-based and node-based counting abstractions.

7   Conclusions and Future Work

This work developed first steps in a unification of the analysis of stochastic spreading processes on networks and Markov population models. Since the so obtained MPM can become very large in terms of species, it is important to be able to control the trade-off between state space size and accuracy.

However, there are still many open research problems ahead. Most evidently, it remains to be determined which of the many techniques developed for the analysis of MPMs (e.g. linear noise, moment closure) work best on our proposed epidemic-type MPMs and how they scale with increasing size of the contact network. We expect also that these reduction methods can provide a good starting point for deriving advanced mean-field equations, similar to ones in [10]. Moreover, literature is very rich in proposed moment-closure-based approximation techniques for MPMs, which can now be utilized [19,44]. We also plan to investigate the relationship between lumped mean-field equations [21,30] and coarse-grained counting abstractions further.

Future work can additionally explore counting abstraction of different types, for instance, a neighborhood-based abstraction like the one proposed by Gleeson in [13,14].

Finally, we expect that there are many more possibilities of partitioning the contact network that remain to be investigated and which might have a significant impact on the final accuracy of the abstraction.

Acknowledgements. This research has been partially funded by the German Research Council (DFG) as part of the Collaborative Research Center "Methods and Tools for Understanding and Controlling Privacy". We thank Verena Wolf for helpful discussions and provision of expertise.

References

1. Allen, G.E., Dytham, C.: An efficient method for stochastic simulation of biological populations in continuous time. Biosystems **98**(1), 37–42 (2009)
2. Barabási, A.-L.: Network Science. Cambridge University Press, Cambridge (2016)
3. Bobbio, A., Cerotti, D., Gribaudo, M., Iacono, M., Manini, D.: Markovian agent models: a dynamic population of interdependent Markovian agents. In: Al-Begain, K., Bargiela, A. (eds.) Seminal Contributions to Modelling and Simulation. SFMA, pp. 185–203. Springer, Cham (2016). https://doi.org/10.1007/978-3-319-33786-9_13
4. Bortolussi, L.: Hybrid behaviour of Markov population models. Inf. Comput. **247**, 37–86 (2016)

5. Bortolussi, L., Hillston, J., Latella, D., Massink, M.: Continuous approximation of collective system behaviour: a tutorial. Perform. Eval. **70**(5), 317–349 (2013)
6. Buchholz, P.: Exact and ordinary lumpability in finite Markov chains. J. Appl. Probab. **31**(1), 59–75 (1994)
7. Cao, Y., Gillespie, D.T., Petzold, L.R.: Efficient step size selection for the tau-leaping simulation method. J. Chem. Phys. **124**(4) (2006)
8. Cardelli, L., Tribastone, M., Tschaikowski, M., Vandin, A.: ERODE: a tool for the evaluation and reduction of ordinary differential equations. In: Legay, A., Margaria, T. (eds.) TACAS 2017. LNCS, vol. 10206, pp. 310–328. Springer, Heidelberg (2017). https://doi.org/10.1007/978-3-662-54580-5_19
9. Cota, W., Ferreira, S.C.: Optimized gillespie algorithms for the simulation of markovian epidemic processes on large and heterogeneous networks. Comput. Phys. Commun. **219**, 303–312 (2017)
10. Devriendt, K., Van Mieghem, P.: Unified mean-field framework for susceptible-infected-susceptible epidemics on networks, based on graph partitioning and the isoperimetric inequality. Phys. Rev. E **96**(5), 052314 (2017)
11. Gan, C., Yang, X., Liu, W., Zhu, Q., Zhang, X.: Propagation of computer virus under human intervention: a dynamicalmodel. Discrete Dyn. Nature Soc. **2012**, 8 (2012)
12. Girvan, M., Newman, M.E.: Community structure in social and biological networks. Proc. Natl. Acad. Sci. **99**(12), 7821–7826 (2002)
13. Gleeson, J.P.: High-accuracy approximation of binary-state dynamics on networks. Phys. Rev. Lett. **107**(6), 068701 (2011)
14. Gleeson, J.P.: Binary-state dynamics on complex networks: pair approximation and beyond. Phys. Rev. X **3**(2), 021004 (2013)
15. Goltsev, A., De Abreu, F., Dorogovtsev, S., Mendes, J.: Stochastic cellular automata model of neural networks. Phys. Rev. E **81**(6), 061921 (2010)
16. Goutsias, J., Jenkinson, G.: Markovian dynamics on complex reaction networks. Phys. Rep. **529**(2), 199–264 (2013)
17. Goyal, P., Ferrara, E.: Graph embedding techniques, applications, and performance: a survey. Knowl.-Based Syst. **151**, 78–94 (2018)
18. Grima, R.: An effective rate equation approach to reaction kinetics in small volumes: theory and application to biochemical reactions in nonequilibrium steady-state conditions. J. Chem. Phys. **133**(3), 035101 (2010)
19. Grima, R.: A study of the accuracy of moment-closure approximations for stochastic chemical kinetics. J. Chem. Phys. **136**(15), 04B616 (2012)
20. Großmann, G., Bortolussi, L.: Reducing spreading processes on networks to Markov population models. arXiv preprint arXiv:1906.11508 (2019)
21. Großmann, G., Kyriakopoulos, C., Bortolussi, L., Wolf, V.: Lumping the approximate master equation for multistate processes on complex networks. In: McIver, A., Horvath, A. (eds.) QEST 2018. LNCS, vol. 11024, pp. 157–172. Springer, Cham (2018). https://doi.org/10.1007/978-3-319-99154-2_10
22. Großmann, G., Wolf, V.: Rejection-based simulation of stochastic spreading processes on complex networks. arXiv preprint arXiv:1812.10845 (2018)
23. Hagberg, A., Swart, P., Chult, D.S.: Exploring network structure, dynamics, and function using networkx. Technical report, Los Alamos National Lab. (LANL), Los Alamos, NM, United States (2008)
24. Henzinger, T.A., Mateescu, M., Wolf, V.: Sliding window abstraction for infinite Markov chains. In: Bouajjani, A., Maler, O. (eds.) CAV 2009. LNCS, vol. 5643, pp. 337–352. Springer, Heidelberg (2009). https://doi.org/10.1007/978-3-642-02658-4_27

25. Holme, P.: Shadows of the susceptible-infectious-susceptible immortality transition in small networks. Phys. Rev. E **92**(1), 012804 (2015)
26. Keeling, M.J., Rohani, P.: Modeling Infectious Diseases in Humans and Animals. Princeton University Press, Princeton (2011)
27. KhudaBukhsh, W.R., Auddy, A., Disser, Y., Koeppl, H.: Approximate lumpability for Markovian agent-based models using local symmetries. arXiv:1804.00910
28. Kiss, I.Z., Miller, J.C., Simon, P.L.: Mathematics of epidemics on networks: from exact to approximate models. Forthcoming in Springer TAM series (2016)
29. Kitsak, M., et al.: Identification of influential spreaders in complex networks. Nat. Phys. **6**(11), 888 (2010)
30. Kyriakopoulos, C., Grossmann, G., Wolf, V., Bortolussi, L.: Lumping of degree-based mean-field and pair-approximation equations for multistate contact processes. Phys. Rev. E **97**(1), 012301 (2018)
31. Li, G., Rabitz, H.: A general analysis of approximate lumping in chemical kinetics. Chem. Eng. Sci. **45**(4), 977–1002 (1990)
32. López-García, M.: Stochastic descriptors in an sir epidemic model for heterogeneous individuals in small networks. Math. Biosci. **271**, 42–61 (2016)
33. Mateescu, M., Wolf, V., Didier, F., Henzinger, T.: Fast adaptive uniformisation of the chemical master equation. IET Syst. Biol. **4**(6), 441–452 (2010)
34. May, R.M., Arinaminpathy, N.: Systemic risk: the dynamics of model banking systems. J. R. Soc. Interface **7**(46), 823–838 (2009)
35. Moslonka-Lefebvre, M., Pautasso, M., Jeger, M.J.: Disease spread in small-size directed networks: epidemic threshold, correlation between links to and from nodes, and clustering. J. Theor. Biol. **260**(3), 402–411 (2009)
36. Ng, T.W., Turinici, G., Danchin, A.: A double epidemic model for the sars propagation. BMC Infect. Dis. **3**(1), 19 (2003)
37. Pautasso, M., Moslonka-Lefebvre, M., Jeger, M.J.: The number of links to and from the starting node as a predictor of epidemic size in small-size directed networks. Ecol. Complex. **7**(4), 424–432 (2010)
38. Porter, M., Gleeson, J.: Dynamical Systems on Networks: A Tutorial, vol. 4. Springer, Switzerland (2016). https://doi.org/10.1007/978-3-319-26641-1
39. Rodrigues, H.S.: Application of sir epidemiological model: new trends. arXiv:1611.02565 (2016)
40. Rodrigues, H.S., Monteiro, M.T.T., Torres, D.F.: Dynamics of dengue epidemics when using optimal control. Math. Comput. Modell. **52**(9–10), 1667–1673 (2010)
41. Schnoerr, D., Sanguinetti, G., Grima, R.: Approximation and inference methods for stochastic biochemical kinetics - a tutorial review. J. Phys. A **51**, 169501 (2018)
42. Simon, P.L., Taylor, M., Kiss, I.Z.: Exact epidemic models on graphs using graph-automorphism driven lumping. J. Math. Biol. **62**(4), 479–508 (2011)
43. Singh, A., Hespanha, J.P.: Stochastic hybrid systems for studying biochemical processes. Roy. Soc. A **368**(1930), 4995–5011 (2010)
44. Soltani, M., Vargas-Garcia, C.A., Singh, A.: Conditional moment closure schemes for studying stochastic dynamics of genetic circuits. IEEE Trans. Biomed. Circuits Syst. **9**(4), 518–526 (2015)
45. St-Onge, G., Young, J.-G., Hébert-Dufresne, L., Dubé, L.J.: Efficient sampling of spreading processes on complex networks using acomposition and rejection algorithm. Comput. Phys. Commun. **240**, 30–37 (2019)
46. Van Kampen, N.G.: Stochastic Processes in Physics and Chemistry, vol. 1. Elsevier, Amsterdam (1992)
47. Van Mieghem, P., Omic, J., Kooij, R.: Virus spread in networks. IEEE/ACM Trans. Networking **17**(1), 1–14 (2009)

48. Ward, J.A., Evans, J.: A general model of dynamics on networks with graph automorphism lumping. In: Aiello, L.M., Cherifi, C., Cherifi, H., Lambiotte, R., Lió, P., Rocha, L.M. (eds.) COMPLEX NETWORKS 2018. SCI, vol. 812, pp. 445–456. Springer, Cham (2019). https://doi.org/10.1007/978-3-030-05411-3_36
49. Watts, D.J., Dodds, P.S.: Influentials, networks, and public opinion formation. J. Consum. Res. **34**(4), 441–458 (2007)
50. Wei, J., Kuo, J.C.: Lumping analysis in monomolecular reaction systems: analysis of the exactly lumpable system. Ind. Eng. Chem. Fundam. **8**(1), 114–123 (1969)
51. Wei, X., Valler, N.C., Prakash, B.A., Neamtiu, I., Faloutsos, M., Faloutsos, C.: Competing memes propagation on networks: a network science perspective. IEEE J. Sel. Areas Commun. **31**(6), 1049–1060 (2013)
52. Zhao, L., Cui, H., Qiu, X., Wang, X., Wang, J.: Sir rumor spreading model in the new media age. Phys. A **392**(4), 995–1003 (2013)
53. Zhao, L., Wang, J., Chen, Y., Wang, Q., Cheng, J., Cui, H.: Sihr rumor spreading model in social networks. Phys. A **391**(7), 2444–2453 (2012)

Applications and Tools

Doping Tests for Cyber-Physical Systems

Sebastian Biewer[1]([☒]), Pedro D'Argenio[1,2,3], and Holger Hermanns[1,4]

[1] Saarland University, Saarland Informatics Campus, Saarbücken, Germany
biewer@depend.uni-saarland.de
[2] FAMAF, Universidad Nacional de Córdoba, Córdoba, Argentina
[3] CONICET, Córdoba, Argentina
[4] Institute of Intelligent Software, Guangzhou, China

Abstract. The software running in embedded or cyber-physical systems (CPS) is typically of proprietary nature, so users do not know precisely what the systems they own are (in)capable of doing. Most malfunctionings of such systems are not intended by the manufacturer, but some are, which means these cannot be classified as bugs or security loopholes. The most prominent examples have become public in the diesel emissions scandal, where millions of cars were found to be equipped with software violating the law, altogether polluting the environment and putting human health at risk. The behaviour of the software embedded in these cars was intended by the manufacturer, but it was not in the interest of society, a phenomenon that has been called *software doping*. Doped software is significantly different from buggy or insecure software and hence it is not possible to use classical verification and testing techniques to discover and mitigate software doping.

The work presented in this paper builds on existing definitions of software doping and lays the theoretical foundations for conducting software doping tests, so as to enable attacking evil manufacturers. The complex nature of software doping makes it very hard to effectuate doping tests in practice. We explain the biggest challenges and provide efficient solutions to realise doping tests despite this complexity.

1 Introduction

Embedded and cyber-physical systems are becoming more and more widespread as part of our daily life. Printers, mobile phones, smart watches, smart home equipment, virtual assistants, drones and batteries are just a few examples. Modern cars are even composed of a multitude of such systems. These systems can have a huge impact on our lives, especially if they do not work as expected. As a result, numerous approaches exist to assure quality of a system. The classical and most common type of malfunctioning is what is widely called "bug". Usually, a bug is a very small mistake in the software or hardware that causes a behaviour that is not intended or expected. Other types of malfunctioning are caused by incorrect or wrongly interpreted sensor data, physical deficiencies of a component, or are simply radiation-induced.

Another interesting kind of malfunction (also from an ethical perspective [4]) arises if the expectation of how the system should behave is different for two (or

© Springer Nature Switzerland AG 2019
Parker and V. Wolf (Eds.): QEST 2019, LNCS 11785, pp. 313–331, 2019.
ps://doi.org/10.1007/978-3-030-30281-8_18

more) parties. Examples for such scenarios are widespread in the context of personal data privacy, where product manufacturers and data protection agencies have notoriously different opinions about how a software is supposed to handle personal data. Another example is the usage of third-party cartridges in printers. Manufacturers and users do not agree on whether their printer should work with third-party cartridges (the user's opinion) or only with those sold by the manufacturer (the manufacturer's opinion). Lastly, an example that received very high media attention are emission cleaning systems in diesel cars. There are regulations for dangerous particles and gases like CO_2 and NO_2 defining how much of these substances are allowed to be emitted during car operation. Part of these regulations are emissions tests, precisely defined test cycles that a car has to undergo on a chassis dynamometer [28]. Car manufacturers have to obey to these regulations in order to get admission to sell a new car model. The central weakness of these regulations is that the relevant behaviour of the car is only a trickle of the possible behaviour on the road. Indeed, several manufacturers equipped their cars with defeat devices that recognise if the car is undergoing an official emissions test. During the test, the car obeys the regulation, but outside test conditions, the emissions extruded are often significantly higher than allowed. Generally speaking, the phenomena described above are considered as incorrect software behaviour by one party, but as intended software behaviour by the other party (usually the manufacturer). In the literature, such phenomena are called software doping [3,10].

The difference between software doping and bugs is threefold: (1) There is a disagreement of intentions about what the software should do. (2) While a bug is most often a small coding error, software doping can be present in a considerable portion of the implementation. (3) Bugs can potentially be detected during production by the manufacturer, whereas software doping needs to be uncovered after production, by the other party facing the final product. Embedded software is typically proprietary, so (unless one finds a way to breach into the intellectual property [9]) it is only possible to detect software doping by observation of the behaviour of the product, i.e., by black-box testing.

This paper develops the foundations for black-box testing approaches geared towards uncovering doped software in concrete cases. We will start off from an established formal notion of robust cleanness (which is the negation of software doping) [10]. Essentially, the idea of robust cleanness is based on a succinct specification (called a "contract") that all involved parties agree on and which captures the intended behaviour of a system with respect to all inputs to the system. Inputs are considered to be user inputs or environmental inputs given by sensors. The contract is defined by input and output distances on standard system trajectories supplemented by input and output thresholds. Simply put, the input distance and threshold induce a tube around the standard inputs, and similar for outputs. For any input in the tube around some standard input the system must be able to react with an output that is in the tube around the output possible according to the standard.

Example 1. For a diesel car the standard trajectory is the behaviour exhibited during the official emissions test cycle. The input distance measures the deviation in car speed from the standard. The input threshold is a small number larger than the acceptable error tolerance of the cycle limiting the inputs considered of interest. The output distance then is the difference between (the total amount of) NO_x extruded by the car facing inputs of interest and that extruded if on the standard test cycle. For cars with an active defeat device we expect to see a violation of the contract even for relatively large output thresholds.

A cyber-physical system (CPS) is influenced by physical or chemical dynamics. Some of this can be observed by the sensors the CPS is equipped with, but some portion might remain unknown, making proper analysis difficult. Nondeterminism is a powerful way of representing such uncertainty faithfully, and indeed the notion of robust cleanness supports non-deterministic reactive systems [10]. Furthermore, the analysis needs to consider (at least) two trajectories simultaneously, namely the standard trajectory and another that stays within the input tube. In the presence of nondeterminism it might even become necessary to consider infinitely many trajectories at the same time. Properties over multiple traces are called hyperproperties [8]. In this respect, expressing robust cleanness as a hyperproperty needs both $\forall$ and $\exists$ trajectory quantifiers. Formulas containing only one type of quantifier can be analysed efficiently, e.g., using model-checking techniques, but checking properties with alternating quantifiers is known to be very complex [7,16]. Even more, testing of such problems is in general not possible. Assume, for example, a property requiring for a (nondeterministic) system that for every input i, there exists the output $o = i$, i.e., one of the system's possible behaviours computes the identity function. For black-box systems with infinite input and output domains the property can neither be verified nor falsified through testing. In order to verify the property, it is necessary to iterate over the infinite input set. For falsification one must show that for some i the system can not produce i as output. However, not observing an output in finitely many steps does not rule out that this output can be generated. As a result, there is no prior work (we are aware of) that targets the automatic generation of test cases for hyperproperties, let alone robust cleanness.

The contribution of this paper is three-fold. (1) We observe that standard behaviour, in particular when derived by common standardisation procedures, can be represented by finite models, and we identify under which conditions the resulting contracts are (un)satisfiable. (2) For a given satisfiable contract we construct the largest non-deterministic model that is robustly clean w.r.t. this contract. We integrate this model into a model-based testing theory, which can provide a non-deterministic algorithm to derive sound test suites. (3) We develop a testing algorithm for bounded-length tests and discretised input/output values. We present test cases for the diesel emissions scandal and execute these tests with a real car on a chassis dynamometer.

2   Software Doping on Reactive Programs

Embedded software is reactive, it reacts to inputs received from sensors by producing outputs that are meant to control the device functionality. We consider a reactive program as a function $P : \mathsf{In}^\omega \to 2^{(\mathsf{Out}^\omega)}$ on infinite sequences of inputs so that the program reacts to the k-th input in the input sequence by producing non-deterministically the k-th output in each respective output sequence. Thus, the program can be seen, for instance, as a (non-deterministic) Mealy or Moore machine. Moreover, we consider an equivalence relation $\approx \subseteq \mathsf{In}^\omega \times \mathsf{In}^\omega$ that equates sequences of inputs. To illustrate this, think of the program embedded in a printer. Here $\approx$ would for instance equate input sequences that agree with respect to submitting the same documents regardless of the cartridge brand, the level of the toner (as long as there is sufficient), etc. We furthermore consider the set $\mathsf{StdIn} \subseteq \mathsf{In}^\omega$ of inputs of interest or *standard inputs*. In the previous example, StdIn contains all the input sequences with compatible cartridges and printable documents. The definitions given below are simple adaptations of those given in [10] (but where parameters are instead treated as parts of the inputs).

Definition 1. *A reactive program P is* clean *if for all inputs* $\mathsf{i}, \mathsf{i}' \in \mathsf{StdIn}$ *such that* $\mathsf{i} \approx \mathsf{i}'$, $P(\mathsf{i}) = P(\mathsf{i}')$. *Otherwise it is* doped.

This definition states that a program is *clean* if its execution exhibits the same visible sequence of output when supplied with two equivalent inputs, provided such inputs comply with the given standard StdIn. Notice that the behaviour outside StdIn is deemed immediately clean since it is of no interest.

In the context of the printer example, a program that would fail to print a document when provided with an ink cartridge from a third-party manufacturer, but would otherwise succeed to print would be considered doped, since this difference in output behaviour is captured by the above definition. For this, the inputs (being pairs of document and printer cartridge) must be considered equivalent (not identical), which comes down to ink cartridges being compatible.

However, the above definition is not very helpful for cases that need to preserve certain intended behaviour *outside* of the standard inputs StdIn. This is clearly the case in the diesel emissions scandal where the standard inputs are given precisely by the emissions test, but the behaviour observed there is assumed to generalise beyond the singularity of this test setup. It is meant to ensure that the amount of NO_2 and NO (abbreviated as NO_x) in the car exhaust gas does not deviate considerably *in general*, and comes with a legal prohibition of defeat mechanisms that simply turn off the cleaning mechanism. This legal framework is obviously a bit short sighted, since it can be circumvented by mechanisms that alter the behaviour gradually in a continuous manner, but in effect drastically. In a nutshell, one expects that if the input values observed by the electronic control unit (ECU) of a diesel vehicle deviate within "reasonable distance" from the *standard* input values provided during the lab emission test, the amount of NO_x found in the exhaust gas is still within the regulated threshold, or at least it does not exceed it more than a "reasonable amount".

This motivates the need to introduce the notion of distances on inputs and outputs. More precisely, we consider distances on finite traces: $d_{\text{In}} : (\text{In}^* \times \text{In}^*) \rightarrow \mathbb{R}_{\geq 0}$ and $d_{\text{Out}} : (\text{Out}^* \times \text{Out}^*) \rightarrow \mathbb{R}_{\geq 0}$. Such distances are required to be pseudometrics. (d is a pseudometric if $d(x,x) = 0$, $d(x,y) = d(y,x)$ and $d(x,y) \leq d(x,z) + d(z,y)$ for all x, y, and z.) With this, D'Argenio et al. [10] provide a definition of robust cleanness that considers two parameters: parameter κ_i refers to the acceptable distance an input may deviate from the norm to be still considered, and parameter κ_o that tells how far apart outputs are allowed to be in case their respective inputs are within κ_i distance (Definition 2 spells out the Hausdorff distance used in [10]).

Definition 2. *Let $\sigma[..k]$ denote the k-th prefix of the sequence σ. A reactive program P is robustly clean if for all input sequences $i, i' \in \text{In}^\omega$ with $i \in \text{StdIn}$, it holds for arbitrary $k \geq 0$ that whenever $d_{\text{In}}(i[..j], i'[..j]) \leq \kappa_i$ for all $j \leq k$, then*

1. for all $o \in P(i)$ there exists $o' \in P(i')$ such that $d_{\text{Out}}(o[..k], o'[..k]) \leq \kappa_o$, and
2. for all $o' \in P(i')$ there exists $o \in P(i)$ such that $d_{\text{Out}}(o[..k], o'[..k]) \leq \kappa_o$.

Notice that this is what we actually need for the non-deterministic case: each possible output generated along one of the executions of the program should be matched within "reasonable distance" by some output generated by the other execution of the program. Also notice that i' does not need to satisfy StdIn, but it will be considered as long as it is within κ_i distance of any input satisfying StdIn. In such a case, outputs generated by $P(i')$ will be requested to be within κ_o distance of some output generated by the respective execution induced by a standard input.

We remark that Definition 2 entails the existence of a *contract* which defines the set of standard inputs StdIn, the tolerance parameters κ_i and κ_o as well as the distances d_{In} and d_{Out}. In the context of diesel engines, one might imagine that the values to be considered, especially the tolerance parameters κ_i and κ_o for a particular car model are made publicly available (or are even advertised by the car manufacturer), so as to enable potential customers to discriminate between different car models according to the robustness they reach in being clean. It is also imaginable that the tolerances and distances are fixed by the legal authorities as part of environmental regulations.

3   Robustly Clean Labelled Transition Systems

This section develops the framework needed for an effective theory of black-box doping tests based on the above concepts. In this, the standard behaviour (e.g. as defined by the emission tests) and the robust cleanness definitions together will induce a set of reference behaviours that then serve as a model in a model-based conformance testing approach. To set the stage for this, we recall the definitions of labelled transition systems (LTS) and input-output transitions systems (IOTS) together with Tretmans' notion on model-based conformance testing [25]. We then recast the characterisation of robust cleanness (Definition 2) in terms of LTS.

Definition 3. *A labelled transition system (LTS) with inputs and outputs is a tuple* $\langle Q, \mathsf{In}, \mathsf{Out}, \rightarrow, q_0 \rangle$ *where (i)* Q *is a (possibly uncountable) non-empty set of states; (ii)* $L = \mathsf{In} \uplus \mathsf{Out}$ *is a (possibly uncountable) set of labels; (iii)* $\rightarrow \subseteq Q \times L \times Q$ *is the transition relation; (iv)* $q_0 \in Q$ *is the initial state. We say that a LTS is an* input-output transition system (IOTS) *if it is input-enabled in any state, i.e., for all* $s \in Q$ *and* $a \in \mathsf{In}$ *there is some* $s' \in Q$ *such that* $s \xrightarrow{a} s'$.

For ease of presentation, we do not consider internal transitions. The following definitions will be used throughout the paper. A *finite path* p in an LTS $\mathcal{L}$ is a sequence $s_1 a_1 s_2 a_2 \ldots a_{n-1} s_n$ with $s_i \xrightarrow{a_i} s_{i+1}$ for all $1 \leq i < n$. Similarly, an *infinite path* p in $\mathcal{L}$ is a sequence $s_1 a_1 s_2 a_2 \ldots$ with $s_i \xrightarrow{a_i} s_{i+1}$ for all $i \in \mathbb{N}$. Let $\mathsf{paths}_*(\mathcal{L})$ and $\mathsf{paths}_\omega(\mathcal{L})$ be the sets of all finite and infinite paths of $\mathcal{L}$ beginning in the initial states, respectively. The sequence $a_1 a_2 \cdots a_n$ is a *finite trace* of $\mathcal{L}$ if there is a finite path $s_1 a_1 s_2 a_2 \ldots a_n s_{n+1} \in \mathsf{paths}_*(\mathcal{L})$, and $a_1 a_2 \cdots$ is an *infinite trace* if there is an infinite path $s_1 a_1 s_2 a_2 \ldots \in \mathsf{paths}_\omega(\mathcal{L})$. If p is a path, we let $\mathsf{trace}(p)$ denote the trace defined by p. Let $\mathsf{traces}_*(\mathcal{L})$ and $\mathsf{traces}_\omega(\mathcal{L})$ be the sets of all finite and infinite traces of $\mathcal{L}$, respectively. We will use $\mathcal{L}_1 \subseteq \mathcal{L}_2$ to denote that $\mathsf{traces}_\omega(\mathcal{L}_1) \subseteq \mathsf{traces}_\omega(\mathcal{L}_2)$.

Model-Based Conformance Tests. In the following we recall the basic notions of *input-output conformance* (**ioco**) testing [25–27], and refer to the mentioned literature for more details. In this setting, it is assumed that the implemented system under test (IUT) $\mathcal{I}$ can be modelled as an IOTS while the specification of the required behaviour is given in terms of a LTS *Spec*. The idea of whether the IUT $\mathcal{I}$ *conforms to* the specification *Spec* is formalized by means of the **ioco** relation which we define in the following.

We first need to identify the *quiescent* (or *suspended*) states. A state is quiescent whenever it cannot proceed autonomously, i.e., it cannot produce an output. We will make each such state identifiable by adding a quiescence transition to it, in the form of a loop with the distinct label δ.

Definition 4. *Let* $\mathcal{L} = \langle Q, \mathsf{In}, \mathsf{Out}, \rightarrow, q_0 \rangle$ *be an LTS. The* quiescence closure *(or δ-closure) of* $\mathcal{L}$ *is the LTS* $\mathcal{L}_\delta := \langle Q, \mathsf{In}, \mathsf{Out} \cup \{\delta\}, \rightarrow_\delta, q_0 \rangle$ *with* $\rightarrow_\delta := \rightarrow \cup \{s \xrightarrow{\delta}_\delta s \mid \forall o \in \mathsf{Out}, t \in Q : s \xcancel{\xrightarrow{o}} t\}$. *Using this we define the* suspension traces *of* $\mathcal{L}$ *by* $\mathsf{traces}_*(\mathcal{L}_\delta)$.

Let $\mathcal{L}$ be an LTS with initial state q_0 and $\sigma = a_1 a_2 \ldots a_n \in \mathsf{traces}_*(\mathcal{L})$. We define $\mathcal{L}$ after σ as the set $\{q_n \mid q_0 a_1 q_1 a_2 \ldots a_n q_n \in \mathsf{paths}_*(\mathcal{L})\}$. For a state q, let $\mathsf{out}(q) = \{o \in \mathsf{Out} \cup \{\delta\} \mid \exists q' : q \xrightarrow{o} q'\}$ and for a set of states $Q' \subseteq Q$, let $\mathsf{out}(Q') = \bigcup_{q \in Q'} \mathsf{out}(q)$.

The idea behind the **ioco** relation is that any output produced by the IUT must have been foreseen by its specification, and moreover, any input in the IUT not foreseen in the specification may introduce new functionality. **ioco** captures this by harvesting concepts from refusal testing. As a result, $\mathcal{I}$ **ioco** *Spec* is defined to hold whenever $\mathsf{out}(\mathcal{I}_\delta$ after $\sigma) \subseteq \mathsf{out}(Spec_\delta$ after $\sigma)$ for all $\sigma \in \mathsf{traces}_*(Spec_\delta)$.

The base principle of *conformance testing* now is to assess by means of testing whether the IUT conforms to its specification w.r.t. **ioco**. An algorithm to derive a corresponding test suite T_{Spec} is available [26,27], so that for any IUT $\mathcal{I}$, $\mathcal{I}$ **ioco** $Spec$ iff $\mathcal{I}$ passes all tests in T_{Spec}.

It is important to remark that the specification in the setting considered here is missing. Instead, we need to construct the specification from the standard inputs and the respective observed outputs, together with the distances and the thresholds given by the contract. Furthermore, this needs to respect the $\forall - \exists$ interaction required by the cleanness property (Definition 2).

Software Doping on LTS. To capture the notion of software doping in the context of LTS, we provide two projections of a trace, projecting to a sequence of the appearing inputs, respectively outputs. To do this, we extend the set of labels by adding the input $-_i$, that indicates that in the respective step some output (or quiescence) was produced (but masking the precise output), and the output $-_o$ that indicates that in this step some (masked) input was given.

The *projection on inputs* $\downarrow_i : L^\omega \to (\text{In} \cup \{-_i\})^\omega$ and the *projection on outputs* $\downarrow_o : L^\omega \to (\text{Out} \cup \{-_o\})^\omega$ are defined for all traces σ and $k \in \mathbb{N}$ as follows: $\sigma \downarrow_i [k] :=$ **if** $\sigma[k] \in \text{In}$ **then** $\sigma[k]$ **else** $-_i$ and $\sigma \downarrow_o [k] :=$ **if** $\sigma[k] \in \text{Out}$ **then** $\sigma[k]$ **else** $-_o$. They are lifted to sets of traces in the usual elementwise way.

Definition 5. *A LTS S is a* standard *for a LTS $\mathcal{L}$, if* $\text{traces}_\omega(S_\delta) \subseteq \text{traces}_\omega(\mathcal{L}_\delta)$.

The above definition provides an interpretation of the notion of StdIn for a given program P modelled in terms of LTS $\mathcal{L}$. This interpretation relaxes the original definition of StdIn, because it requires to fix only a subset of the behaviour that $\mathcal{L}$ exhibits when executed with standard inputs. This corresponds to a testing context, in which recordings of the system executing standard inputs are the baseline for testing. StdIn can then be considered as implicitly determined as the input sequences $\text{traces}_\omega(S) \downarrow_i$ occurring in S. If instead $\mathcal{L}$ and StdIn $\subseteq (\text{In} \cup -_i)^\omega$ are given, we denote by $S^{(\mathcal{L}, \text{StdIn})}$ a standard LTS which is maximal w.r.t. StdIn and $\mathcal{L}$, i.e., for all $\sigma \in \text{traces}_\omega(S_\delta^{(\mathcal{L}, \text{StdIn})})$ iff $\sigma \downarrow_i \in \text{StdIn}$ and $\sigma \in \text{traces}_\omega(\mathcal{L}_\delta)$.

In this new setting, we assume that the distance functions d_{In} and d_{Out} run on traces containing labels $-_i$ and $-_o$, i.e. they are pseudometrics in $(\text{In} \cup \{-_i\})^* \times (\text{In} \cup \{-_i\})^* \to \mathbb{R}_{\geq 0}$ and $(\text{Out} \cup \{-_o\})^* \times (\text{Out} \cup \{-_o\})^* \to \mathbb{R}_{\geq 0}$, respectively. We will denote a contract explicitly by a 5-tuple $\mathcal{C} = \langle S, d_{\text{In}}, d_{\text{Out}}, \kappa_i, \kappa_o \rangle$, which contains a LTS S representing some standard behaviour, the distances and thresholds (the domains In and Out are captured implicitly as the domains of d_{In}, respectively d_{Out}). With this, robust cleanness can be restated in terms of LTS as follows.

Definition 6. *Let $\mathcal{L}$ be an IOTS and $\mathcal{C} = \langle S, d_{\text{In}}, d_{\text{Out}}, \kappa_i, \kappa_o \rangle$ a contract so that S is standard for $\mathcal{L}$. This $\mathcal{L}$ is* robustly clean *w.r.t. $\mathcal{C}$ if for all $\sigma \in \text{traces}_\omega(S_\delta)$ and $\sigma' \in \text{traces}_\omega(\mathcal{L}_\delta)$ it holds for arbitrary $k \geq 0$ that whenever $d_{\text{In}}(\sigma[..j] \downarrow_i, \sigma'[..j] \downarrow_i) \leq \kappa_i$ for all $j \leq k$ then*

1. *there exists $\sigma'' \in \text{traces}_\omega(\mathcal{L}_\delta)$ s.t. $\sigma' \downarrow_i = \sigma'' \downarrow_i$ and $d_{\text{Out}_\delta}(\sigma[..k] \downarrow_o, \sigma''[..k] \downarrow_o) \leq \kappa_o$,*
2. *there exists $\sigma'' \in \text{traces}_\omega(S_\delta)$ s.t. $\sigma \downarrow_i = \sigma'' \downarrow_i$ and $d_{\text{Out}_\delta}(\sigma'[..k] \downarrow_o, \sigma''[..k] \downarrow_o) \leq \kappa_o$.*

In the spirit of model-based testing with **ioco**, Definition 6 takes specific care of quiescence in a system. In order to properly integrate quiescence into the context of robust cleanness it must be considered as a unique output. As a consequence, in the presence of a contract $\mathcal{C} = \langle \mathcal{S}, d_{\mathsf{In}}, d_{\mathsf{Out}}, \kappa_i, \kappa_o \rangle$, we use – instead of $\mathcal{S}$, Out and d_{Out} – the quiescence closure $\mathcal{S}_\delta$ of $\mathcal{S}$, $\mathsf{Out}_\delta = \mathsf{Out} \cup \{\delta\}$ and an extended output distance defined as $d_{\mathsf{Out}_\delta}(\sigma_1, \sigma_2) := d_{\mathsf{Out}}(\sigma_{1\backslash\delta}, \sigma_{2\backslash\delta})$ if $\sigma_1[i] = \delta \Leftrightarrow \sigma_2[i] = \delta$ for all i, and $d_{\mathsf{Out}_\delta}(\sigma_1, \sigma_2) := \infty$ otherwise, where $\sigma_{\backslash\delta}$ is the same as σ with all δ removed.

For the maximal standard LTS $\mathcal{S}^{(\mathcal{L}, \mathsf{StdIn})}$, Definition 6 echoes the semantics of the HyperLTL interpretation appearing in Proposition 19 of [10] restricted to programs with no parameters. Thus, the proof showing that Definition 6 is the correct interpretation of Definition 2 in terms of LTS, can be obtained in a way similar to that of Proposition 19 in [10].

In the sequel, we will at some places need to refer to Definition 6 only considering the second condition (but not the first one). We denote this as Definition 6.2.

4   Reference Implementation for Contracts

As mentioned before, doping tests need to be based on a contract $\mathcal{C}$, which we assume given. $\mathcal{C}$ specifies the domains In, Out, a standard LTS $\mathcal{S}$, the distances d_{In} and d_{Out} and the bounds κ_i and κ_o. We intuitively expect the contract to be satisfiable in the sense that it never enforces a single input sequence of the implementation to keep outputs close enough to two different executions of the specification while their outputs stretch too far apart. We show such a problematic case in the following example.

Example 2. On the right a quiescence-closed standard LTS $\mathcal{S}_\delta$ for an implementation $\mathcal{L}$ (shown below) is depicted. For simplicity some input transitions are omitted. Assume $\mathsf{Out} = \{o\}$ and $\mathsf{In} = \{i, i - \kappa_i, i + \kappa_i\}$. Consider the transition labelled x of $\mathcal{L}$. This must be one of either o or δ, but we will see that either choice leads to a contradiction w.r.t. the output distances induced. The input projection of the middle path in $\mathcal{L}$ is $i -_i$ and the input distance to $(i - \kappa_i) -_i$ and $(i + \kappa_i) -_i$ exactly κ_i, so both branches $(i + \kappa_i)\, o$ and $(i - \kappa_i)\, \delta$ of $\mathcal{S}_\delta$ must be considered to determine x. For $x = o$, the output distance of $-_o x$ to $-_o o$ in the right branch of $\mathcal{S}_\delta$ is 0, i.e. less than κ_o. However, $d_{\mathsf{Out}_\delta}(-_o \delta, -_o o) = \infty > \kappa_o$. Thus the output distance to the left branch of $\mathcal{S}_\delta$ is too high if picking o. Instead picking $x = \delta$ does not work either, for the symmetric reasons, the problem switches sides. Thus, neither picking o nor δ for x satisfies robust cleanness here. Indeed, no implementation satisfying robust cleanness exists for the given contract.

We would expect that a correct implementation fully entails the standard behaviour. So, to satisfy a contract, the standard behaviour itself must be robustly clean. This and the need for satisfiability of particular inputs lead to Definition 7.

Definition 7 (Satisfiable Contract). *Let* $\mathcal{C} = \langle \mathcal{S}, d_{\mathsf{In}}, d_{\mathsf{Out}}, \kappa_i, \kappa_o \rangle$ *be a contract. Let input* $\sigma_i \in (\mathsf{In} \cup \{-_i\})^\omega$ *be the input projection of some trace.* σ_i *is satisfiable for* $\mathcal{C}$ *if and only if for every standard trace* $\sigma_S \in \mathsf{traces}_\omega(\mathcal{S}_\delta)$ *and* $k > 0$ *such that for all* $j \leq k$ $d_{\mathsf{In}}(\sigma_i[..j], \sigma_S[..j]\!\downarrow_i) \leq \kappa_i$ *there is some implementation* $\mathcal{L}$ *that satisfies Definition 6.2 w.r.t.* $\mathcal{C}$ *and has some trace* $\sigma \in \mathsf{traces}_\omega(\mathcal{L}_\delta)$ *with* $\sigma\!\downarrow_i = \sigma_i$ *and* $d_{\mathsf{Out}_\delta}(\sigma[..k]\!\downarrow_o, \sigma_S[..k]\!\downarrow_o) \leq \kappa_o$.
$\mathcal{C}$ *is* satisfiable *if and only if all inputs* $\sigma_i \in (\mathsf{In} \cup \{-_i\})^\omega$ *are satisfiable for* $\mathcal{C}$ *and if* $\mathcal{S}$ *is robustly clean w.r.t.* $\mathcal{C}$. *A contract that is not satisfiable is called* unsatisfiable.

Given a satisfiable contract it is always possible to construct an implementation that is robustly clean w.r.t. to this contract. Furthermore, for every contract there is exactly one implementation (modulo trace equivalence) that contains all possible outputs that satisfy robust cleanness. Such an implementation is called the *largest implementation*.

Definition 8 (Largest Implementation). *Let* $\mathcal{C}$ *be a contract and* $\mathcal{L}$ *an implementation that is robustly clean w.r.t.* $\mathcal{C}$. $\mathcal{L}$ *is the* largest *implementation within* $\mathcal{C}$ *if and only if for every* $\mathcal{L}'$ *that is robustly clean w.r.t.* $\mathcal{C}$ *it holds that* $\mathsf{traces}_\omega(\mathcal{L}'_\delta) \subseteq \mathsf{traces}_\omega(\mathcal{L}_\delta)$.

In the following, we will focus on the fragment of satisfiable contracts with standard behaviour defined by finite LTS. For unsatisfiable contracts, testing is not necessary, because every implementation is not robustly clean w.r.t. to $\mathcal{C}$. Finiteness of $\mathcal{S}$ will be necessary to make testing feasible in practice. For simplicity we will further assume *past-forgetful* output distance functions. That is, $d_{\mathsf{Out}}(\sigma_1, \sigma_2) = d_{\mathsf{Out}}(\sigma'_1, \sigma'_2)$ whenever $\mathsf{last}(\sigma_1) = \mathsf{last}(\sigma'_1)$ and $\mathsf{last}(\sigma_2) = \mathsf{last}(\sigma'_2)$ (where $\mathsf{last}(a_1 \, a_2 \ldots a_n) = a_n$.) Thus, we simply assume that $d_{\mathsf{Out}} : (\mathsf{Out} \cup \{-_o\} \times \mathsf{Out} \cup \{-_o\}) \to \mathbb{R}_{\geq 0}$, i.e., the output distances are determined by the last output only. We remark that $d_{\mathsf{Out}_\delta}(\delta, o) = \infty$ for all $o \neq \delta$.

We will now show how to construct the largest implementation for any contract (of the fragment we consider), which we name *reference implementation* $\mathcal{R}$. It is derived from $\mathcal{S}_\delta$ by adding inputs and outputs in such a way that whenever the input sequence leading to a particular state is within κ_i distance of an input sequence σ_i of $\mathcal{S}_\delta$, then the outputs possible in such a state should be at most κ_o distant from those outputs possible in the unique state on $\mathcal{S}_\delta$ reached through σ_i. This ensures that $\mathcal{R}$ will satisfy condition (2) in Definition 6.

Reference Implementation. To construct the reference implementation $\mathcal{R}$ we decide to model the quiescence transitions explicitly instead of using the quiescence closure. We preserve the property, that in each state of the LTS it is possible to do an output or a quiescence transition. The construction of $\mathcal{R}$ proceeds by adding all transitions that satisfy the second condition of Definition 6.

Definition 9. *Let* $\mathcal{C} = \langle \mathcal{S}, d_{\mathsf{In}}, d_{\mathsf{Out}}, \kappa_i, \kappa_o \rangle$ *be a contract. The* reference implementation $\mathcal{R}$ *for* $\mathcal{C}$ *is the LTS* $\langle (\mathsf{In} \cup \mathsf{Out})^*, \mathsf{In}, \mathsf{Out}, \to_\mathcal{R}, \epsilon \rangle$ *where* $\to_\mathcal{R}$ *is defined by*

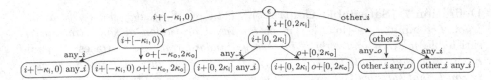

Fig. 1. The reference implementation $\mathcal{R}$ of $\mathcal{S}$ in Example 3.

$$\forall \sigma_i \in \mathsf{traces}_\omega(\mathcal{S}_\delta) \downarrow_i :$$
$$(\forall j \leq |\sigma| + 1 : d_{\mathsf{In}}((\sigma \cdot a)\downarrow_i[..j], \sigma_i[..j]) \leq \kappa_i)$$
$$\Rightarrow \exists \sigma_S \in \mathsf{traces}_\omega(\mathcal{S}_\delta) : \sigma_S \downarrow_i = \sigma_i \wedge d_{\mathsf{Out}_\delta}(a\downarrow_o, \sigma_S[|\sigma| + 1]\downarrow_o) \leq \kappa_o$$

$$\overline{\sigma \xrightarrow{a}_{\mathcal{R}} \sigma \cdot a}$$

Notably, $\mathcal{R}$ is deterministic, since only transitions of the form $\sigma \xrightarrow{a}_{\mathcal{R}} \sigma \cdot a$ are added. As a consequence of this determinism, outputs and quiescence may coexist as options in a state, i.e. they are not mutually exclusive.

Example 3. Fig. 1 gives a schematic representation of the reference implementation $\mathcal{R}$ for the LTS $\mathcal{S}$ on the right. Input (output) actions are denoted with letter i (o, respectively), quiescence transitions are omitted. We use the absolute difference of the values, so that $d_{\mathsf{In}}(i, i') := |i - i'|$ and $d_{\mathsf{Out}}(o, o') := |o - o'|$. For this example, the quiescence closure $\mathcal{S}_\delta$ looks like $\mathcal{S}$ but with δ-loops in states s_0, s_4,

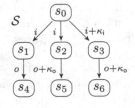

s_5, and s_6. Label $r+[a, b]$ should be interpreted as any value $r' \in [a + r, b + r]$ and similarly $r+[a, b)$ and $r+(a, b]$, appropriately considering closed and open boundaries; "other$\_i$" represents any other input not explicitly considered leaving the same state; and "any$\_i$" and "any$\_o$" represent any possible input and output (including δ), respectively. In any case $-_i$ and $-_o$ are not considered since they are not part of the alphabet of the LTS. Also, we note that any possible sequence of inputs becomes enabled in the last states (omitted in the picture).

Robust Cleanness of Reference Implementation. In the following, the aim is to show that $\mathcal{R}$ is robustly clean. By construction, each state in $\mathcal{R}$ equals the trace that leads to that state. In other words, $\mathsf{last}(p) = \mathsf{trace}(p)$ for any $p \in \mathsf{paths}_*(\mathcal{R})$ can be shown by induction. As a consequence, a path in $\mathcal{R}$ can be completely identified by the trace it defines. The following lemma states that $\mathcal{R}$ preserves all traces of the standard $\mathcal{S}_\delta$ it is constructed from. This can be proven by using that $\mathcal{S}_\delta$ is robustly clean w.r.t. the (satisfiable) contract $\mathcal{C}$ (see Definition 7).

Lemma 1. *Let $\mathcal{R}$ be the reference implementation for $\mathcal{C} = \langle \mathcal{S}, d_{\mathsf{In}}, d_{\mathsf{Out}}, \kappa_i, \kappa_o \rangle$. Then $\mathcal{S}$ is standard for $\mathcal{R}$.*

The following theorem states that the reference implementation $\mathcal{R}$ is robustly clean w.r.t. the contract it was constructed from.

Theorem 1. *Let $\mathcal{R}$ be the reference implementation for some contract $\mathcal{C}$. Then $\mathcal{R}$ is robustly clean w.r.t. $\mathcal{C}$.*

Furthermore, it is not difficult to show that $\mathcal{R}$ is indeed the largest implementation within the contract it was constructed from.

Theorem 2. *Let $\mathcal{R}$ be the reference implementation for some contract $\mathcal{C}$. Then $\mathcal{R}$ is the largest implementation within $\mathcal{C}$.*

5   Model-Based Doping Tests

Following the conceptual ideas behind **ioco**, we need to construct a specification that is compatible with our notion of robust cleanness in such a way that a test suite can be derived. Intuitively, such a specification must be able to foresee every behaviour of the system that is allowed by the contract. We will take the reference implementation from the previous section as this specification. Indeed we claim that $\mathcal{R}$ is constructed in such a way that whenever an IUT $\mathcal{I}$ is robustly clean, $\mathcal{I}$ **ioco** $\mathcal{R}$ holds. The latter translates to

$$\forall \sigma \in \mathsf{traces}_*(\mathcal{R}_\delta) : \mathsf{out}(\mathcal{I}_\delta \text{ after } \sigma) \subseteq \mathsf{out}(\mathcal{R}_\delta \text{ after } \sigma). \tag{1}$$

Theorem 3. *Let $\mathcal{C}$ be a contract with standard $\mathcal{S}$ and let IOTS $\mathcal{I}$ be robustly clean w.r.t. $\mathcal{C}$. If $\mathcal{R}$ is the reference implementation for $\mathcal{C}$, then $\mathcal{I}$ **ioco** $\mathcal{R}$.*

The key observations to prove this theorem are: (i) the reference implementation is the largest implementation within the contract, i.e. if the IUT is robustly clean, then all its traces are covered by $\mathcal{R}$, and (ii) by construction of $\mathcal{R}$ and satisfiability of $\mathcal{C}$, the suspension traces of $\mathcal{R}$ are exactly its finite traces.

Test Algorithm. An important element of the model-based testing theory is a non-deterministic algorithm to generate test cases. It is, however, not guaranteed that this algorithm, even if existing, is implementable, a problem which we will tackle in this section. A set of test cases is called a *test suite*. It is shown elsewhere [27], that there is an algorithm that can produce a (possibly infinitely large) test suite T, for which a system $\mathcal{I}$ passes T if $\mathcal{I}$ is correct w.r.t. **ioco** and, conversely, $\mathcal{I}$ is correct w.r.t. **ioco** if $\mathcal{I}$ passes T. The former property is called *soundness* and the latter is called *exhaustiveness*. Algorithm 1 shows a tail-recursive algorithm to test for robust cleanness. This DT algorithm takes as an argument the history h of the test currently running. Every doping test is initialized by $\mathsf{DT}(\epsilon)$. Several runs of the algorithm constitute a test suite. Each test can either **pass** or **fail**, which is reported by the output of the algorithm. In each call DT picks one of three choices: (i) it either terminates the test by returning **pass** (line 3), (ii) if there is no pending output that has to be read from the system under test, the algorithm may pick a new input and pass it to the system (lines 5–6), or (iii) DT reads and checks the next output (or quiescence) that the system produces (lines 9–10). Quiescence can be recognized by using a timeout mechanism that returns δ if no output has been received in a given amount of time. In the original algorithm, the case and the next input are determined non-deterministically. Our algorithm is parameterized by Ω_{case}

Algorithm 1. Doping Test (DT)

Input: history $h \in (\text{In} \cup \text{Out} \cup \{\delta\})^*$
Output: pass or **fail**

1   $c \leftarrow \Omega_{\text{case}}(h)$ /* Pick from one of three cases */
2   **if** $c = 1$ **then**
3   **return pass**   /* Finish test generation */
4   **else if** $c = 2$ **and** no output from $\mathcal{I}$ is available **then**
5     $i \leftarrow \Omega_{\text{In}}(h)$ /* Pick next input */
6     $i \twoheadrightarrow \mathcal{I}$ /* Forward input to IUT */
7   **return** $\text{DT}(h \cdot i)$ /* Continue with next step */
8   **else if** $c = 3$ **or** output from $\mathcal{I}$ is available **then**
9     $o \twoheadleftarrow \mathcal{I}$ /* Receive output from IUT */
10   **if** $o \in \text{acc}(h)$ **then**
11   **return** $\text{DT}(h \cdot o)$ /* If o is foreseen by oracle continue with next step */
12   **else**
13   **return fail**   /* Otherwise, report test failure */
14   **end if**
15 **end if**

---

and Ω_{In}, which can be instantiated by either non-determinism or some optimized test-case selection. Until further notice we assume non-deterministic selection. An output or quiescence that has been produced by the IUT is checked by means of an oracle acc (line 10). The oracle reflects the reference implementation $\mathcal{R}$, that is used as the specification for the **ioco** relation and is defined in Eq. (2).

$$\text{acc}(h) := \{o \in \text{Out}_\delta \mid \tag{2}$$
$$\forall \sigma_i \in \text{traces}_\omega(\mathcal{S}_\delta){\downarrow}_i : (\forall j \leq |h|+1 : d_{\text{In}}(\sigma_i[..j]{\downarrow}_i, (h \cdot o)[..j]{\downarrow}_i) \leq \kappa_i\})$$
$$\Rightarrow \exists \sigma \in \text{traces}_\omega(\mathcal{S}_\delta) : \sigma{\downarrow}_i = \sigma_i{\downarrow}_i \wedge d_{\text{Out}_\delta}(o, \sigma[|h|+1]{\downarrow}_o) \leq \kappa_o$$

Given a finite execution, acc returns the set of acceptable outputs (after such an execution) which corresponds exactly to the set of outputs in $\mathcal{R}$ (after such an execution). Thus $\text{acc}(h)$ is precisely the set of outputs that satisfies the premise in the definition of $\mathcal{R}$ after the trace h, as stipulated in Definition 9.

We refer to acc as an oracle, because it cannot be computed in general due to the infinite traces of $\mathcal{S}_\delta$ in the definition. However, we get the following theorem stating that the algorithm is sound and exhaustive with respect to **ioco** (and we present a computable algorithm in the next section). The theorem follows from the soundness and exhaustiveness of the original test generation algorithm for model-based testing and Definition 9.

Theorem 4. *Let $\mathcal{C}$ be a contract with standard $\mathcal{S}$. Let $\mathcal{I}$ be an implementation with $\mathcal{S}_\delta \subseteq \mathcal{I}_\delta$ and let $\mathcal{R}$ be the largest implementation within $\mathcal{C}$. Then, $\mathcal{I}$ **ioco** $\mathcal{R}$ if and only if for every test execution $t = \text{DT}(\epsilon)$ it holds that $\mathcal{I}$ **passes** t.*

Corollary 1. *Let $\mathcal{C}$ be a contract with standard $\mathcal{S}$. Let $\mathcal{I}$ be an implementation with $\mathcal{S}_\delta \subseteq \mathcal{I}_\delta$. If $\mathcal{I}$ is robustly clean w.r.t. $\mathcal{C}$, then for every test execution $t = \text{DT}(\epsilon)$ it holds that $\mathcal{I}$ **passes** t.*

This corollary is derived from Theorem 3 and the satisfiability of $\mathcal{C}$. It is worth noting that in this corollary we do not get that $\mathcal{I}$ is robustly clean if $\mathcal{I}$ always passes DT. This is due the intricacies of genuine hyperproperties. By testing, we will never be able to verify the first condition of Definition 6, because this needs a simultaneous view on all possible execution traces of $\mathcal{I}$. During testing, however, we always can observe only one trace.

Finite Doping Tests. As mentioned before, the execution of DT is not possible, because the oracle acc is not computable. There is, however, a computable version acc_b of acc for executions up to some test length b for bounded and discretised In and Out. Even for infinite executions, b can be seen as a limit of interest and testing is still sound. acc_b is shown in Eq. (3). The only variation w.r.t. acc lies in the use of the set $traces_b(\mathcal{S}_\delta)$, instead of $traces_\omega(\mathcal{S}_\delta)$, so as to return all traces of $\mathcal{S}_\delta$ whose length is exactly b. Since $\mathcal{S}_\delta$ is finite, function acc_b can be implemented.

$$acc_b(h) := \{o \in \mathsf{Out}_\delta \mid \tag{3}$$
$$\forall \sigma_i \in \mathsf{traces}_b(\mathcal{S}_\delta){\downarrow}_i : (\forall j \leq |h|+1 : d_{\mathsf{In}}(\sigma_i[..j]{\downarrow}_i, (h \cdot o)[..j]{\downarrow}_i) \leq \kappa_i)$$
$$\Rightarrow \exists \sigma \in \mathsf{traces}_b(\mathcal{S}_\delta) : \sigma{\downarrow}_i = \sigma_i{\downarrow}_i \wedge d_{\mathsf{Out}_\delta}(o, \sigma[|h|+1]{\downarrow}_o) \leq \kappa_o\}$$

Now we get a new algorithm DT_b by replacing acc by acc_b in DT and by forcing case 1 when and only when $|h| = b$. We get a similar soundness theorem for DT_b as in Corollary 1.

Theorem 5. *Let $\mathcal{C}$ be a contract with standard $\mathcal{S}$. Let $\mathcal{I}$ be an implementation with $\mathcal{S}_\delta \subseteq \mathcal{I}_\delta$. If $\mathcal{I}$ is robustly clean w.r.t. $\mathcal{C}$, then for every boundary b and every test execution $t = DT_b(\epsilon)$ it holds that $\mathcal{I}$ passes t.*

Since $\mathcal{I}$ passes $DT_b(\epsilon)$ implies $\mathcal{I}$ passes $DT_a(\epsilon)$ for any $a \leq b$, we have in summary arrived at an on-the-fly algorithm DT_b that for sufficiently large b (corresponding to the length of the test) will be able to conduct a "convicting" doping test for any IUT $\mathcal{I}$ that is not robustly clean w.r.t. a given contract $\mathcal{C}$. The bounded-depth algorithm effectively circumvents the fact that, except for $\mathcal{S}$ and $\mathcal{S}_\delta$, all other objects we need to deal with are countably or uncountably infinite and that the property we check is a hyperproperty.

We implemented a prototype of a testing framework using the bounded-depth algorithm. The specification of distances, value domains and test case selection are parameters of the algorithm that can be set specific for a concrete test scenario. This flexibility enables us to use the framework in a two-step approach for cyber-physical systems not equipped with a digital interface to forward the inputs to: first, the tool can generate test inputs, that are executed by a human or a robot on the CPS under test. The actual inputs (possibly deviating from the generated inputs) and outputs from the system are recorded so that in the second step our tool determines if the (actual) test is passed or failed.

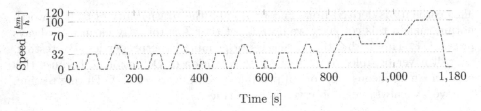

Time [s]

Fig. 2. NEDC speed profile.

6 Evaluation

The normed emission test NEDC (New European Driving Cycle) (see Fig. 2) is the legally binding framework in Europe [28] (at the time the scandal surfaced). It is to be carried out on a chassis dynamometer and all relevant parameters are fixed by the norm, including for instance the outside temperature at which it is run.

For a given car model, the normed test induces a standard LTS $\mathcal{S}$ as follows. The input dimensions of $\mathcal{S}$ are spanned by the sensors the car model is equipped with (including e.g. temperature of the exhaust, outside temperature, vertical and lateral acceleration, throttle position, time after engine start, engine rpm, possibly height above ground level etc.) which are accessible via the standardized OBD-2 interface [24]. The output is the amount of NO_x per kilometre that has been extruded since engine start. Inputs are sampled at equidistant times (once per second). The standard LTS $\mathcal{S}$ is obtained from the trace representing the observations of running NEDC on the chassis dynamometer, say $\sigma_S := i_1 \cdots i_{1180} \, o_S \, \delta \delta \delta \cdots$ with inputs $i_1, \cdots i_{1180}$ given by the NEDC over its 20 min (1180 s) duration, and o_S is the amount of NO_x gases accumulated during the test procedure. This σ_S is the only standard trace of our experiments. The trace ends with an infinite suffix δ^ω of quiescence steps.

The input space, In is a vector space spanned by all possible input parameter dimensions. For $a \in$ In we distinguish the speed dimension as $v(a) \in \mathbb{R}$ (measured in km/h). We can use past-forgetful distances with $d_{In}(a, b) := |v(a) - v(b)|$ if $a, b \in$ In, $d_{In}(-_i, -_i) = 0$ and $d_{In}(a, b) = \infty$ otherwise. The speed is the decisive quantity defined to vary along the NEDC (cf. Fig. 2). Hence $d_{In}(a, b) = 0$ if $v(a) = v(b)$ regardless of the values of other parameters. We also take Out $= \mathbb{R}$ for the average amount of NO_x gases per kilometre since engine start (in mg/km). We define $d_{Out}(a, b) = |a - b|$ if $a, b \in$ Out, and $d_{Out}(a, b) = \infty$ otherwise.

Doping Tests in Practice. For the purpose of practically exercising doping tests, we picked a Renault 1.5 dci (110*hp*) (Diesel) engine. This engine runs, among others, inside a Nissan NV200 Evalia which is classified as a Euro 6 car. The test cycle used in the original type approval of the car was NEDC (which corresponds to Euro 6b). Emissions are cleaned using *exhaust gas recirculation* (EGR). The technical core of EGR is a valve between the exhaust and intake pipe, controlled by a software. EGR is known to possibly cause performance losses, especially at

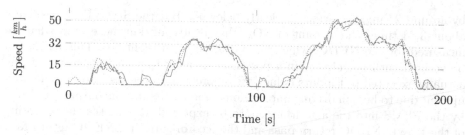

Fig. 3. Initial 200 s of a SINENEDC (red, dotted), its test drive (green) and the NEDC driven (blue, dashed). (Color figure online)

higher speed. Car manufacturers might be tempted to optimize EGR usage for engine performance unless facing a known test cycle such as the NEDC.

We fixed a contract with $\kappa_i = 15\,km/h$, $\kappa_o = 180\,mg/km$. We report here on two of the tests we executed apart from the NEDC reference: *(i)* POWERNEDC is a variation of the NEDC, where acceleration is increased from $0.94\frac{m}{s^2}$ to $1.5\frac{m}{s^2}$ in phase 6 of the NEDC elementary urban cycle (i.e. after 56 s, 251 s, 446 s and 641 s) and *(ii)* SINENEDC defines the speed at time t to be the speed of the NEDC at time t plus $5 \cdot sin(0.5t)$ (but capped at 0). Both can be generated by $DT_{1181}(\epsilon)$ for specific deterministic Ω_{case} and Ω_{ln}. For instance, SINENEDC is given below. Fig. 3 shows the initial 200 s of SINENEDC (red, dotted).

$$\Omega_{case}(h) = \begin{cases} 2 & \text{, if } |h| \leq 1179 \\ 3 & \text{, if } |h| = 1180 \end{cases} \quad \Omega_{ln}(h) = \max \left\{ \begin{matrix} 0, \\ NEDC(|h|) + 5 \cdot sin(0.5|h|) \end{matrix} \right\}$$

The car was fixed on a *Maha LPS 2000* dynamometer and attached to an *AVL M.O.V.E iS* portable emissions measurement system (PEMS, see Fig. 4) with speed data sampling at a rate of 20 Hz, averaged to match the 1 Hz rate of the NEDC. The human driver effectuated the NEDC with a deviation of at most 9 km/h relative to the reference (notably, the result obtained for NEDC are not consistent with the car data sheet, likely caused by lacking calibration and absence of any further manufacturer-side optimisations).

Fig. 4. Nissan NV200 Evalia on a dynamometer

Table 1. Dynamometer measurements (sample rate: 1 Hz)

	NEDC	Power	Sine
Distance $[m]$	11,029	11,081	11,171
Avg. Speed $\left[\frac{km}{h}\right]$	33	29	34
CO_2 $\left[\frac{g}{km}\right]$	189	186	182
NO_x $\left[\frac{mg}{km}\right]$	180	204	**584**

The POWERNEDC test drive differed by less than 15 km/h and the SINENEDC by less than 14 km/h from the NEDC test drive, so both inputs deviate by less than κ_i. The green line in Fig. 3 shows SINENEDC driven. The test outcomes are summarised in Table 1. They show that the amount of CO_2 for the two tests is lower than for the NEDC driven. The NO_x emissions of POWERNEDC deviate by around 24 mg/km, which is clearly below κ_o. But the SINENEDC produces about 3.24 times the amount of NO_x, that is 404 mg/km more than what we measured for the NEDC, which is a violation of the contract. This result can be verified with our algorithm a posteriori, namely by using Ω_{In} to replay the actually executed test inputs (which are different from the test inputs generated upfront due to human driving imprecisions) and by feeding the outputs recorded by the PEMS into the algorithm. As to be expected, this makes the recording of the POWERNEDC return **pass** and the recording of SINENEDC return **fail**.

Our algorithm is powerful enough to detect other kinds of defeat devices like those uncovered in investigations of the Volkswagen or the Audi case. Due to lack of space, we cannot present the concrete Ω_{case} and Ω_{In} for these examples.

7   Discussion

Related Work. The present work complements white-box approaches to software doping, like model-checking [10] or static code analysis [9] by a black-box testing approach, for which the specification is given implicitly by a contract, and usable for on-the-fly testing. Existing test frameworks like TGV [18] or TorX [29] provide support for the last step, however they fall short on scenarios where (i) the specification is not at hand and, among others, (ii) the test input is distorted in the testing process, e.g., by a human driving a car under test.

Our work is based on the definition of robust cleanness [10] which has conceptual similarities to continuity properties [6,17] of programs. However, continuity itself does not provide a reasonably good guarantee of cleanness. This is because physical outputs (e.g. the amount of NO_x gas in the exhaust) usually do change continuously. For instance, a doped car may alter its emission cleaning in a discrete way, but that induces a (rapid but) continuous change of NO_x gas concentrations. Established notions of stability and robustness [13,19,21,23] differ from robust cleanness in that the former assure the outputs (of a white-box system model) to stabilize despite transient input disturbances. Robust cleanness does not consider perturbations but (intentionally) different inputs, and needs a hyperproperty formulation.

Concluding Remarks. This work lays the theoretical foundations for black-box testing approaches geared towards uncovering doped software. As in the diesel emissions scandal – where manufacturers were forced to pay excessive fines [22] and where executive managers are facing lawsuits or indeed went to prison [5,14] – doped behaviour is typically strongly related to illegal behaviour.

As we have discussed, software doping analysis comes with several challenges. It can be performed (i) only after production time on the final embedded or cyber-physical product, (ii) notoriously without support by the manufacturer, and (iii) the property belongs to the class of hyperproperties with alternating quantifiers. (iv) Non-determinism and imprecision caused by a human in-the-loop complicate doping analysis of CPS even further.

Conceptually central to the approach is a contract that is assumed to be explicitly offered by the manufacturer. The contract itself is defined by very few parameters making it easy to form an opinion about a concrete contract. And even if a manufacturer is not willing to provide such contractual guarantees, instead a contract with very generous parameters can provide convincing evidence of doping if a test uncovers the contract violation. We showed this in a real automotive example demonstrating how a legally binding reference behaviour and a contract altogether induce a finite state LTS enabling to harvest input-output conformance testing for doping tests. We developed an algorithm that can be attached directly to a system under test or in a three-step process, first generating a valid test case, afterwards used to guide a human interacting with the system, possibly adding distortions, followed by an a-posteriori validation of the recorded trajectory. For more effective test case selection [11,15] we are exploring different guiding techniques [1,2,12] for cyber-physical systems.

Acknowledgements. We gratefully acknowledge Thomas Heinze, Michael Fries, and Peter Birtel (Automotive Powertrain Institute of HTW Saar) for sharing their automotive engineering expertise with us, and for providing the automotive test infrastructure. This work is partly supported by the ERC Grant 695614 (POWVER), by the Deutsche Forschungsgemeinschaft (DFG, German Research Foundation) grant 389792660 as part of TRR 248, see https://perspicuous-computing.science, by the Saarbrücken Graduate School of Computer Science, by the Sino-German CDZ project 1023 (CAP), by ANPCyT PICT-2017-3894 (RAFTSys), and by SeCyT-UNC 33620180100354CB (ARES).

References

1. Adimoolam, A., Dang, T., Donzé, A., Kapinski, J., Jin, X.: Classification and coverage-based falsification for embedded control systems. In: Majumdar, R., Kunčak, V. (eds.) CAV 2017. LNCS, vol. 10426, pp. 483–503. Springer, Cham (2017). https://doi.org/10.1007/978-3-319-63387-9_24
2. Annpureddy, Y., Liu, C., Fainekos, G., Sankaranarayanan, S.: S-TaLiRo: a tool for temporal logic falsification for hybrid systems. In: Abdulla, P.A., Leino, K.R.M. (eds.) TACAS 2011. LNCS, vol. 6605, pp. 254–257. Springer, Heidelberg (2011). https://doi.org/10.1007/978-3-642-19835-9_21
3. Barthe, G., D'Argenio, P.R., Finkbeiner, B., Hermanns, H.: Facets of software doping. In: Margaria and Steffen [20], pp. 601–608. https://doi.org/10.1007/978-3-319-47169-3_46
4. Baum, K.: What the hack is wrong with software doping? In: Margaria and Steffen [20], pp. 633–647. https://doi.org/10.1007/978-3-319-47169-3_49
5. BBC: Audi chief Rupert Stadler arrested in diesel emissions probe. BBC (2018). https://www.bbc.com/news/business-44517753. Accessed 28 Jan 2019
6. Chaudhuri, S., Gulwani, S., Lublinerman, R.: Continuity analysis of programs. In: Proceedings of the 37th ACM SIGPLAN-SIGACT Symposium on Principles of Programming Languages, POPL 2010, Madrid, Spain, 17–23 January 2010, pp. 57–70. ACM (2010). http://doi.acm.org/10.1145/1706299.1706308
7. Clarkson, M.R., Finkbeiner, B., Koleini, M., Micinski, K.K., Rabe, M.N., Sánchez, C.: Temporal logics for hyperproperties. In: Abadi, M., Kremer, S. (eds.) POST 2014. LNCS, vol. 8414, pp. 265–284. Springer, Heidelberg (2014). https://doi.org/10.1007/978-3-642-54792-8_15

8. Clarkson, M.R., Schneider, F.B.: Hyperproperties. In: CSF 2008, pp. 51–65 (2008). http://dx.doi.org/10.1109/CSF.2008.7

9. Contag, M., et al.: How they did it: an analysis of emission defeat devices in modern automobiles. In: 2017 IEEE Symposium on Security and Privacy, SP 2017, San Jose, CA, USA, 22–26 May 2017, pp. 231–250. IEEE Computer Society (2017). https://doi.org/10.1109/SP.2017.66

10. D'Argenio, P.R., Barthe, G., Biewer, S., Finkbeiner, B., Hermanns, H.: Is your software on dope? In: Yang, H. (ed.) ESOP 2017. LNCS, vol. 10201, pp. 83–110. Springer, Heidelberg (2017). https://doi.org/10.1007/978-3-662-54434-1_4

11. de Vries, R.: Towards formal test purposes. In: Formal Approaches to Testing of Software 2001 (FATES 2001). BRICS Notes Series, No. NS-01-4, pp. 61–76. BRICS, University of Aarhus, August 2001

12. Deshmukh, J., Jin, X., Kapinski, J., Maler, O.: Stochastic local search for falsification of hybrid systems. In: Finkbeiner, B., Pu, G., Zhang, L. (eds.) ATVA 2015. LNCS, vol. 9364, pp. 500–517. Springer, Cham (2015). https://doi.org/10.1007/978-3-319-24953-7_35

13. Doyen, L., Henzinger, T.A., Legay, A., Nickovic, D.: Robustness of sequential circuits. In: 10th International Conference on Application of Concurrency to System Design, ACSD 2010, Braga, Portugal, 21–25 June 2010, pp. 77–84. IEEE Computer Society (2010). https://doi.org/10.1109/ACSD.2010.26

14. Ewing, J.: Ex-Volkswagen C.E.O. Charged With Fraud Over Diesel Emissions. New York Times (2018). https://www.nytimes.com/2018/05/03/business/volkswagen-ceo-diesel-fraud.html. Accessed 28 Jan 2019

15. Feijs, L.M.G., Goga, N., Mauw, S., Tretmans, J.: Test selection, trace distance and heuristics. In: Testing of Communicating Systems XIV, Applications to Internet Technologies and Services, Proceedings of the IFIP 14th International Conference on Testing Communicating Systems - TestCom 2002, Berlin, Germany, 19–22 March 2002. IFIP Conference Proceedings, vol. 210, pp. 267–282. Kluwer (2002)

16. Finkbeiner, B., Rabe, M.N., Sánchez, C.: Algorithms for model checking Hyper-LTL and HyperCTL*. In: Kroening, D., Păsăreanu, C.S. (eds.) CAV 2015. LNCS, vol. 9206, pp. 30–48. Springer, Cham (2015). https://doi.org/10.1007/978-3-319-21690-4_3

17. Hamlet, D.: Continuity in sofware systems. In: Proceedings of the International Symposium on Software Testing and Analysis, ISSTA 2002, Roma, Italy, 22–24 July 2002, pp. 196–200. ACM (2002). https://doi.org/10.1145/566172.566203

18. Jard, C., Jéron, T.: TGV: theory, principles and algorithms. STTT 7(4), 297–315 (2005)

19. Majumdar, R., Saha, I.: Symbolic robustness analysis. In: Proceedings of the 30th IEEE Real-Time Systems Symposium, RTSS 2009, Washington, DC, USA, 1–4 December 2009, pp. 355–363. IEEE Computer Society (2009). https://doi.org/10.1109/RTSS.2009.17

20. Margaria, T., Steffen, B. (eds.): ISoLA 2016, Part II. LNCS, vol. 9953. Springer, Cham (2016). https://doi.org/10.1007/978-3-319-47169-3

21. Pettersson, S., Lennartson, B.: Stability and robustness for hybrid systems. In: Proceedings of 35th IEEE Conference on Decision and Control, vol. 2, pp. 1202–1207, December 1996

22. Riley, C.: Volkswagen's diesel scandal costs hit $30 billion. CNN Business (2018). https://money.cnn.com/2017/09/29/investing/volkswagen-diesel-cost-30-billion/index.html. Accessed 28 Jan 2019

23. Tabuada, P., Balkan, A., Caliskan, S.Y., Shoukry, Y., Majumdar, R.: Input-output robustness for discrete systems. In: Proceedings of the 12th International Conference on Embedded Software, EMSOFT 2012, Part of the Eighth Embedded Systems Week, ESWeek 2012, Tampere, Finland, 7–12 October 2012, pp. 217–226. ACM (2012). http://doi.acm.org/10.1145/2380356.2380396

24. The European Parliament and the Council of the European Union: Directive 98/69/ec of the european parliament and of the council. Official Journal of the European Communities (1998). http://eur-lex.europa.eu/LexUriServ/LexUriServ.do?uri=CELEX:31998L0069:EN:HTML

25. Tretmans, J.: A formal approach to conformance testing. Ph.D. thesis, University of Twente, Enschede, Netherlands (1992). http://purl.utwente.nl/publications/58114

26. Tretmans, J.: Conformance testing with labelled transition systems: implementation relations and test generation. Comput. Netw. ISDN Syst. **29**(1), 49–79 (1996). https://doi.org/10.1016/S0169-7552(96)00017-7

27. Tretmans, J.: Model based testing with labelled transition systems. In: Hierons, R.M., Bowen, J.P., Harman, M. (eds.) Formal Methods and Testing. LNCS, vol. 4949, pp. 1–38. Springer, Heidelberg (2008). https://doi.org/10.1007/978-3-540-78917-8_1

28. United Nations: UN Vehicle Regulations - 1958 Agreement, Revision 2, Addendum 100, Regulation No. 101, Revision 3 – E/ECE/324/Rev.2/Add.100/Rev.3 (2013). http://www.unece.org/trans/main/wp29/wp29regs101-120.html

29. de Vries, R.G., Tretmans, J.: On-the-fly conformance testing using SPIN. STTT **2**(4), 382–393 (2000). https://doi.org/10.1007/s100090050044

Safety Guarantees for the Electricity Grid with Significant Renewables Generation

Andrea Peruffo[1]([⊠]) (iD), Emeline Guiu[2], Patrick Panciatici[2],
and Alessandro Abate[1] (iD)

[1] Department of Computer Science, University of Oxford, Oxford, UK
{andrea.peruffo,alessandro.abate}@cs.ox.ac.uk
[2] Réseau de Transport d'Électricité, Paris, France
{emeline.guiu,patrick.panciatici}@rte-france.com

Abstract. This work presents a study of the frequency dynamics of the electricity grid under significant presence of generation from renewable sources. A safety requirement, namely ensuring that frequency does not deviate excessively from a reference level, is formally studied by means of probabilistic model checking of a finite-state abstraction of the grid dynamics. The dynamics of the electric network comprise a model of the frequency evolution, which is in a feedback connection with a model of renewable power generation by a heterogeneous population of solar panels. Each panel switches independently between two states (ON and OFF) in response to frequency deviations, and the power generated by the population of solar panels affects the network frequency response. A power generation loss scenario is analysed and its consequences on the overall network are formally quantified in terms of probabilistic safety. We thus provide guarantees on the grid frequency dynamics under several scenarios of solar penetration and population heterogeneity.

Keywords: Population models · Aggregated models ·
Formal abstractions · Quantitative model checking ·
Probabilistic safety

1 Introduction

Renewable energy sources have shown potential to revolutionise power systems, not only on the generation side but also for demand-response programs [3], for fast frequency response [4], and for ancillary services [5]. Energy generation from a large population of photovoltaic (PV, or solar) panels, resulting from either an industrial setting (e.g., large PV farms) or numerous single households, can have economically and environmentally relevant consequences for energy providers and consumers alike. In this work we focus on PV populations composed of predominantly household devices. Such populations are naturally heterogeneous, in view of diverse weather conditions, of different panel sizes, makes and ages, and of the actual ratio between power generated and consumed.

© Springer Nature Switzerland AG 2019
D. Parker and V. Wolf (Eds.): QEST 2019, LNCS 11785, pp. 332–349, 2019.
https://doi.org/10.1007/978-3-030-30281-8_19

A rich literature on models of solar panels encompasses several features, such as their electrical characteristics [22] (where a panel comprises its components and their inter-connections), their power output generation [23], or their role in the larger economy of renewable power production [18]. A discrete-time Markov chain (dtMC) model for a population of PV panels is presented in [15], where an analysis on the effect of heterogeneity (as different disconnection/reconnection rules) is discussed as a function of the dynamics of the frequency in the electric network. The relationship between the panels working interval (to be discussed shortly) and the stability of the electric network is further addressed in [17]. In particular, the consequences of generation- and load-loss incidents are studied, under several scenarios of network load and of population dynamics. This paper expands earlier results by newly employing techniques from formal verification: we tailor a formal abstraction procedure [2] to generate finite probabilistic models (i.e., Markov chains) from the population models above, which are then analysed by means of probabilistic model checking.

Cognate to this work, [10] presents models of power grids with a significant penetration of solar: these models are employed to investigate runtime control algorithms, introducing control designs from randomised distributed algorithms, for photovoltaic micro-generators to assess grid stability. In [11] the authors study the German regulation framework exploiting ideas from communication protocol design. These works study a 50.2-Hz-disconnect/reconnect mechanism as well as the emergency switch-off procedure. A reachable set computation is presented in [12] to assess the stability of networked micro-grids in the presence of uncertainties induced by penetration of distributed energy resources: this results in bounds for systems dynamics and in its stability margins.

Technically, the models in this work are partially-degenerate discrete-time stochastic processes [19], for which formal abstractions can be computed. However, the abstraction procedure in this work is different from [19] and following work, as detailed next. In [21] a Markov model is constructed as the aggregation of the temperature dynamics of an inhomogeneous population of thermostatically controlled loads (TCLs): the population model is based on Markov chains obtained as abstractions of each TCL model. In this work, unlike [21], the formal abstraction is applied *after* the aggregation procedure. As discussed in [20], the aggregation of population models from earlier work [21] introduces two kinds of errors: the abstraction error (over a single device) and a population heterogeneity error. Instead, in this work we directly abstract the model of a heterogeneous population of PV panels, thus removing the second error term. Similar to [19–21], refining the abstraction improves the accuracy: the error converges to zero as the number of generated abstract states increases.

Models in this work (cf. Sect. 2.2) are derived from the following description of the workings of a solar panel. An inverter-panel device is equipped with a sensor to sample the network frequency, and with an internal counter. Two quantities are key to model the panel behaviour: (1) $\mathcal{I}_f$, the working interval for the grid frequency (only when the frequency lies within $\mathcal{I}_f$ can a panel inject power into the grid); and (2) τ_r, the time delay required for a safe connection to the network (the network frequency needs to remain inside $\mathcal{I}_f$ long enough before the panel connects back to the grid). Each device, in principle, can have different admissible frequency range and time delay. The behaviour of solar-inverter devices affects the grid and can lead [7] to *load-shedding*. This is a process activated to prevent frequency imbalance and subsequent blackouts, by means of an engineered stop of electricity delivery in order to avoid a complete shut-down of the electricity grid. In order to secure a network with no frequency imbalance, power generation and consumption must be matched: this is attained by Load-Frequency Control [6], which is distinguished in primary, secondary, and tertiary control, each activated at different timescales and with different goals. This study focusses on few instants after an incident, when primary control is relevant, which is thus included within the network model. We leverage model abstractions to formally quantify the absence of load-shedding, by probabilistic model checking a safety specification. In the end, we are able to provide certificates on the safe and reliable operation of the grid under penetration of solar generation.

This work is organised as follows. Section 2 introduces the solar panel behaviour, its description as a dynamical system and the electric network model. Section 3 discusses the formal abstraction techniques and computes the introduced error. Section 4 presents the generation-loss incident scenario, and shows experimental results in terms of probability of load-shedding under several parameter configurations, ranging over population heterogeneity and solar penetration level. Finally, conclusions are drawn in Sect. 5.

2   A Model of the Electricity Grid with Solar Generation

In this Section we present a description of the behaviour of a physical device, of its corresponding Markov model, and a model of the electric network.

2.1   Operation of a PV Panel

We briefly describe the workings of a photovoltaic panel that is connected to the electric network [15]. A panel-inverter device is connected to the electricity grid and samples it with a fixed sampling time - we'll work with discrete-time models indexed by $k \in \mathbb{N}$. The panel can be either ON (connected) or OFF (disconnected), and its activation/deactivation depends on two quantities: the network frequency $f(k)$ and a time delay τ_r. Table 1 summarises the behaviour of a PV panel, considering the value of the network frequency and a requirement on the time delay: regulations impose the panel to produce electricity, i.e. being in the ON state, exclusively when the frequency $f(k)$ belongs to a predefined interval $\mathcal{I}_f$, a neighbourhood of the nominal frequency $f_0 = 50$ Hz. If the frequency exits $\mathcal{I}_f$, the panel must disconnect, i.e. switch to the OFF state.

Whilst we assume the ON-to-OFF transition to be instantaneous, this does not hold for the OFF-to-ON transition. The reconnection happens if the frequency remains within $\mathcal{I}_f$ for τ_r time steps: this requirement forces the network frequency to be "stable" for a sufficient amount of time before allowing a safe connection of the panel to the grid. The panel is thus equipped with an internal counter $\tau(k)$ that increases when $f(k) \in \mathcal{I}_f$ and is reset as soon as $f(k) \notin \mathcal{I}_f$. When $\tau(k) \geq \tau_r$ the panel reconnects to the grid.

Note that the values of $\mathcal{I}_f$ and τ_r are not homogeneous across a population of panels. Beyond the intrinsic differences due to the small panels size that we have mentioned above, our network setting – a continental grid or part of it – is geographically wide enough to comprise different norms from several countries, across many years of installation. Moreover, digital systems are sensitive to noise in the measurements and suffer from ageing of its components: these elements make the system under consideration highly heterogeneous.

Table 1. Switching behaviour of a single PV panel. The network frequency is $f(\cdot)$, $\tau(\cdot)$ is the internal counter, τ_r the re-connection delay, and k the time index.

State $s(k)$	Frequency measurement	Delay	State $s(k+1)$
OFF	$f(k) \in \mathcal{I}_f$	$\tau(k) \geq \tau_r$	ON
ON	$f(k) \in \mathcal{I}_f$	–	ON
ON	$f(k) \notin \mathcal{I}_f$	–	OFF
OFF	$f(k) \in \mathcal{I}_f$	$\tau(k) < \tau_r$	OFF

2.2   Markov Model of a Heterogeneous Population of Solar Panels

We introduce a model of a heterogeneous population of solar panels, which is originally developed in [15]. Heterogeneity stems from differences between solar panels, and globally translates to the use of different $\mathcal{I}_f$ intervals and reconnection settings. In order to aggregate this heterogeneity at the population level, we assume to know a distribution function describing the panel intervals $\mathcal{I}_f$, and to know how the delays are distributed across the population.

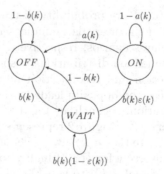

Fig. 1. A (time-varying) Markov model for the aggregated dynamics of a heterogeneous population of solar panels.

At a population level, we describe the portion of panels engaged in either of the following states: active (ON, for panels sampling $f(k) \in \mathcal{I}_f$ and $\tau \geq \tau_r$); inactive (OFF, for panels sampling $f(k) \notin \mathcal{I}_f$); and in between these conditions (WAIT, for panels sampling $f(k) \in \mathcal{I}_f$ but where $\tau < \tau_r$). The pictorial representation of such model is shown in Fig. 1. The values attached to the transition edges depend on the grid frequency: ideally, when $f(k) = f_0$ every panel can (eventually) connect (back) to the network, thus switching to the ON state, whereas if $f(k)$ is different than f_0, then solar panels might disconnect. As seen shortly, the transition values are probabilities, and characterise a Markov chain model with the following dynamics:

$$\begin{cases} x(k+1) = (1 - a(k))x(k) + b(k)\varepsilon(k)y(k) \\ y(k+1) = b(k)(1 - x(k) - \varepsilon(k)y(k)). \end{cases} \tag{1}$$

Here $x(k)$ and $y(k)$ represent the probability (that is, the portion of panels) of being in the ON and WAIT state at time k, respectively. Note that the probability of being OFF can be obtained as $1 - x(k) - y(k)$, $\forall k$. The function $\varepsilon(k)$ is a time-varying term accounting for the probabilistic description of the delay [15]: in this work, we assume to know its value at any k. The quantities $a(k)$ and $b(k)$ are functions of $f(k)$ as

$$a(k) = \begin{cases} \displaystyle\int_{-\infty}^{f(k)} p_o^d(u)du & \text{if } f(k) > f_0 \\ \displaystyle\int_{f(k)}^{+\infty} p_u^d(u)du & \text{otherwise,} \end{cases}$$

$$b(k) = \begin{cases} \displaystyle\int_{f(k)}^{\infty} p_o^r(u)du & \text{if } f(k) > f_0 \\ \displaystyle\int_{-\infty}^{f(k)} p_u^r(u)du & \text{otherwise,} \end{cases}$$

where p_i^j, $i = \{u, o\}$, $j = \{d, r\}$ are probability distributions encompassing the population heterogeneity over the intervals $\mathcal{I}_f$: indices u and o indicate the underfrequency and over-frequency scenarios, respectively, whereas d and r indicate the disconnection and reconnection distributions, respectively. As their variance increases, panels disconnection and reconnection become more scattered over the frequency range, whereas the opposite leads to the synchronisation of panels switching their configuration. As such, $a(\cdot)$ and $b(\cdot)$ describe the population heterogeneity over the interval $\mathcal{I}_f$. Figure 2 represents function $a(k)$ in underfrequency and overfrequency. Note that $a(k)$ always denotes the part of the integral that is closer to f_0, and conversely for $b(k)$. Finally, notice that $a(\cdot)$ and $b(\cdot)$ are functions of the frequency signal $f(k)$: to ease the notation we denote them as $a(k)$, instead of $a(f(k))$.

Remark 1 (On the modelling assumptions). [15] has shown that the introduced three-state Markov model has an almost identical frequency response to that of a

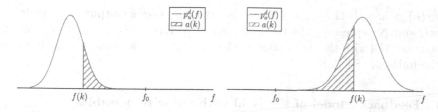

Fig. 2. Pictorial representation of $a(k)$ in over-frequency, i.e. $f(k) > f_0$ (right) and in under-frequency, i.e. $f(k) < f_0$ (left). The value of $f(k)$ is indicated as a red vertical line, which defines the upper or lower integration extrema in over- and under-frequency, respectively. In general p_u and p_o might not be symmetric nor belong to the same distribution family.

population of devices modelled individually: in experiments, given threshold and delay distributions, heterogeneous panels in the population are modelled with values extracted from the distributions. The presented modelling framework is tuneable to real data: the distributions can be interpolated from the behaviour of real devices that are measured across the population under study.   □

2.3   Model of the Grid Dynamics

The electricity grid reference model is derived from the ENTSO-E report in [6]. It consists of a discrete-time model of the electric network in the form of a second-order transfer function, $G(z, C_P)$ (z is the variable of the Z-transform and denotes a one-step time difference in the signal), which depends on the amount of conventional power (C_P) feeding the network, the total load of which is denoted as S. Note that $C_P \leq S$, where $C_P = S$ in a network without renewable energy sources. The model relates the photovoltaic power deviation, $\Delta P_{PV}(k)$ (its input) to the frequency deviation $\Delta f(k)$ (output) as

$$\Delta f(k) = G(z, C_P)[P_{PV}(k) - P_{PV,0}],$$

where $\Delta f(k) = f(k) - f_0$, and $\Delta P_{PV} = P_{PV}(k) - P_{PV,0}$ represents the deviation from $P_{PV,0}$, the power output at the equilibrium. Finally, $G(z, C_P)$ is the transfer function

$$G(z, C_P) = \frac{\beta_1 z + \beta_2}{z^2 + \alpha_1(C_P)z + \alpha_2(C_P)}. \tag{2}$$

Here $\alpha_1(C_P)$, $\alpha_2(C_P)$, β_1, β_2 are parameters that are selected to render the transfer function stable around the equilibrium, in accordance with values in [6]; in particular $\alpha_1(C_P)$ and $\alpha_2(C_P)$ depend on the conventional power C_P in the network [16]. Further, the structure of $G(z, C_P)$ encompasses the network primary control. Each scenario analysed in Sect. 4 includes a different network transfer function, depending on the solar penetration considered.

The total power output of the solar population $P_{PV}(k)$ is directly proportional to the portion of panels in the ON mode (variable $x(k)$), as $P_{PV}(k) \sim$

$\bar{P}Nx(k)$, where $\bar{P}$ is assumed to be the constant power output of a single PV panel, and N represents the total number of panels. This quantity couples the network model with the population model: their feedback connection is discussed in the following Section.

2.4   Feedback Model of the Grid with Solar Renewables

We now place in feedback the (time-varying) Markov chain modelling the solar panels dynamics in Eq. (1), with the model of the electric network in Eq. (2), expressing the transfer function $G(z, C_P)$ as a difference equation, resulting in:

$$\begin{cases} \Delta f(k+1) = \alpha_1 \Delta f(k) + \alpha_2 \Delta f(k-1) + \\ \qquad\qquad + \beta_1 \Delta P_{PV}(k) + \beta_2 \Delta P_{PV}(k-1) + \omega_f(k) \\ x(k+1) = (1 - a(k))x(k) + b(k)\varepsilon(k)y(k) \\ y(k+1) = b(k)(1 - x(k) - \varepsilon(k)y(k)), \end{cases} \tag{3}$$

where
$$P_{PV}(k) = \bar{P}Nx(k) + \omega_P(k).$$

Notice that we have added a frequency noise term $\omega_f(k) \in \mathcal{N}(0, \sigma_f)$, which represents the imperfect balance of the electric network; $P_{PV}(k)$ represents the solar power injected in the grid at time k; and $\omega_P(k) \in \mathcal{N}(0, \sigma_P)$ is the noise over the solar power generation at time k. ω_P represents the unpredictability of solar panels: their power output depends on characteristics as weather conditions, occlusions, temperature, that allow a stochastic description. The process noises $\omega_f(k)$ and $\omega_P(k)$, are made up by i.i.d. random variables, characterised by density functions $t_f(\cdot)$ and $t_P(\cdot)$, to be used below. We assume also that $\omega_f(\cdot)$ and $\omega_P(\cdot)$ are independent of each other.

Note that the dynamics of $\Delta P_{PV}(k+1)$ can be formed simply by operating a change of variable as $P_{PV}(k) = \Delta P_{PV}(k) + P_{PV,0}$. Note also that $a(k)$, $b(k)$, $x(k)$, $y(k)$ by construction belong to the interval $[0, 1]$ $\forall k \in \mathbb{N}$.

The equations in (3) represent a so called partially-degenerate stochastic model [19]. It comprises two stochastic equations (the dynamics of $\Delta f(k+1)$ and $P_{PV}(k+1)$) and two deterministic equations (for the Markovian dynamics of $x(k+1)$ and $y(k+1)$). The stochastic nature of the reconnection is embedded into the $\varepsilon(k)$ term, so an additional noise is not necessary.

3   Formal Abstractions

The dynamics of variables $x(k)$ and $y(k)$ represent the portion of panels in the population that are in state ON and in state WAIT at time k, respectively (cf. Fig. 1). Both $x(k)$ and $y(k)$, as well as $P_{PV}(k)$, are continuous variables, which makes their formal verification tricky.

Further, the dynamics in (3) also include state variables with delays (i.e. $f(k-1)$ and $P_{PV}(k-1)$). This issue is handled by variable renaming, namely we introduce two new state variables

$$\phi(k) = f(k-1), \quad \xi(k) = P_{PV}(k-1),$$

so that

$$\Delta\phi(k) = f(k-1) - f_0 = \Delta f(k-1), \quad \Delta\xi(k) = P_{PV}(k-1) - P_{PV,0} = \Delta P_{PV}(k-1).$$

Recall that f_0 and $P_{PV,0}$ denote fixed quantities at the equilibrium points. The model in (3) becomes

$$\begin{cases} \Delta f(k+1) = \alpha_1 \Delta f(k) + \alpha_2 \Delta\phi(k) + \beta_1 \Delta P_{PV}(k) + \beta_2 \Delta\xi(k) + \omega_f(k) \\ \Delta\phi(k+1) = \Delta f(k) \\ x(k+1) = (1 - a(k))x(k) + b(k)\varepsilon(k)y(k) \\ y(k+1) = b(k)(1 - x(k) - \varepsilon(k)y(k)) \\ P_{PV}(k) = \bar{P}Nx(k) + \omega_P(k) \\ \xi(k+1) = P_{PV}(k), \end{cases} \tag{4}$$

Let us focus on the domain of the six state-space variables: $x(k)$, and $y(k)$ belong to the interval $[0,1]$, whereas by definition, $\Delta f(k)$, $\Delta\phi(k)$, $P_{PV}(k)$ and $\xi(k)$ range over $\mathbb{R}$. However, as mentioned above, whenever $f(k)$ exits its operational range (for instance, as shall be seen in the experiments, because of a generation loss), primary control mechanisms act to restore the frequency to its nominal value. As such, we can limit our models to values of frequency within the operational range $\mathbb{F} = [f_u, f_o] = [-0.8, +0.8]$ Hz, which corresponds to the frequency interval $[49.2, 50.8]$ Hz. Similarly, we restrict dynamics of $P_{PV}(k)$ to belong to the interval $\mathbb{P} = [0, \bar{P}N]$ to model the physical limitation of real devices. Finally, introduce the interval $\mathbb{X} = [0,1]$ for variables x, y.

The state space of the model is thus characterised by a vector variable $q = (\Delta f, \Delta\phi, x, y, P_{PV}, \xi) \in \mathbb{F}^2 \times \mathbb{X}^2 \times \mathbb{P}^2 := \mathcal{Q}$, with six continuous components. Let us also introduce a noise vector $\omega(k) = (\omega_f(k), \omega_P(k))$.

We now discuss the one-step transition density kernel $t_\omega(\cdot|q)$, a function that defines the transition from state q to state q', derived from the model in (4) as per [2]. Conditional on point $q \in \mathcal{Q}$, it can be written as

$$\begin{aligned} t_\omega(q'|q) = {} & t_f(\Delta f' - \alpha_1 \Delta f - \alpha_2 \Delta\phi - \beta_1 \Delta P_{PV} - \beta_2 \Delta\xi) \cdot \\ & \cdot \delta(\Delta\phi' - \Delta f) \cdot \delta(x' - (1-a)x - b\varepsilon y) \cdot \\ & \cdot \delta(y' - b(1 - x - \varepsilon y)) \cdot t_P(P'_{PV} - \bar{P}Nx) \cdot \delta(\xi' - P_{PV}), \end{aligned} \tag{5}$$

where primed variables indicate the next value in time, and $\delta(p)$ is the Dirac delta function pointed at p (namely it assumes value 1 if $p = 0$, or value 0 otherwise) that characterises the dynamics of deterministic vector fields for variables $\Delta\phi$ x, y, ξ. Stochastic vector fields are instead characterised by the two densities $t_f(p)$ and $t_P(p)$, centred at point p, which are decoupled in view of the independence of the two corresponding noise processes.

3.1   Finite Abstraction via State-Space Partitioning

We introduce a formal abstraction technique, proposed in [2], aimed at reducing a discrete-time, uncountable state-space Markov process to a discrete-time finite-state Markov chain, for the purpose of probabilistic model checking. The abstraction is based on a state-space partitioning procedure: consider an arbitrary and finite partition of the continuous domain $\mathbb{F} = \bigcup_{i=1}^{n} \mathcal{F}_i$, where $\mathcal{F}_i$ are non-overlapping, and a set of representative points within the partitions $\{\bar{f}_i \in \mathcal{F}_i, \ i = 1, \ldots n\}$, which in practice are taken to be their middle points. This partition intervals represent the values of Δf. Similarly, we introduce a partition of the other variables and respective domains, and define representative points $\{\bar{\phi}_i \in \Phi_i, \ i = 1, \ldots n\}$, $\{\bar{x}_j \in \mathcal{X}_j, \ j = 1, \ldots m\}$, $\{\bar{y}_j \in \mathcal{Y}_j, \ j = 1, \ldots m\}$, $\{\bar{p}_j \in \mathcal{P}_j, \ j = 1, \ldots m\}$, $\{\bar{\xi}_j \in \Xi_j, \ j = 1, \ldots m\}$, for variables $\Delta\phi$, x, y, ΔP_{PV} and $\Delta\xi$, respectively[1].

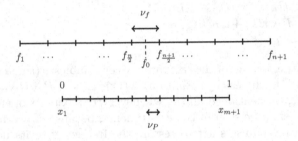

Fig. 3. Partition intervals for frequency (top) and active panels (bottom).

Let us now provide details on the selection of the intervals resulting in the partitions of $\mathcal{S}$. Let us select a partition size ν_f and quantity $n = \frac{f_o - f_u}{\nu_f}$, representing the number of partitions created in the frequency domain. Note that the symmetry of the interval $[f_u, f_o]$ with respect to f_0 implies that f_0 becomes the reference point of the $\frac{n}{2}$-th partition, i.e. $\bar{f}_{\frac{n}{2}} = f_0$. Analogously, denote ν_P as the second partition size and $m = \frac{1}{\nu_P}$ as the number of partitions in the active panels domain. We denote the boundary points of the partitions as (Fig. 3)

$$f_{i+1} = f_i + \nu_f, \quad i = 1, \ldots n, \quad \mathcal{F}_i = [f_i, f_{i+1}), \quad \mathbb{F} = \bigcup_{i=1}^{n} \mathcal{F}_i,$$
$$x_{j+1} = x_j + \nu_P, \quad j = 1, \ldots m, \quad \mathcal{X}_j = [x_j, x_{j+1}), \quad \mathbb{X} = \bigcup_{j=1}^{m} \mathcal{X}_j, \tag{6}$$

and analogously for $\Delta\phi$, y, and P_{PV}, ξ. Let us remark that interval $\mathcal{F}_1$ represents frequency values just above f_u: introduce the unsafe interval $\mathcal{F}_0 = (-\infty, f_1)$ and render $\mathcal{F}_0$ absorbing – this allows evaluating the cumulative load-shedding probability over time (as further detailed shortly).

---

[1] In principle, we could employ different partitioning intervals for different variables, however to ease the notation we have used n intervals for frequency-related variables and m intervals for x, y, P_{PV}, ξ.

Introduce now a discrete-time and finite-state Markov chain $\mathcal{M}$, composed by $n^2 \times m^4$ abstract states $s = (\bar{f}_{i1}, \bar{\phi}_{i2}, \bar{x}_{j1}, \bar{y}_{j2}, \bar{p}_{j3}, \bar{\xi}_{j4})$, where $i1, i2 \in [1, n]$ and $j1, j2, j3, j4 \in [1, m]$. Denote by $\mathcal{S}$ the finite state space of $\mathcal{M}$ and by $\mathcal{S}_{i1,i2,j1,j2,j3,j4} \in \mathcal{S}$ one of its states, which corresponds to a hyper-rectangle centred at $(\bar{f}_{i1}, \bar{\phi}_{i2}, \bar{x}_{j1}, \bar{y}_{j2}, \bar{p}_{j3}, \bar{\xi}_{j4})$ and with bounds (see Eq. (6)) defined by the intervals $\mathcal{F}_i$, $\mathcal{P}_j$ and $\mathcal{X}_j$ and corresponding copies. Denote $\mu : \mathcal{S} \rightarrow \mathcal{Q}$ the one-to-one mapping between the abstract state s and the corresponding region of the state-space q.

The transition probability matrix of $\mathcal{M}$ comprises the probabilies obtained by marginalising the kernel t_ω over the hyper-rectangular partitions, as

$$P(s, s') = \int_{\mu(\mathcal{S}_{i1',i2',j1',j2',j3',j4'})} t_\omega((df', d\phi', dx', dy', dP'_{PV}, d\xi')|q). \tag{7}$$

The abstraction procedure applied to the model in (4) carries a discretisation error: in the following, we formally derive a bound for this error as a function of the discretisation steps ν_f and ν_P. As argued in [2], a finer grid results in a smaller abstraction error, however it generates a larger state space.

In view of the presence of non-probabilistic dynamics in the degenerate stochastic model, the abstraction results in a Markov chain structured as the following example.

Example 1. Consider, as an illustrative example, the following model:

$$\begin{cases} r_1(k + 1) = \zeta_1 r_1(k) + \zeta_2 r_2(k) + \omega_z(k) \\ r_2(k + 1) = r_1(k), \end{cases} \tag{8}$$

where ζ_1, ζ_2 are constants and $\omega_z(k)$ is a Gaussian noise term at time $k \in \mathbb{N}$. These models are typical in control engineering, as they derive from autoregressive systems, such as

$$r_1(k + 1) = \zeta_1 r_1(k) + \zeta_2 r_1(k - 1) + \omega_z(k),$$

where a new variable is introduced $(r_2(k))$ to replace the delayed variable of interest. Let us introduce a 2-set partition of $\mathbb{R} = A \cup B$ with reference points $\bar{r}_1 = r_A, \bar{r}_2 = r_B$. Both variables $r_1(\cdot)$ and $r_2(\cdot)$ can take value r_A or r_B. The state-space of (8) is $q = (\bar{r}_1, \bar{r}_2) \in \{r_A, r_B\}^2 = \{r_A r_A, r_A r_B, r_B r_A, r_B r_B\}$. The dynamics of r_2 allow only for deterministic transitions, as the next value of r_2 must be the current value of r_1. As an example, if the current state is $q = (r_A, r_A)$, the next state must be $q' = (*, r_A)$. Thus, adding a auxiliary variable r_2 expands the state-space while forbidding several transitions.   □

3.2   Quantification of Safety Probability and of Abstraction Error

Let us now formally characterise the load-shedding probability. Consider the model in (4) with initial state q_0 and select a discrete time horizon H. We assume that the electric network activates the load-shedding procedure whenever

$f(k) \leq 49.2$ Hz, namely if $q(k) \in \mathcal{L}$, where $\mathcal{L} := \{\Delta f \leq -0.8\}^2$. The aim of this work is the computation of

$$p_{q_0}(\mathcal{L}) := Prob(q(i) \in \mathcal{L}, i \in [1, H] \mid q_0), \tag{9}$$

where q_0 is the initial state of the continuous model. This probability can be formally characterised via value functions $V_k : \mathcal{Q} \to [0, 1]$, $k = 1, \ldots H$, which can be computed recursively as

$$V_k(q) = \mathbf{1}_{\mathcal{L}}(q) \int_{\mathcal{Q}} V_{k+1}(u) t_\omega(u|q) du, \quad \text{with } V_H(q) = \mathbf{1}_{\mathcal{L}}(q), \tag{10}$$

so the initial value function $V_1(q_0) = p_{q_0}(\mathcal{L})$ is the quantity of interest. We recall a procedure presented in [1] to approximate the model in Eq. (4) by a finite-state dtMC. We therefore define the discrete version of Eq. (9), as $p_{s_0}(\mathcal{L}_s) := Prob(s(i) \in \mathcal{L}_s, i \in [1, H] \mid s_0) = V_1^s(s_0)$, where $\mathcal{L}_s := \{\bar{f} \in \mathcal{F}_0\}$ (consider it the dtMC-equivalent of $\mathcal{L}$), $V_1^s(\cdot)$ is the value function computed over $\mathcal{S}$ similarly to Eq. (10), and s_0 represents the initial state of the dtMC according to the procedure in Sect. 3.1.

In the dtMC model, functions $a(k)$ and $b(k)$ are approximated and assume a finite number of values (one for each of the $\bar{f}_i$). This introduces an error term: let us define a_{max} as

$$a_{max} = \max_{\substack{i \in [1, n] \\ f \in \mathcal{F}_i}} \left\| \int_{f_i}^{\bar{f}_i} p^d(u) du \right\|,$$

where p^d represents the probability distribution for disconnection. This quantity defines the maximum approximation error introduced with the discretisation in the computation of $a(k)$.

Note that the presence of $\delta(\cdot)$ functions in Eq. (5) introduces discontinuities within the domain of the kernel: continuity regions of the kernel (density) are parts of the state space where the $\delta(\cdot)$ functions are equal to one. Within such regions (which are formally defined in Appendix A) the value functions are continuous, and the following holds over pairs of points $q, \tilde{q}$ (cf. Appendix C):

$$|V_k(q) - V_k(\tilde{q})| \leq \frac{2\alpha_1}{\sigma_f \sqrt{2\pi}} |\Delta f - \Delta \tilde{f}| + \frac{2a_{max}}{\sigma_P \sqrt{2\pi}} |\Delta P_{PV} - \Delta \tilde{P}_{PV}|,$$

where α_1 is a term introduced in Eq. (2).

We now abstract the aggregated population of solar panels as a Markov chain based on the procedure of Sect. 3. Computing the solution of (10) over the Markov chain, the overall approximation error can be upper-bounded [1] as follows

$$|p_{q_0}(\mathcal{L}) - p_{s_0}(\mathcal{L}_s)| \leq (H - 1) \left[\frac{2\alpha_1}{\sigma_f \sqrt{2\pi}} \nu_f + \frac{2a_{max}}{\sigma_P \sqrt{2\pi}} \nu_P \right].$$

---

[2] We argue in Appendix B that the delayed variables are not necessary for the characterisation of the load-shedding probability.

This error allows to refine the outcomes of the model checking procedure (obtained from $p_{s_0}(\mathcal{L}_s)$) over the concrete population model (corresponding to the unknown quantity $p_{q_0}(\mathcal{L})$).

Remark 2 (On the population heterogeneity). The model in Sect. 3 allows for a crisp expression of the population heterogeneity in terms of p^d and p^r distributions: the working intervals are encapsulated by the integrals $a(k)$ and $b(k)$. These quantities can easily be extended to encompass a population made up of diverse parts: assume that $a(k)$ is the sum of various integrals, each of them encompassing a portion of the whole population, as

$$a(k) = \lambda_1 \int_{-\infty}^{f(k)} p_1^d(u)du + \ldots \lambda_r \int_{-\infty}^{f(k)} p_r^d(u)du,$$

where $\lambda_i \in (0,1)$, $i = 1, \ldots r$, $\sum_{i=1}^r \lambda_i = 1$, are weights representing the contribution of power production for the i-th portion with respect to the total population. A similar setup can be made for $b(k)$. $\square$

4   Experimental Results

In this section we use the abstract Markov chain to compute the load-shedding probability after a sudden generation loss, under several scenarios.

In line with the ENTSO-E requirements [6], we assume an infeed loss of 3 GW in a global network with a demand of $S = 220$ GW. Power and frequency values are normalised (per unit) relative to S and to 50 Hz. Power production of a single panel $\bar{P}$ is set to 3 kW. The variance σ_P is set to 1% of $\bar{P}$. The variance σ_f is set to 0.05. Time delays are modelled in accordance with [13,14]: the minimum reconnection delay is set to 20 seconds, whereas the maximum to 40 seconds. Whilst these two quantities are handled deterministically, the delays are modelled via a geometric distribution. The probabilistic model checking tests are implemented using the MATLAB software. Due to the large state space, ν_f and ν_P are set to 0.01 and 0.05 respectively. The grid frequency is sampled at a rate of 0.2 s, consistently with the requirements introduced in [8]. The discussion is focused on the consequences of an incident after a few seconds: the time interval considered is 20 s. After this time interval, we assume the frequency control would stabilise $f(\cdot)$ around its nominal value. The discrete time horizon is thus composed of 100 steps: results shown in the following Section carry an abstraction error of 0.1, as quantified in Sect. 3.2. This error ought to be attached to the certificates on safety probability derived in Sect. 4.2.

4.1   Study of Generation-Loss Incidents - Setup

As anticipated above, Transmission Systems Operators are tasked with ensuring the safe operation of the grid, and are thus interested in formal guarantees on its dynamics, and in reliable forecasting of potentially problematic situations, such as issues related to frequency responses after a generation loss incident.

Our study concerns the so-called normal incidents, classified as a loss of up to 2 GW of load, and as a loss of up to 3 GW of power generation. We assume the initial condition to be $f(0) = f_0$, with the population of panels in active (ON) mode $(x(0) = 1)$. The generation-loss incident is modelled as a negative step injected into the dynamics in Eq. (2). Assuming that an incident of magnitude M occurs at time $k = \bar{k}$, the dynamics of $f(\bar{k} + 1)$ become

$$f(\bar{k} + 1) = \alpha_1 \Delta f(\bar{k}) + \alpha_2 \Delta\phi(\bar{k}) + \beta_1(\Delta P_{PV}(\bar{k}) - M) + \beta_2 \Delta\xi(\bar{k}) + \omega_f(\bar{k}), \quad (11)$$

and then evolve from time $(\bar{k} + 2)$ on as

$$\begin{aligned} f(\bar{k} + 2) = &\alpha_1 \Delta f(\bar{k} + 1) + \alpha_2 \Delta\phi(\bar{k} + 1) + \beta_1(\Delta P_{PV}(\bar{k} + 1) - M) + \\ &+ \beta_2(\Delta\xi(\bar{k} + 1) - M) + \omega_f(\bar{k} + 1). \end{aligned} \quad (12)$$

Equations (11) and (12) display two different deterministic drifts, which lead to two different transition matrices P_1 and P_2 defined over the same state space. We further assume that $\bar{k} = 0$, namely the incident occurs at the beginning of the time horizon. This results in a time-varying safety verification problem: given the initial probability distribution vector π_0, the dynamics evolve as

$$\pi_1 = \pi_0 \cdot P_1, \qquad \pi_2 = \pi_1 \cdot (P_2)^{H-1},$$

where π_2 is a vector with the probabilities of being in each state after H steps.

4.2   Computation of Load-Shedding Probability

Our tests encompass several scenarios, in which we vary: (a.) the choice of the distributions p^d and p^r; (b.) the associated variance of p^d and p^r; and (c.) the total solar penetration in the network. Recall from Sect. 2.3 that the solar penetration modifies the network transfer function.

We obtain results with solar penetration from 10% to 40% of the network load (for brevity we show only the results with a 10% load) which represents current values for solar power contribution (e.g. Germany's 2017 average production is around 7%, reaching peaks of 60% in Summer [9]). The threshold distributions for $\mathcal{I}_f$ are either Gaussian or χ^2: these are notably dissimilar and are here used to model different panel dynamics. The use of different distributions denotes different *modelling* choices: a Gaussian distribution models the inverter measurement noise, whereas a χ^2 can be used to define a minimum performance setting, namely a minimum working interval. Whilst the Gaussian distribution results in a more realistic choice, the χ^2 offers interesting outlooks on how distributions affect the safety property. In practice, the selection of a distribution must depend on data measurements coming from real devices. Ultimately, increasing variance of p^d and p^r reflects a more heterogeneous population (more diverse thresholds characterising $\mathcal{I}_f$). As discussed shortly, a larger variance can have opposite consequences on stability, depending on which threshold distribution is used.

(1) $\mathcal{I}_f$ thresholds distributed as a Gaussian. Figure 4 depicts the load-shedding probability in presence of 10% solar penetration, varying values of the mean and

variance of p^d. A Gaussian distribution has a symmetric shape around its mean: when its variance increases, the tails on both sides spread out. As such, increasing their variance cause a higher number of panels (represented by the tails of the distribution) to have thresholds closer to f_0. Consequently, we observe more panels with a narrow working interval around the nominal frequency. Therefore, a greater portion of the population is likely to disconnect under frequency deviations, causing the network frequency to decrease.

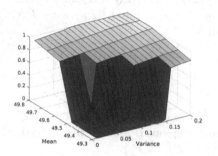

Fig. 4. Load-shedding probability with 10% solar penetration, with Gaussian distribution of thresholds. Variance within $[0.01, 0.20]$ and mean within $[49.3, 49.8]$ Hz.

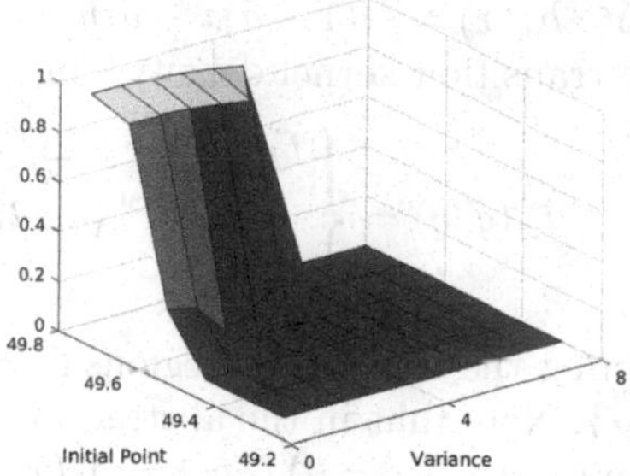

Fig. 5. Load-shedding probability with 10% solar penetration, with χ^2 distribution of thresholds. Variance within $[1, 8]$ and initial point within $[49.3, 49.8]$ Hz.

(2) $\mathcal{I}_f$ thresholds distributed as a χ^2. Figure 5 depicts the load-shedding probability under 10% solar penetration, with varying values of the initial point of the support and of the variance of p^d. Note that, due to the nature of the χ^2 distribution, instead of the average value we denote an initial point of the support. Unlike the Gaussian case, increasing the variance of a χ^2 distribution results in larger thresholds. As expected, the experiments show that, in this scenario, an increased heterogeneity guarantees a more reliable network.

Experiments with a higher penetration of solar contributions (20%, 30%, 40% of the total), under either Gaussian or χ^2 scenarios, indicate that the probability of load-shedding increases when a larger PV population is connected to the grid.

5 Conclusions

We have introduced a formal procedure to abstract the dynamics of a heterogeneous population of solar panels, embedded within the frequency dynamics of the grid. The computation of error bounds on the abstraction guarantees the correctness of the outcomes of a formal verification procedure run on the obtained abstract model. The focus of the verification procedure has been on a grid safety property, under significant energy generation from renewables via formal abstractions: we have assessed the load-shedding probability of the network, under several scenarios of population heterogeneity. Operators can use these certificates to monitor the distribution of solar panels over the grid and to assess its reliability in case of incidents.

A   Definition of Kernel Continuous Regions

We want to underline the discontinuity of the kernel density $t_\omega(\cdot|q)$ caused by the presence of the $\delta(\cdot)$ functions. Let us define $g(\Delta f) = -\alpha_1\Delta f - \alpha_2\Delta\phi - \beta_1\Delta x - \beta_2\Delta\xi$, $h_1(x) = -(1-a)x - b\varepsilon y$, $h_2(y) = -b(1 - x - \varepsilon y)$ and $l(P_{PV}) = -\bar{P}Nx$. The transition kernel density can be written as

$$t_\omega(q'|q) = \begin{cases} t_f(\Delta f' - g(\Delta f))\cdot & \text{if } \Delta\phi' = \Delta f \wedge x' = h_1(x) \\ \quad \cdot t_P(P'_{PV} - l(P_{PV})) & \wedge\, y' = h_2(y) \wedge \xi' = P_{PV} \\ 0 & \text{otherwise} \end{cases}$$

defining the continuous regions $\mathcal{C} = \{\Delta\phi' = \Delta f \wedge x = h_1(x) \wedge y' = h_2(y) \wedge \xi' = P_{PV}\}$. Note that in the abstraction framework, regions $\mathcal{C}$ assume the discretised form $\mathcal{C}_d = \{\Delta\bar{\phi}' = \Delta\bar{f}_i \wedge \bar{x} = h_1(\bar{x}) \wedge \bar{y}' = h_2(\bar{y}) \wedge \bar{\xi}' = \bar{P}_j\}$.

B   Probabilistic Safety for Partially Degenerate Models

Let us show that for a partially degenerate stochastic model the safety probability computation depends only on the stochastic state. Consider the model

$$\begin{cases} x(k + 1) = f(z(k)) + \omega(k) \\ y(k + 1) = x(k), \end{cases}$$

where $\omega(k) \sim \mathcal{N}(0, \sigma)$ and where $z = (x, y)^T$ denotes the complete state vector. Let us denote with $t_\omega(\cdot)$ the density of the Gaussian kernel. The one-step transition probability kernel can be split as follows:

$$P(x(k+1)|z(k)) = t_\omega(f(z(k))),$$
$$P(y(k+1)|z(k)) = P(y(k+1)|x(k)) = \delta(y(k+1) - x(k)),$$

where $\delta(z-p)$ represent the Dirac delta function of variable z, centred at point p. Let us consider a safe set $A = A_x \times A_y$, where A_x and A_y denote its projections on variables x and y, respectively. Define the value function at time step H as $V_H(z) = \mathbf{1}_A(z)$ and compute the one-step backward recursion:

$$V_{H-1}(z) = \int_A V_H(z')P(z'|z)dz' = \int_A P(z'|z)dz' =$$
$$= \int_{A_y}\int_{A_x} t_\omega(dx'|f(z))\delta(dy' - x) = \mathbf{1}_{A_y}(z)\int_{A_x} t_\omega(dx'|f(z)) =$$
$$= \int_{A_x} t_\omega(dx'|f(z)),$$

showing that the computation of the safety probability depends solely on the stochastic kernel affecting the dynamics of variable x.

C   Value Function Continuity for Probabilistic Safety

In the following, we consider a generation-loss incident scenario; the load-loss case can be derived analogously. Recall the value function definition from Sect. 3 and compute the backward Bellman equation as

$$V_k(q) = \mathbf{1}_{\mathcal{L}}(q) \int_Q V_{k+1}(\tilde{q}) t_s(\tilde{q}|q) d\tilde{q}, \quad \text{with } V_H(q) = \mathbf{1}_{\mathcal{L}}(q).$$

We show that the value functions are continuous within the continuity regions of the state space, thus there must exist a constant γ so that

$$|V_k(q) - V_k(\tilde{q})| \leq \gamma \|q - \tilde{q}\|. \tag{13}$$

To enhance the readability let us define $g(\Delta f) = -\alpha_1 \Delta f - \alpha_2 \Delta \phi - \beta_1 \Delta x - \beta_2 \Delta \xi$, $h(P_{PV}) = -\bar{P} N x$ and $\Delta f = \rho$, $P_{PV} = \psi$. We now show the validity of Equation (13) by finding a value for γ. From the definition of $V_k(q)$, we obtain:

$$\left| \int_Q V_{k+1}(q) t_f(\underline{\rho} - g(\rho)) t_P(\underline{\psi} - h(\psi)) d(\underline{\rho}) d(\underline{\psi}) \right.$$
$$\left. - \int_Q V_{k+1}(\tilde{q}) t_f(\underline{\rho} - g(\tilde{\rho})) t_P(\underline{\psi} - h(\tilde{\psi})) d(\underline{\rho}) d(\underline{\psi}) \right| \leq$$
$$\left| \int_{\mathcal{F}} V_{k+1}(q) t_f(\underline{\rho} - g(\rho)) d(\underline{\rho}) \cdot \int_{\mathcal{P}} V_{k+1}(q) t_P(\underline{\psi} - h(\psi)) d(\underline{\psi}) \right.$$
$$\left. - \int_{\mathcal{F}} V_{k+1}(\tilde{q}) t_f(\underline{\rho} - g(\tilde{\rho})) d(\underline{\rho}) \cdot \int_{\mathcal{P}} V_{k+1}(\tilde{q}) t_P(\underline{\psi} - h(\tilde{\psi})) d(\underline{\psi}) \right|,$$

where $\mathcal{F}$ and $\mathcal{P}$ denote the domain of frequency and power, respectively. In order to continue, we introduce a useful lemma.

Lemma 1. *Assume* $A, B, C, D \in [0, 1]$, *then* $|AB - CD| \leq |A - C| + |B - D|$.

Proof. Assume $A > C$, then
if $AB - CD > 0$,
$|AB - CD| \leq |CB - CD| = C|B - D| \leq |B - D| \leq |B - D| + |A - C|$.
if $AB - CD < 0$,
$|AB - CD| \leq |AB - AD| = A|B - D| \leq |B - D| \leq |B - D| + |A - C|$.
Analogously for $A \leq C$.   □

Thanks to this Lemma, we can write

$$\left| \int_{\mathcal{F}} V_{k+1}(q) t_f(\underline{\rho} - g(\rho)) d(\underline{\rho}) \cdot \int_{\mathcal{P}} V_{k+1}(q) t_P(\underline{\psi} - h(\psi)) d(\underline{\psi}) \right.$$
$$\left. - \int_{\mathcal{F}} V_{k+1}(\tilde{q}) t_f(\underline{\rho} - g(\tilde{\rho})) d(\underline{\rho}) \cdot \int_{\mathcal{P}} V_{k+1}(\tilde{q}) t_P(\underline{\psi} - h(\tilde{\psi})) d(\underline{\psi}) \right| \leq$$
$$\left| \int_{\mathcal{F}} t_f(\underline{\rho} - g(\rho)) d(\underline{\rho}) \cdot \int_{\mathcal{P}} t_P(\underline{\psi} - h(\psi)) d(\underline{\psi}) \right.$$
$$\left. - \int_{\mathcal{F}} t_f(\underline{\rho} - g(\tilde{\rho})) d(\underline{\rho}) \cdot \int_{\mathcal{P}} t_P(\underline{\psi} - h(\tilde{\psi})) d(\underline{\psi}) \right| \leq$$
$$\int_{\mathcal{F}} \left| t_f(\underline{\rho} - g(\rho)) - t_f(\underline{\rho} - g(\tilde{\rho})) \right| d(\underline{\rho}) +$$
$$+ \int_{\mathcal{P}} \left| t_P(\underline{\psi} - h(\psi)) - t_P(\underline{\psi} - h(\tilde{\psi})) \right| d(\underline{\psi}).$$

Let us focus on the first integral:

$$\int_{\mathcal{F}} |t_f(\underline{\rho} - g(\rho)) - t_f(\underline{\rho} - g(\tilde{\rho}))| d(\underline{\rho}) = \frac{1}{\sigma_f} \int_{\mathcal{F}} \left| \Phi\left(\frac{\underline{\rho} - g(\rho)}{\sigma_f}\right) - \Phi\left(\frac{\underline{\rho} - g(\tilde{\rho})}{\sigma_f}\right) \right| d\underline{\rho}$$
$$= \int_{\mathcal{F}} \left| \Phi\left(u - \frac{\alpha_1(\rho - \tilde{\rho})}{2\sigma_f}\right) - \Phi\left(u + \frac{\alpha_1(\rho - \tilde{\rho})}{2\sigma_f}\right) \right| d\underline{\rho} \leq \frac{2\alpha_1}{\sqrt{2\pi}\sigma_f} |\rho - \tilde{\rho}|,$$

and similarly for the second integral. Therefore,

$$|V_k(q) - V_k(\tilde{q})| \leq \frac{2\alpha_1}{\sqrt{2\pi}\sigma_f} |\rho - \tilde{\rho}| + \frac{2a_{max}}{\sqrt{2\pi}\sigma_P} |\psi - \tilde{\psi}|.$$

References

1. Abate, A., Katoen, J.P., Lygeros, J., Prandini, M.: Approximate model checking of stochastic hybrid systems. Eur. J. Control **16**(6), 624–641 (2010)
2. Abate, A., Soudjani, S.E.Z.: Quantitative approximation of the probability distribution of a Markov process by formal abstractions. Logical Methods Comput. Sci. **11** (2015)
3. Aghaei, J., Alizadeh, M.I.: Demand response in smart electricity grids equipped with renewable energy eources: a review. Renew. Sustain. Energy Rev. **18**, 64–72 (2013)
4. Banks, J., Bruce, A., Macgill, I.: Fast frequency response markets for high renewable energy penetrations in the future Australian NEM. In: Proceedings of the Asia Pacific Solar Research Conference 2017. Australian PV Institute, December 2017
5. Banshwar, A., Sharma, N.K., Sood, Y.R., Shrivastava, R.: Renewable energy sources as a new participant in ancillary service markets. Energy Strategy Rev. **18**, 106–120 (2017)
6. ENTSO-E: Policy 1: Load-frequency Control and Performance. Technical report (2009)
7. ENTSO-E: Dispersed Generation Impact on CE Region, Dynamic Study. Technical report (2014)
8. European Commission: Commission regulation (EU) 2016/631 of 14th April 2016. Technical report (2016)
9. Wirth, H.: Recent Facts about Photovoltaics in Germany. Technical report, Fraunhofer ISE (2018)
10. Hartmanns, A., Hermanns, H.: Modelling and decentralised runtime control of self-stabilising power micro grids. In: Margaria, T., Steffen, B. (eds.) ISoLA 2012. LNCS, vol. 7609, pp. 420–439. Springer, Heidelberg (2012). https://doi.org/10.1007/978-3-642-34026-0_31
11. Hartmanns, A., Hermanns, H., Berrang, P.: A comparative analysis of decentralized power grid stabilization strategies. In: Proceedings of the Winter Simulation Conference, p. 158. Winter Simulation Conference (2012)
12. Li, Y., Zhang, P., Luh, P.B.: Formal analysis of networked microgrids dynamics. IEEE Trans. Power Syst. **33**(3), 3418–3427 (2017)
13. Jung, M., Wiss, O., Lazpita, B.: Analyses et Conclusions - Tests en sous-frequence. Technical report, RTE (2016)
14. Jung, M., Wiss, O., Lazpita, B.: Analyses et Conclusions - Tests en sur-frequence. Technical report, RTE (2016)

15. Peruffo, A., Guiu, E., Panciatici, P., Abate, A.: Aggregated Markov models of a heterogeneous population of photovoltaic panels. In: Bertrand, N., Bortolussi, L. (eds.) QEST 2017. LNCS, vol. 10503, pp. 72–87. Springer, Cham (2017). https://doi.org/10.1007/978-3-319-66335-7_5

16. Peruffo, A., Guiu, E., Panciatici, P., Abate, A.: Impact of solar panels and cooling devices on frequency control after a generation loss incident. In: Decision and Control (CDC) 2018 IEEE 57th Annual Conference Proceedings. IEEE (2018)

17. Peruffo, A., Guiu, E., Panciatici, P., Abate, A.: Synchronous frequency grid dynamics in the presence of a large-scale population of photovoltaic panels (2018)

18. Rehman, S., Bader, M.A., Al-Moallem, S.A.: Cost of solar energy generated using PV panels. Renew. Sustain. Energy Rev. **11**(8), 1843–1857 (2007)

19. Soudjani, S.E.Z., Abate, A.: Probabilistic reach-avoid computation for partially degenerate stochastic processes. IEEE Trans. Autom. Control **59**(2), 528–534 (2014)

20. Soudjani, S.E.Z., Abate, A.: Aggregation and control of populations of thermostatically controlled loads by formal abstractions. IEEE Trans. Control Syst. Technol. **23**(3), 975–990 (2015)

21. Esmaeil Zadeh Soudjani, S., Gerwinn, S., Ellen, C., Fränzle, M., Abate, A.: Formal synthesis and validation of inhomogeneous thermostatically controlled loads. In: Norman, G., Sanders, W. (eds.) QEST 2014. LNCS, vol. 8657, pp. 57–73. Springer, Cham (2014). https://doi.org/10.1007/978-3-319-10696-0_6

22. Tiam, H., Mancilla-David, F., Ellis, K.: A Detailed Performance Model for Photovoltaic Systems. Technical report, NREL (2012)

23. Tse, K., Ho, M., Chung, H.H., Hui, S.: A novel maximum power point tracker for PV panels using switching frequency modulation. IEEE Trans. Power Electron. **17**(6), 980–989 (2002)

WiseMove: A Framework to Investigate Safe Deep Reinforcement Learning for Autonomous Driving

Jaeyoung Lee, Aravind Balakrishnan, Ashish Gaurav, Krzysztof Czarnecki[✉],
and Sean Sedwards

University of Waterloo, Waterloo, Canada
kczarnec@gsd.uwaterloo.ca

Abstract. WiseMove is a platform to investigate safe deep reinforcement learning (DRL) in the context of motion planning for autonomous driving. It adopts a modular architecture that mirrors our autonomous vehicle software stack and can interleave learned and programmed components. Our initial investigation focuses on a state-of-the-art DRL approach from the literature, to quantify its safety and scalability in simulation, and thus evaluate its potential use on our vehicle.

1 Introduction

Ensuring the safety of learned components is of interest in many contexts and particularly in autonomous driving, which is the concern of our group.[1] We have hand-coded an autonomous driving motion planner that has already been used to drive autonomously for 100 km,[2] but we observe that further extensions by hand will be very labour-intensive. The success of deep reinforcement learning (DRL) in playing Go [5], and its success with other applications having intractable state space [1], suggests DRL as a more scalable way to implement motion planning. A recent DRL-based approach [4] seems particularly plausible, since it incorporates temporal logic (safety) constraints and its architecture is broadly similar to our existing software stack. The claimed results are promising, but the authors provide no means of verifying them and there is apparently no other platform in which to test their ideas. We have thus devised WiseMove, to quantify the trade-offs between safety, performance and scalability of both learned and programmed motion planning components.

Below we describe the key features of WiseMove and briefly present results of experiments that corroborate some of the claimed quantitative results of [4]. In contrast to that work, our results can be reproduced by installing our publicly-available code.[3]

---

[1] uwaterloo.ca/waterloo-intelligent-systems-engineering-lab/.
[2] therecord.com/news-story/8859691-waterloo-s-autonomoose-hits-100-kilometre-milestone/.
[3] git.uwaterloo.ca/wise-lab/wise-move/.

J. Lee, A. Balakrishnan, A. Gaurav and S. Sedwards—Contributed equally.

© Springer Nature Switzerland AG 2019
D. Parker and V. Wolf (Eds.): QEST 2019, LNCS 11785, pp. 350–354, 2019.
https://doi.org/10.1007/978-3-030-30281-8_20

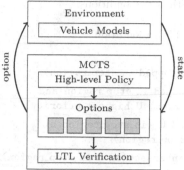

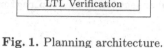

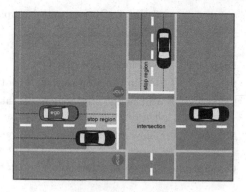

Fig. 1. Planning architecture.   **Fig. 2.** Experimental environment.

2   Features and Architecture

WiseMove is an options-based modular DRL framework, written in Python, with a hierarchical structure designed to mirror the architecture of our autonomous driving software stack. Options [6, Chap. 17] are intended to model primitive manoeuvres, to which are associated low-level policies that implement them. A learned high-level policy over options decides which option to take in any given situation, while Monte Carlo tree search (MCTS [6, Chap. 8]) is used to improve overall performance during deployment (planning). High-level policies correspond to the *behaviour planner* in our software stack, while low-level policies correspond to the *local planner*. These standard concepts are discussed in, e.g., [3]. To define correct behaviour and option termination conditions, Wise-Move incorporates "learntime" verification to validate individual simulation traces and assign rewards during both learning and planning. This typically improves safety, but does not guarantee it, given finite training and function approximation [1,2].

When an option is chosen by the decision maker (the high-level policy or MCTS), a sequence of actions is generated according to the option's low-level policy. An option terminates if there is a violation of a logical requirement, a collision, a timeout, or successful completion. In the latter case, the decision maker then chooses the next option to execute, and so on until the whole episode ends. Fig. 1 gives a diagrammatic overview of WiseMove's planning architecture. The current state is provided by the environment. The planning algorithm (MCTS) explores and verifies hypothesized future trajectories using the learned high-level policy as a baseline. MCTS chooses the best next option it discovers, which is then used to update the environment.

WiseMove comprises four high-level Python modules: `worlds`, `options`, `backends` and `verifier`. The `worlds` module provides support for environments that adhere to the OpenAI Gym[4] interface, which includes methods to initialize, update and visualize the environment, among others. The `options` module

---

[4] gym.openai.com.

Table 1. Options and examples of their preconditions.

Option	Description	Example LTL precondition
`KeepLane`	Keep lane while driving	-
`Stop`	Stop in the stop region	`G(not has_stopped_in_stop_region)`
`Wait`	Wait in the stop region then drive forward	`G((has_stopped_in_stop_region and in_stop_region) U highest_priority)`
`Follow`	Follow vehicle ahead	`G(veh_ahead U (in_stop_region or close_to_stop_region))`
`ChangeLane`	Change to other lane	`G(not(in_intersection or in_stop_region))`

defines the hierarchical decision-making structure. The `backends` module provides the code that implements the learned or possibly programmed components of the hierarchy. WISEMOVE currently uses `keras`[5] and `keras-rl`[6] for DRL training. The training hierarchy can be specified through a `json` file.

The `verifier` module provides methods for checking LTL-like properties constructed according to the following syntax:

$$\varphi = F\,\varphi \mid G\,\varphi \mid X\,\varphi \mid \varphi => \varphi \mid \varphi \text{ or } \varphi \mid \varphi \text{ and } \varphi \mid \text{not } \varphi \mid \varphi\,U\,\varphi \mid (\varphi) \mid \alpha \qquad (1)$$

Atomic propositions, α, are functions of the global state, represented by human-readable strings. In what follows we use the term LTL to mean properties written according to (1). The verifier decides during learning and planning when various LTL properties are satisfied or violated, in order to assign the appropriate reward. Learning proceeds one step at a time, so the verifier works incrementally, without revisiting the prefix of a trace. WISEMOVE uses LTL to express the preconditions and terminal conditions of each option, as well as to encode traffic rules. E.g.,

`G(in_stop_region => (in_stop_region U has_stopped_in_stop_region))`.

Some options and preconditions are listed in Table 1.

3   Experiments

Our experiments reproduce the architecture and some of the results of [4],[7] using the scenario illustrated in Fig. 2. We learned the low-level policies for each option first, then learned the high-level policy that determines which option to use at each decision instant. We used the DDPG [1] and DQN [2] algorithms to learn the low- and high-level polices, respectively. Each episode is initialized with the

---

[5] keras.io.

[6] github.com/keras-rl/keras-rl.

[7] Details and scripts to reproduce our results can be found in our repository (see Footnote 3).

ego vehicle placed at the left hand side, and up to six other randomly placed vehicles driving "aggressively" [4]. The goal of the ego is to reach the right hand side with no collisions and no LTL violations.

Table 2. Performance of low-level policies trained for 10^5 steps, with and without additional LTL: mean (std) % success, averaged over 100 trials of 100 episodes.

Add'l LTL	KeepLane	Stop	Wait	Follow	ChangeLane
Without	7.7 (21.4)	53.9 (32.0)	36.7 (43.3)	52.0 (19.6)	60.9 (34.0)
With	78.1 (29.4)	87.6 (20.4)	78.3 (28.8)	81.0 (15.4)	92.8 (14.3)

Table 3. Overall performance with and without MCTS, using low-level policies trained for 10^6 steps: mean (std) %, averaged over 1000 episodes.

Without MCTS			With MCTS		
Success	LTL violation	Collision	Success	LTL violation	Collision
92.0 (2.0)	5.40 (1.9)	2.60 (1.6)	98.5 (1.5)	0.9 (0.9)	0.6 (0.8)

We found that training low-level policies only according to the information given in [4] is unreliable; training would often not converge and good policies had to be selected from multiple attempts. We thus introduced additional LTL to give more information to the agent during training, including liveness constraints (e.g., G(not stopped_now)) to promote exploration, and safety-related properties (e.g., G(not veh_ahead_too_close)). Table 2 reports typical performance gains for 10^5 training steps. Note in particular the sharp increase in performance for KeepLane, which is principally due to the addition of a liveness constraint. Without this, the agent avoids the high penalty of collisions by simply waiting, thus not completing the option.

Having trained good low-level policies with DDPG using 10^6 steps, we trained high-level policies with DQN using 2×10^5 steps. We then tested the policies with and without MCTS. Table 3 reports the results, which suggest a ca. 7% improvement using MCTS.

4   Conclusion and Prospects

We have constructed WiseMove to investigate safe deep reinforcement learning in the context of autonomous driving. Learning is via options, whose low- and high-level policies broadly mirror the behaviour planner and local planner in our autonomous driving stack. The learned policies are deployed using a Monte Carlo tree search planning algorithm, which adapts the policies to situations that may not have been encountered during training. During both learning and planning, WiseMove uses linear temporal logic to enable and terminate options, and to specify safe and desirable behaviour.

Our initial investigation using WISEMOVE has reproduced some of the quantitative results of [4]. To achieve these we found it necessary to use additional logical constraints that are not mentioned in [4]. These enhance training by promoting exploration and generally encouraging good behaviour. We leave a detailed analysis for future work.

Our ongoing research will use WISEMOVE with different scenarios and more complex vehicle dynamics. We will also use different types of non-ego vehicles (aggressive, passive, learned, programmed, etc.) and interleave learned components with programmed components from our autonomous driving stack.

Acknowledgment. This work is supported by the Japanese Science and Technology agency (JST) ERATO project JPMJER1603: HASUO Metamathematics for Systems Design, and by the Natural Sciences and Engineering Research Council of Canada (NSERC) Discovery Grant: Model-Based Synthesis and Safety Assurance of Intelligent Controllers for Autonomous Vehicles.

References

1. Lillicrap, T.P., et al.: Continuous control with deep reinforcement learning (2015). http://arxiv.org/abs/1509.02971
2. Mnih, V., et al.: Playing Atari with deep reinforcement learning (2013). http://arxiv.org/abs/11312.5602
3. Paden, B., Čáp, M., Yong, S.Z., Yershov, D., Frazzoli, E.: A survey of motion planning and control techniques for self-driving urban vehicles. IEEE Trans. Intell. Veh. 1(1), 33–55 (2016)
4. Paxton, C., Raman, V., Hager, G.D., Kobilarov, M.: Combining neural networks and tree search for task and motion planning in challenging environments. In: 2017 IEEE/RSJ International Conference on Intelligent Robots and Systems (IROS), IEEE, pp. 6059–6066 (2017)
5. Silver, D., et al.: Mastering the game of Go with deep neural networks and tree search. Nature **529**, 484–489 (2016)
6. Sutton, R.S., Barto, A.G.: Reinforcement Learning: An Introduction. MIT Press, Cambridge (2018)

Great-Nsolve: A Tool Integration for (Markov Regenerative) Stochastic Petri Nets

Elvio Gilberto Amparore[1(✉)], Peter Buchholz[2(✉)], and Susanna Donatelli[1(✉)]

[1] Dipartimento di Informatica, Università degli Studi di Torino, Torino, Italy
{amparore,donatelli}@di.unito.it
[2] Department of Computer Science, TU Dortmund, Dortmund, Germany
peter.buchholz@cs.tu-dortmund.de

Abstract. This paper presents *Great-Nsolve*, the integration of Great-SPN (with its user-friendly graphical interface and its numerous possibilities of stochastic Petri net analysis) and *Nsolve* (with its very efficient numerical solution methods) aimed at solving large Markov Regenerative Stochastic Petri Nets (MRSPN). The support for general distribution is provided by the alphaFactory library.

1 The Baseline

Generalized stochastic Petri nets (GSPN) [1] are a stochastic extension of place/transition to associate exponentially distributed delays to transitions. The stochastic process described by a GSPN is a continuous time Markov chain (CTMC). Markov Regenerative Stochastic Petri Nets (MRSPN) [16] are an extension of GSPNs to allow transitions to have a generally distributed delay, given that, in each state, at most one non-exponential transition is enabled. The solution of a MRSPN is based on the solution of its underlying Markov Regenerative Processes (MRgP) which is typically based on the construction and solution of the embedded discrete time Markov chain (DTMC) at regeneration points (the so-called global kernel) and of the CTMC that describes the stochastic behavior of the net in-between regeneration points (the so-called local kernel). It is well known that the embedded DTMC matrix can be very dense (even if the MRgP transition matrix is sparse) and therefore it can be built, stored and solved only for small systems (thousands of states). This approach is called explicit, in contrast to the implicit, matrix-free technique [6,18] which does not require to build and store the embedded DTMC, but works with the (usually sparse) transition matrix.

When a net model is formulated as a set of components, the state space and transition matrix (and even the solution vector [14]) can be effectively represented in Kronecker form [11,20] allowing us to treat much larger state spaces.

A previous paper [5] has introduced the combination of matrix-free, as in [6,18], and Kronecker representation to solve MRSPN. This demo tool paper describes how the matrix-free solutions provided by GreatSPN [4] and the advanced Kronecker-based techniques provided by *Nsolve* [12] have been integrated to implement the theory presented in [5] and have been made accessible in an easy manner through the graphical interface of GreatSPN [3]. The support

© Springer Nature Switzerland AG 2019
D. Parker and V. Wolf (Eds.): QEST 2019, LNCS 11785, pp. 355–360, 2019.
https://doi.org/10.1007/978-3-030-30281-8_21

for general distributions is provided by the alphaFactory library [8]. The resulting *Great-Nsolve* tool allows the user to experiment and compare solutions of MRSPN based on a large variety of techniques: implicit and matrix-free approaches [2], already present in GreatSPN, and Kronecker-based approaches, thanks to the tool integration presented in this paper.

The *Nsolve* Tool. *Nsolve* [12] is a collection of advanced numerical solution methods for the computation of stationary distributions of large structured Markov chains. The tool is written in C and uses a two level hierarchical Kronecker structure to represent generator matrices which can be built from a flat description of a model as a network of synchronized components [13]. The matrix structure is described in [15]. Due to the modular structure and the availability of basic functions for numerical operations it is possible to easily integrate new numerical solution methods at the level of C functions. Different interfaces exist to combine the numerical methods provided by *Nsolve* with other tools as front- and/or back-ends. First, a file interface using small component matrices stored in sparse format has been defined [12], then the internal C data structures for sparse matrices can be used and the solution functions can be called directly. Finally, models can be provided in APNN format [10], an XML based format to describe extended stochastic Petri nets. At the back-end *Nsolve* provides the stationary vector and results computed from an appropriate reward structure. *Nsolve* cannot deal with MRSPN models directly.

The GreatSPN Tool. GreatSPN [4] is a tool for the qualitative and stochastic analysis of (stochastic) Petri nets and various extensions of them. Through the support on a Java-based, highly portable graphical interface (GUI) [3] that allows one to draw and compose nets and to play the (colored) token game, GreatSPN offers a qualitative analysis that includes state space construction, model-checking of CTL properties based on decision diagrams, various structural analysis techniques (like P- and T- invariants) and a stochastic analysis with a rich variety of numerical solutions for ergodic and non-ergodic models, as well as stochastic model checking of the CSL^{TA} logic [17]. GreatSPN has been tailored for teaching, but it is also a tool with advanced solution techniques that regularly participates in the Petri net model checking competition [19]. For the integration with *Nsolve* the most relevant features of the tool are the solvers [2] for MRSPNs, that include an explicit and a matrix-free solver as well as a component-based method [7] for non-ergodic MRSPN.

The alphaFactory Library. This library [8] supports the definition of general distributions through their PDF $f(x)$, and the computation of the α factors to be used inside any uniformization method implementation. Examples of $f(x)$ are $I[\delta]$ for a Dirac impulse at time δ, i.e. a deterministic event, $R(a, b)$ for a uniform rectangular signal in the $[a, b]$ range, or more sophisticated functions like $f(x) = \frac{\lambda^r}{(r-1)!} \cdot x^{r-1} \cdot e^{-\lambda x}$ for the Erlang distribution with r phases of rate λ. Syntactic sugar for common distributions (like *Uniform*, *Erlang* or *Pareto*) is available too.

2   *Great-Nsolve*: The Integration

Changes to the GreatSPN GUI. GreatSPN already has support for MRSPN: the GUI supports the labelling of general transitions with a PDF function $f(x)$ in *alphaFactory* form, and the execution of different forms of MRSPN solutions, including a matrix-free approach. In *Great-Nsolve* the net model of the GUI has been extended to include the module information (a partition of places) to be exploited by *Nsolve*. The same net model is then shared by both the GreatSPN solvers (that ignores it) and the Kronecker-based solvers of *Nsolve*. The extensions of the GUI include: (1) Places can be *partitioned* by the use of labels: places with the same label are in the same partition; Such separation is compulsory for *Nsolve*, and it is ignored by other solvers. (2) The GUI outputs the model in the internal representation of *Nsolve*, i.e. the APNN model. This format requires the place bounds to be known in advance, hence place bounds are first computed from the place invariants. Nets not fully covered by P-invariants are currently not supported. (3) the results computed by *Nsolve* are shown back in the GUI (directly on the net or in tabular form), as for any other GreatSPN solver. These changes allow any user to exploit the efficient techniques presented in [5] in a fully transparent, 1-click approach manner. As a by-product of the integration, Kronecker based solution for GSPNs are also 1-click available.

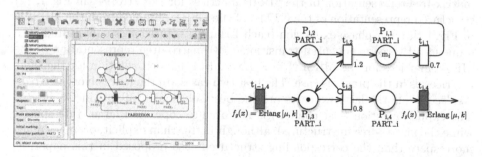

Fig. 1. Net partition in the GUI and a component of a Moving Server System model.

Extension of *Nsolve* for MRP Solutions. Two main modifications were needed: (1) Computation of the hierarchical state space structure cannot be done with the standard algorithm from [13] since the hierarchical matrix structure for MRgPs differs from the matrix structure of CTMCs and requires the implementation of the algorithm in [5]. (2) The steady-state solution for MRgPs has been implemented to work using matrix in Kronecker form inside a matrix-free iterative method, similar to the one in [6]. Currently the iterative methods supported are the Power method, GMRES and BiCG-Stab. The matrix-free solution requires the computation of the local kernels using Uniformization w.r.t. the PDF $f(x)$ of the general transition, and the computation of the Uniformization coefficients is delegated to the alphaFactory library.

3   *Great-Nsolve* in Action

Great-Nsolve is available as a virtual machine with all software preinstalled and ready to use that can be downloaded at: http://www.di.unito.it/~greatspn/VBox/Ubuntu64_with_Great-nsolve.ova. The code of GreatSPN with the *Nsolve* integration can be found at https://github.com/greatspn/SOURCES.

AlphaFactory can be found at https://github.com/amparore/alphaFactory.

The Kronecker-based solution of MRSPN in *Great-Nsolve* is limited to ergodic systems. If the system is non-ergodic, but finite, the model can still be solved using the matrix-free techniques and the component-based method [2,7] of GreatSPN.

To show the use of the integrated tool consider the Moving Server System (MSS) obtained by concatenating 4 stations (components) modeled by the net in Fig. 1(right). Places are partitioned according to stations. The black and thick transitions have a general distribution. This is an Erlang-3 with rate 1. Experiments are performed for a variable number m_i of requests arriving to the station's queues (in the reported experiments the same value for all queues). The number of macro states (top states in the hierarchical structure of the MRSPN) is constant, and equal to 12. Macro states are built automatically. Table 1 summarizes the comparison in space based on the number of states and of non-zeros of the matrices used. The following cases are considered (left to right): (1) structured matrix-free representation of the MRgP matrices for the MRSPN in Fig. 1, (2) structured representation of the CTMC of the GSPN obtained from the MRSPN in Fig. 1 through phase-expansion (each Erlang-3 transition is substituted by a sequence of three exponential transitions) (3) matrix-free representation of the MRgP matrices for the MRSPN in Fig. 1 and (4) explicit solution for the same matrices as in the previous case. The first two cases are the new solvers of *Great-Nsolve*, while the last two cases are the best solutions available in GreatSPN. The explicit solution (last column) is not able to cope with large state spaces, while the matrix-free unstructured, although better than explicit, occupies much more space than the corresponding structured case proposed in this paper, as an example: 840 entries against more than 7 millions for the $m_i = 20$ case. Note

Table 1. MSS: State space sizes and memory occupations.

	Structured representation (*Great-Nsolve*)						Unstructured representation				
	General (MRgP)			Phase-expansion (CTMC)			General (MRgP)				
	Matrix-free			–			Matrix-free			Explicit	
m_i	states	nnz	Time	states	nnz	Time	states	nnz	Time	nnz	Time
1	128	80	0.1	320	416	0.2	128	384	0.6	544	1.1
10	117128	440	1.8	244904	1856	2.5	117128	543048	11.8	7452116	67.7
20	1555848	840	47.5	3185784	3456	64.3	1555848	7482888	606.8	88318068	12202.1
25	3655808	1040	140.5	7452224	4256	208.5	3655808	17716608	1687.8	–	–
30	7388168	1240	369.3	15014664	5056	613.4	7388168	35987528	4048.4	–	–

also that the structured approach does not suffer much for the substitution of general distribution with their phase expansion.

Future Work. From a theoretical point of view, we are currently working on an optimization of the solution presented in [5]. We are also considering whether the component-based MRgP solution techniques presented in [7] can profit of a Kronecker-based approach for the solution of single components. From a tool point of view we plan to lift the requirement that all places are covered by a p-invariant. If this is not the case, indeed the place bounds cannot be computed through structural analysis, but, if the state space is finite, they can be computed on the actual state space, that can be efficiently generated in GreatSPN using decision-diagrams [9].

References

1. Ajmone Marsan, M., Conte, G., Balbo, G.: A class of generalized stochastic Petri nets for the performance evaluation of multiprocessor systems. ACM Trans. Comput. Sys. **2**, 93–122 (1984)
2. Amparore, E.G., Donatelli, S.: DSPN-tool: a new DSPN and GSPN solver for GreatSPN. In: Proceedings of the 2010 Seventh International Conference on the Quantitative Evaluation of Systems, QEST 2010, Washington, DC, USA, pp. 79–80. IEEE Computer Society (2010). ISBN: 978-0-7695-4188-4, https://doi.org/10.1109/QEST.2010.17
3. Amparore, E.G.: A new greatSPN GUI for GSPN editing and CSL^{TA} model checking. In: Norman, G., Sanders, W. (eds.) QEST 2014. LNCS, vol. 8657, pp. 170–173. Springer, Cham (2014). https://doi.org/10.1007/978-3-319-10696-0_13
4. Amparore, E.G., Balbo, G., Beccuti, M., Donatelli, S., Franceschinis, G.: 30 years of GreatSPN. In: Fiondella, L., Puliafito, A. (eds.) Principles of Performance and Reliability Modeling and Evaluation. SSRE, pp. 227–254. Springer, Cham (2016). https://doi.org/10.1007/978-3-319-30599-8_9
5. Amparore, E.G., Buchholz, P., Donatelli, S.: A structured solution approach for Markov regenerative processes. In: Norman, G., Sanders, W. (eds.) QEST 2014. LNCS, vol. 8657, pp. 9–24. Springer, Cham (2014). https://doi.org/10.1007/978-3-319-10696-0_3
6. Amparore, E.G., Donatelli, S.: Revisiting the matrix-free solution of Markov regenerative processes. Numer. Linear Algebra Appl. **18**, 1067–1083 (2011)
7. Amparore, E.G., Donatelli, S.: A component-based solution for reducible Markov regenerative processes. Perform. Eval. **70**(6), 400–422 (2013)
8. Amparore, E.G., Donatelli, S.: alphaFactory: a tool for generating the alpha factors of general distributions. In: Bertrand, N., Bortolussi, L. (eds.) QEST 2017. LNCS, vol. 10503, pp. 36–51. Springer, Cham (2017). https://doi.org/10.1007/978-3-319-66335-7_3
9. Babar, J., Beccuti, M., Donatelli, S., Miner, A.: GreatSPN enhanced with decision diagram data structures. In: Lilius, J., Penczek, W. (eds.) PETRI NETS 2010. LNCS, vol. 6128, pp. 308–317. Springer, Heidelberg (2010). https://doi.org/10.1007/978-3-642-13675-7_19
10. Bause, F., Buchholz, P., Kemper, P.: A toolbox for functional and quantitative analysis of DEDS. In: Puigjaner, R., Savino, N.N., Serra, B. (eds.) TOOLS 1998. LNCS, vol. 1469, pp. 356–359. Springer, Heidelberg (1998). https://doi.org/10.1007/3-540-68061-6_32

11. Buchholz, P., Ciardo, G., Donatelli, S., Kemper, P.: Complexity of memory-efficient Kronecker operations with applications to the solution of Markov models. INFORMS J. Comput. **12**(3), 203–222 (2000)
12. Buchholz, P.: Markov matrix market. http://ls4-www.cs.tu-dortmund.de/download/buchholz/struct-matrix-market.html
13. Buchholz, P.: Hierarchical structuring of superposed GSPNs. IEEE Trans. Softw. Eng. **25**(2), 166–181 (1999)
14. Buchholz, P., Dayar, T., Kriege, J., Orhan, M.C.: On compact solution vectors in Kronecker-based Markovian analysis. Perform. Eval. **115**, 132–149 (2017)
15. Buchholz, P., Kemper, P.: Kronecker based matrix representations for large Markov models. In: Baier, C., Haverkort, B.R., Hermanns, H., Katoen, J.-P., Siegle, M. (eds.) Validation of Stochastic Systems. LNCS, vol. 2925, pp. 256–295. Springer, Heidelberg (2004). https://doi.org/10.1007/978-3-540-24611-4_8
16. Choi, H., Kulkarni, V.G., Trivedi, K.S.: Markov regenerative stochastic Petri nets. Perform. Eval. **20**(1–3), 337–357 (1994)
17. Donatelli, S., Haddad, S., Sproston, J.: Model checking timed and stochastic properties with CSL^{TA}. IEEE Trans. Software Eng. **35**(2), 224–240 (2009)
18. German, R.: Iterative analysis of Markov regenerative models. Perform. Eval. **44**, 51–72 (2001)
19. Kordon, F., & all: Complete results for the 2019 edition of the Model Checking Contest (2019). http://mcc.lip6.fr/2019/results.php
20. Plateau, B., Fourneau, J.M.: A methodology for solving Markov models of parallel systems. J. Parallel Distrib. Comput. **12**(4), 370–387 (1991)

Author Index